BIOSOURCES
LAB PROGRAM

LABORATORY TECHNIQUES AND EXPERIMENTAL DESIGN
Teacher's Edition

INCLUDES LABS C1–C48

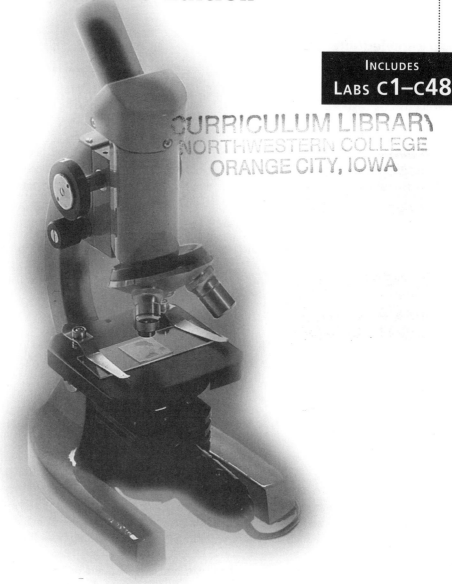

HOLT, RINEHART AND WINSTON
Harcourt Brace & Company

Austin • New York • Orlando • Atlanta • San Francisco • Boston • Dallas • Toronto • London

HOLT BIOSOURCES *LAB PROGRAM*

LABORATORY TECHNIQUES AND EXPERIMENTAL DESIGN

Staff Credits

Editorial Development

Carolyn Biegert
Janis Gadsden
Debbie Hix

Copyediting

Amy Daniewicz
Denise Haney
Steve Oelenberger

Prepress

Rose Degollado

Manufacturing

Mike Roche

Design Development and Page Production

Morgan-Cain & Associates

Acknowledgments

Contributors

David Jaeger
Will C. Wood High School
Vacaville, CA

George Nassis
Kenneth G. Rainis
WARD'S Natural Science Establishment
Rochester, NY

Suzanne Weisker
Science Teacher and Department Chair
Will C. Wood High School
Vacaville, CA

Editorial Development

WordWise, Inc.

Cover

Design—Morgan-Cain & Associates
Photography—Sam Dudgeon

Lab Reviewers

Lab Activities
Ted Parker
Forest Grove, OR

Mark Stallings, Ph.D.
Chair, Science Department
Gilmer High School
Ellijay, GA

George Nassis
Kenneth G. Rainis
Geoffrey Smith
WARD'S Natural Science Establishment
Rochester, NY

Lab Safety
Kenneth G. Rainis
WARD'S Natural Science Establishment
Rochester, NY

Jay Young, Ph.D
Chemical Safety Consultant
Silver Spring, MD

Printed in the United States of America

ISBN 0-03-051404-5

123456 022 00 99 98 97

LABORATORY TECHNIQUES AND EXPERIMENTAL DESIGN

Contents

Contents, *continued*

Maintaining a Safe Lab

Building Safety Partnerships: You're Not Alone

A safe laboratory can only be achieved through a partnership among all parties concerned, not just among students or teachers. Materials for teachers and students need to be thorough, explicit, and persuasive. Teachers must actively boost safety consciousness among students, fellow faculty members, administrators, and parents. For success, everyone must agree to respect the same laboratory rules, to obtain and use the proper safety equipment, and to take appropriate precautions during a lab activity.

An excellent way to start building this safety partnership with your students is to use the Safety Contract on the **Holt BioSources Teaching Resources CD-ROM.** Have each student fill out a contract and return it to you. Keep the contracts on file, in case you need to remind students of their promises.

Where to Start

In each lab activity, safety symbols are included and specific safety procedures are highlighted where appropriate. Detailed descriptions of each safety symbol and hazards and precautions related to each one can be found in your biology textbook and in the "Laboratory Safety" section of the pupil's edition of the *Inquiry Skills Development, Laboratory Techniques and Experimental Design,* and *Biotechnology* manuals. The safety symbol descriptions are included in expanded form in the following section.

The expanded safety symbol descriptions and other information included in the following sections will help you plan and maintain a safe and healthy laboratory environment.

This information is not all-inclusive. Each school's lab situation is different, and no publication could list safe practices for all situations that could possibly arise.

Be sure that you are aware of any federal, state, or local laws that may cover your lab. Although laws and regulations can vary from place to place and from time to time, you can build a safe program suited to your situation using this information.

Safety Symbols

Eye Safety

- **Wear approved chemical safety goggles as directed.** Goggles should always be worn whenever you or your students are working with a chemical or solution, heating substances, using any mechanical device, or observing physical processes. See the "Safety Equipment" section for specific tips on the types of goggles to be worn.

  Teachers should model appropriate behavior by wearing safety goggles when appropriate. Some teachers have success when they turn this into a game: students found without safety goggles must wear them during the next lecture period. If students catch the teacher not wearing safety goggles, the teacher must wear them during the next lecture period.

- **In case of eye contact, go to an eyewash station and flush eyes (including under the eyelids) with running water for at least 15 minutes.** The teacher or other adult in charge must be notified immediately.

- **Wearing contact lenses for cosmetic reasons is prohibited in the laboratory.** Be sure to make this clear to students at the beginning of the year. First, take a poll of contact-lens wearers. Explain the precautions necessary, noting that liquids or gases can be drawn up under a contact lens and onto the eyeball. If a student must wear contact lenses prescribed by a physician, be sure the students wear approved eyecup safety goggles, which are similar to goggles worn for swimming.

- **Never look directly at the sun through any optical device or lens system, and never gather direct sunlight to illuminate a microscope.** Such actions will concentrate light rays that can severely burn the retina, possibly causing blindness. At the beginning of the year, make sure each microscope you will use has an appropriate and functioning light source.

Electrical Supply

- **Be sure you know the location of the master shut-off for all circuits and other utilities in the lab.** If the circuit breakers are locked up, make sure you have a key in case of an emergency. Color code or label the necessary switches. Be sure you will remember what to do under the pressures of an emergency.

- **Be sure all outlets have correct polarity and ground-fault interruption.** Polarity can be tested with an inexpensive (about $5.00) continuity tester available from most electronic hobby shops. Use only electrical equipment with three-prong plugs and three-wire cords. Each electrical socket in the laboratory must be a three-holed socket with a GFI (ground fault interrupter) circuit. In some cases, rewiring the lab may be necessary. Be sure your supervisors understand the potential hazards and costs of leaving the lab in an unsafe configuration.

- **Electrical equipment should be in the "off" position before it is plugged into a socket.** After a lab activity is completed, the equipment should first be turned off and then unplugged. Wiring hookups should not be made or altered except when an apparatus is disconnected from its AC or DC power source and the power switch, if applicable, is off.

- **Do not let electrical cords dangle from work stations.** Dangling cords are a hazard that can cause tripping or electrical shock.

- **Tape electrical cords to work surfaces.** This will prevent falls and decrease the chances of equipment being pulled off the table. If you find you have too many cords to be taped down, that could be a sign of a poorly designed lab, which will be prone to other problems as well. Sometimes simply rearranging the lab desks alleviates some flaws.

- **Never use or allow students to use equipment with frayed or kinked cords.** Check all of the electrical equipment at the beginning and end of each year. It is better to omit an activity because the equipment is unsafe than to proceed with an activity that results in an injury to yourself or a student.

- **Never use electrical equipment around water or with wet hands or clothing.** The area under and around electrical equipment should be dry. Electrical cords should not lie in puddles of spilled liquid.

- **Use dry cells or rechargeable batteries as direct current (DC) sources.** Do not use automobile storage batteries or AC-to-DC converters; these two sources of DC current can present serious electrical shock hazards. When storing dry cells and rechargeable batteries, cover both terminals with insulating tape.

- **Before leaving the laboratory, be sure all electrical equipment has been turned off and is unplugged.**

Clothing Protection

- **Wear a lab apron or lab coat when working in the laboratory to prevent chemicals or chemical solutions from contacting skin or contaminating clothes.** Suggested styles of lab aprons are discussed in the "Safety Equipment" section that follows. Be sure students confine all loose clothing and long jewelry. Open-toed shoes should not be allowed in the laboratory.

Animal Care

- **Do not touch or approach any animal in the wild.** Be sure you and your students are aware of any poisonous or dangerous animals in any area where you will be doing fieldwork.

- **Always insist that students obtain your permission before bringing any animal (or pet) into the school building.** There are legitimate reasons to bring animals to school, but be certain that such an occasion does not present a danger or a distraction to students.

- **Handle all animals with proper caution and respect.** Mishandling or abuse of any animal should not be tolerated. The National Association of Biology Teachers guidelines for the use of live animals, reproduced in the "Animal Care" section, provide a good framework for planning specific procedures.

Sharp Object Safety

- **Use extreme care with all sharp instruments, such as scalpels, sharp probes, and knives.** You may want to consider restricting the use of such objects to lab activities for which there are no substitutes.

- **Never use double-edged razors in the laboratory.**

- **Never cut objects while holding them in your hand.** Place objects on a suitable work surface. Be sure your lab has an adequate supply of dissecting pans and similar surfaces for cutting.

Chemical Safety

More detailed information on chemical hazards, the use of MSDSs (Material Safety Data Sheets), and safe chemical storage can be found in the "Reagents and Storage" and "Chemical Handling and Disposal" sections that follow.

- **Always wear appropriate personal protective equipment.** Safety goggles, gloves, and a lab apron or lab coat should always be worn when working with any chemical or chemical solution.

- **Never taste, touch, or smell any substance or bring it close to your eyes, unless specifically directed to do so.** If students need to note the odor of a substance, have them do so by waving the fumes toward themselves with their hands. Make sure there are enough suction bulbs for any pipetting that needs to be done. Set a good example, and never pipet any substance by mouth.

- **Always handle any chemical or chemical solution with care.** Nothing in the lab should be considered harmless. Even nontoxic substances can easily be contaminated. Check the MSDS for each chemical prior to a lab activity, and observe safe-use procedures. Be sure to have appropriate containers available for unused reagents so that students won't return them to reagent bottles. Store chemicals according to the directions in the "Reagents and Storage" section.

- **Never mix any chemicals unless you are certain about what you are doing and why.** Many common chemicals react violently with each other. Consult section V of a chemical's MSDS for compatibility information.

- **Never pour water into a strong acid or base.** The mixture can produce heat and splatter. Remember this rhyme:

    "Do as you oughta—
    Add acid (or base) to water."

- **Have a spill-control plan and kit ready.** Students should not handle chemical spills. Be sure that your spill-control kit contains neutralizing agents, sand, and other absorbent material.

- **Check for the presence of any source of flames, sparks, or heat (open flame, electrical heating coils, etc.) before working with flammable liquids or gases.**

Plant Safety

- **Do not ingest any plant part used in the laboratory (especially seeds sold commercially).** Commercially sold seeds often are coated with fungicidal agents. Do not rub any sap or plant juice on your eyes, skin, or mucous membranes.

- **Wear protective (disposable polyethylene) gloves when handling any wild plant.**

- **Wash hands thoroughly after handling any plant or plant part (particularly seeds).** Avoid touching your hands to your face and eyes.

- **Do not inhale or expose yourself to the smoke of any burning plant.** Some irritants travel in smoke and can cause inflammation in the throat and lungs.

- **Do not pick wildflowers or other plants unless permission from appropriate authorities has been obtained in advance.**

Proper Waste Disposal

- **Have students clean and decontaminate all work surfaces and personal protective equipment after each lab activity.** Prompt and frequent cleaning helps keep contamination problems to a minimum.

- **Set aside special containers for the disposal of all sharp objects (broken glass and other contaminated sharp objects) and other contaminated materials (biological or chemical).** Make sure these items are disposed of in an environmentally sound way.

Hygienic Care

- **Keep your hands away from your face and mouth.**

- **Wash your hands thoroughly before leaving the laboratory.** Have bactericidal soap available for students to use.

- **Remove contaminated clothing immediately; launder contaminated clothing separately.** Have a few spare T-shirts and shorts or sweatsuits available in case of an emergency involving clothing.

- **Demonstrate the proper techniques when handling bacteria or microorganisms.** Examine microorganism cultures (such as those in petri dishes) without opening them.

- **Collect all stock and experimental cultures for proper disposal.** See the "Safety With Microbes" and "Release and Disposal of Organisms" sections for instructions on materials and cultures used in these lab activities.

Heating Safety

- **When heating chemicals or reagents in a test tube, never point the test tube toward anyone.**

- **Use hot plates, not open flames.** Be sure hot plates have an On-Off switch and indicator light. Never leave hot plates unattended, even for a minute.

  Check all hot plates for malfunctions several times during the school year. Never use alcohol lamps.

- **Know the location of laboratory fire extinguishers and fire blankets.** Have ice readily available in case of burns or scalds. Make certain that your laboratory fire extinguishers are tri-class (A-B-C) and are useful for all types of fires.

- **Use tongs or appropriate insulated holders when heating objects.** Heated objects often do not look hot. Set a good example by using tongs or other holders to handle an object whenever there is a possibility that the object could be warm.

- **Keep combustibles away from heat and other ignition sources.**

Hand Safety

- **Never cut objects while holding them in your hand.**

- **Wear protective gloves when working with stains, chemicals, chemical solutions, or wild (unknown) plants.**

Glassware Safety

- **Inspect glassware before use; never use chipped or cracked glassware.** Use borosilicate glass for heating. Check all glassware several times a year, and discard anything that shows signs of chipping or cracking.

- **Hold glassware firmly, but do not squeeze it.** Glass is fragile and may break if it is not handled carefully. Be sure your hands and the glassware are dry when you are handling glassware.

- **Do not attempt to insert glass tubing into a rubber stopper without taking proper precautions.** Lubricate the stopper and the glass tubing. Use heavy leather gloves to protect your hands from shattering glass. To prevent puncture wounds, be sure your hand is clear of the hole where the glass tubing will emerge.

- **Always clean up broken glass by using tongs and a brush and dustpan.** Discard the pieces in an appropriately labeled "sharps" container.

Safety With Gases

- **Never directly inhale any gas or vapor.** Do not put your nose close to any substance having an odor.

- **Be sure that your lab has excellent ventilation.** Some work will still require a chemical fume hood. If your lab does not have good ventilation, investigate opportunities to improve it. Be certain your supervisors are aware of potential hazards due to ineffective ventilation.

Laboratory Rules

Post the following rules in the laboratory, and discuss them with students. Afterward, give students the "Safety Quiz" on the **Holt BioSources Teaching Resources CD-ROM.**

- **Never work alone in the laboratory.**

- **Never perform any experiment not specifically assigned by your teacher.** Never work with any unauthorized material.

- **Never eat, drink, or apply cosmetics in the laboratory.** Never store food in the laboratory. Keep hands away from faces. Wash your hands at the conclusion of each laboratory investigation and before leaving the laboratory. Remember that some hair products are highly flammable, even after application.

- *NEVER* **taste chemicals.** *NEVER* **touch chemicals.** Even common substances should be considered dangerous, since they can be easily contaminated in the lab.

- **Do not wear contact lenses in the lab.** Chemical vapors can get between the lenses and the eyes and cause permanent eye damage.

- **Know the location of all safety and emergency equipment used in the laboratory.** Examples include eyewash stations, safety blankets, safety shower, fire extinguisher, first-aid kit, and chemical-spill kit.

- **Know fire drill procedures and the locations of exits.**

- **Know the location of the closest telephone,** and be sure there is a posted list of emergency phone numbers, including the poison control center, fire department, police department, and ambulance service.

- **Familiarize yourself with a lab activity—especially safety issues—before entering the lab.** Know the potential hazards of the materials and equipment to be used and the procedures required for the activity. Before you start, ask the teacher to explain any parts you do not understand.

- **Before beginning work: tie back long hair, roll up loose sleeves, and put on any personal protective equipment as required by your teacher.** Avoid wearing loose clothing or confine loose clothing that could knock things over, ignite from flame, or soak up chemical solutions. Do not wear open-toed shoes to the lab. If there is a spill, your feet could be injured.

- **Report any accidents, incidents, or hazards—no matter how trivial—to your teacher immediately.** Any incident involving bleeding, burns, fainting, chemical exposure, or ingestion should also be reported to the school nurse or physician.

- **In case of fire, alert the teacher and leave the laboratory.**

- **Keep your work area neat and uncluttered.** Bring only the books and materials needed to conduct a lab activity. Stay at your work area as much as possible. The less movement in a lab, the fewer spills and other accidents that can occur.

- **Clean your work area at the conclusion of a lab activity as your teacher directs.**

- **Wash your hands with soap and water after each lab activity.**

Safety Equipment

Do You Have What It Takes?

- **Chemical goggles** (meeting ANSI [American National Standards Institute] standard Z87.1): These should be worn when working with any chemical or chemical solution other than water, when heating substances, when using any mechanical device, or when observing physical processes that could eject an object.

  Wearing contact lenses for cosmetic reasons should be prohibited in the laboratory. If a student must wear contact lenses prescribed by a physician, that student should wear eyecup safety goggles meeting ANSI standard Z87.1 (similar to swimmers' goggles).

- **Face shield** (meeting ANSI standard Z87.1): Use in combination with safety goggles when working with corrosives.

- **Eyewash station:** The station must be capable of delivering a copious, gentle flow of water to both eyes for at least 15 minutes. **Portable liquid supply devices are not satisfactory and should not be used.** A plumbed-in fixture or a perforated spray head on the end of a hose attached to a plumbed-in outlet and designed for use as an eyewash fountain is suitable if it meets ANSI standard Z358.1 and is within a 30-second walking distance from any spot in the room.

- **Safety shower** (meeting ANSI standard Z358.1): Location should be within a 30-second walking distance from any spot in the room. Students should be instructed in the use of the safety shower for a fire or chemical splash on their body that cannot be simply washed off.

- **Gloves:** Polyethylene, neoprene, or disposable plastic may be used. Nitrile or butyl rubber gloves are recommended when handling corrosives.

- **Apron:** Gray or black rubber-coated cloth or a nylon-coated vinyl halter is recommended.

Prudent Precautions

What would you do if a student dropped a liter bottle of concentrated sulfuric acid? RIGHT NOW? Are you prepared? Could you have altered your handling and storage methods to prevent or lessen the severity of this incident? PLAN now how to effectively react BEFORE you need to. Planning tips include the following:

1. Post the phone numbers of your regional poison control center, fire department, police department, ambulance service, and hospital ON your telephone.

2. Practice fire and evacuation drills during labs and at all times during the year, not just in the fall. Post an evacuation diagram and an established evacuation procedure by every entrance to the laboratory.

3. Have drills on what students MUST do if they are on fire or experience chemical contact or exposure.

4. Mark the locations of eyewash stations, the safety shower, fire extinguishers (A-B-C tri-class), the chemical-spill kit, the first-aid kit, and fire blankets in the laboratory and storeroom. Make sure you have all of the necessary safety equipment prior to conducting each lab activity, and be certain the equipment is in good working order.

5. Lock your laboratory (and storeroom) when you are not present.

6. Compile an MSDS file for all chemicals. This reference resource should be readily accessible in case of spills or other incidents. (Information about MSDSs is found in the "Reagents and Storage" section.)

7. Develop spill-control procedures. Handle only incidents that you FEEL COMFORTABLE handling. Situations of greater severity should be handled by trained hazardous-material responders.

8. Under no circumstances should students fight fires or handle chemical spills.

9. Be sure to recognize and heed the signal words used on most safety labels for materials, equipment, and procedures:
CAUTION—low level of risk associated with use or misuse
WARNING—moderate level of risk associated with use or misuse
DANGER—high level of risk associated with use or misuse

10. Be trained in first aid and basic life support (CPR) procedures. Have first-aid kits and spill kits readily available.

11. Before the class begins a lab activity, review specific safety rules and demonstrate proper procedures.

12. Never permit students to work in your laboratory without your supervision. No unauthorized investigations should ever be conducted, nor should unauthorized materials be brought into the laboratory.

13. Fully document ANY INCIDENT that occurs. Documentation will provide the best defense in terms of liability, and it is a critical tool in helping to identify area(s) of laboratory safety that need improvement. Remind students that any safety incident, no matter how trivial, must be reported directly to you.

Safety With Microbes

What You Can't See CAN Hurt You

Pathogenic (disease-causing) microorganisms are not appropriate investigation tools in the high school laboratory and should never be used.

Consult with the school nurse to screen students whose immune system may be compromised by illness or who may be receiving immunosuppressive drug therapy. Such individuals are extraordinarily sensitive to potential infection from generally harmless microorganisms and should not participate in laboratory activities unless permitted to do so by a physician. Do not allow students with any open cuts, abrasions, or sores to work with microorganisms.

Aseptic Technique

Demonstrate correct aseptic technique to students PRIOR to conducting a lab activity. Never pipet liquid media by mouth. Wherever possible, use sterile cotton applicator sticks in place of inoculating loops and Bunsen burner flames for culture inoculation. Remember to use appropriate precautions when disposing of cotton applicator sticks: they should be autoclaved or sterilized before disposal.

Treat ALL microbes as pathogenic. Seal with tape all petri dishes containing bacterial cultures. Do not use blood agar plates, and never attempt to cultivate microbes from a human or animal source.

Never dispose of microbe cultures without first sterilizing them. Autoclave or steam-sterilize all used cultures and any materials that have come in contact with them at 120°C and 15 psi for 15–20 minutes. If an autoclave or steam sterilizer is not available, flood or immerse these articles with full-strength household bleach for 30 minutes, and then discard. Use the autoclave or steam sterilizer yourself; do not allow students to use these devices.

Wash all lab surfaces with a disinfectant solution before and after handling bacterial cultures.

Handling Bacteriological Spills

Never allow students to clean up bacteriological spills. Keep on hand a spill kit containing 500 mL of full-strength household bleach, biohazard bags (autoclavable), forceps, and paper towels.

In the event of a bacterial spill, cover the area with a layer of paper towels. Wet the paper towels with the bleach, and allow to stand for 15–20 minutes. Wearing gloves and using forceps, place the residue in the biohazard bag. If broken glass is present, use a brush and dustpan to collect the broken material and place it in a suitably marked container.

Safety in the Field

Your ability to control the laboratory environment is far greater than your ability to foresee consequences that could occur in the field. Extra care should be used when planning and conducting field activities.

- Have in your possession a "Permission Slip" for each student, signed by a parent or legal guardian, that details travel plans as well as field activities for EVERY INDIVIDUAL field trip.

- Check with the school nurse regarding any restrictions or medical conditions that could affect your field trip, such as allergies, asthma, reactions to contact irritants (poison ivy), etc. Know how to contact emergency personnel in case of need.

- "Pre-field" your location, checking any obvious and insidious dangers, such as plants to which students may be allergic, water hazards, steep slopes and cliffs, and animals (ticks, bees, snakes, etc.)

- File a detailed itinerary, including a complete roster of students and adult participants, with your department head or principal.

- Be sure that all participants dress appropriately (i.e., long pants and shirt sleeves in potentially tick-infested areas).

- Bring a first-aid kit and a canteen of water. At least one adult should possess a current first-aid certification from the Red Cross. If engaging in water activities, at least one adult must also hold a current life-saving certificate.

- Use only plastic containers.

- Allow no mercury thermometers; use alcohol.

- Upon your return, have all students wash their hands, arms, and face with soap and water. In certain circumstances, it may be advisable to have students shower and examine themselves for ticks.

- In case of any incident or accident, including those involving other adults, compile a complete report for filing with school administrators.

Reagents and Storage

General Guidelines

- Store bulk quantities of chemicals in a safe and secure storeroom, not in the teaching laboratory. Store them in well-ventilated, dry areas protected from sunlight and localized heat. Store by similar hazard characteristics, not alphabetically. (See "Chemical Hazard Classes" and "Chemical Storage" for additional recommendations.)

- Label student reagent containers with the substance's name and hazard class(es). Be sure to use labeling materials that won't be affected by the reagent or other chemicals that will be stored nearby. (See "Chemical Hazard Classes" for additional recommendations.)

- Dispose of hazardous waste chemicals according to federal, state, and local regulations. Refer to the Material Safety Data Sheets (available through your supplier) for recommended disposal procedures. Some disposal information is also included in the "Chemical Handling and Disposal" section on pages T23–T33. **NEVER ASSUME** that a reagent can be safely poured down the drain.

- Have a chemical-spill kit immediately available. Know the procedures for handling a spill of any chemical used during a lab activity or in preparing reagents. Never allow students to clean up hazardous chemical spills.

- Remove all sources of flames, sparks, and heat from the laboratory when any flammable material is being used.

Chemical Record-Keeping and MSDSs

Maintaining a current inventory of chemicals can help you keep track of purchase dates and amounts. In addition, if you use your inventory to determine what you need for the school year, you can cut down on storage problems.

The purpose of a Material Safety Data Sheet (MSDS) is to provide readily accessible information on chemical substances commonly used in the science laboratory or in industry. MSDSs are available from suppliers of chemicals.

MSDSs should be kept on file and referred to BEFORE handling ANY chemical. The MSDSs can also be used to instruct students on chemical hazards, to evaluate spill and disposal procedures, and to warn of incompatibilities with other chemicals or mixtures.

Each MSDS is divided into the following sections:

    I. Material Identification: includes name, common synonyms, reference codes, and precautionary labeling

    II. Ingredients and Hazards: identifies dangerous components of mixtures

    III. Physical Data: includes information such as melting point, boiling point, appearance, odor, density, etc.

IV. Fire and Explosion Hazard Data: includes flash point, description of fire-extinguishing media and procedures, and information on unusual fire and explosion hazards

V. Health Hazard Data: describes problems associated with inhalation, skin contact, eye contact, skin absorption, and ingestion, along with first-aid procedures

VI. Reactivity Data: includes information on incompatible types of chemicals and likely decomposition products

VII. Spill, Leak, and Disposal Procedures: includes step-by-step information

VIII. Special Protection Information: describes equipment needed for safe use

IX. Special Precautions and Comments: describes storage requirements and other notes

WARD'S has a pocket guide (Ward's Catalog No. 32 T 0002) that explains in greater detail how to use Material Safety Data Sheets.

Chemical Hazard Classes

The hazards presented by any chemical can be grouped into the following categories. (It is important to keep in mind that a particular chemical may have more than one of these hazards.)

- **Flammable**
- **Corrosive**
- **Poisonous (toxic)**
- **Reactive**

A fifth category, those chemicals that do not possess the above properties, are termed "low hazard" materials. Water is an example of a low hazard material. Although these materials may not ordinarily represent a hazard, their presence in the lab requires that they be treated differently than they would be in a kitchen or backyard.

The following precautions should be used with all hazard classes:

- Require the use of safety goggles, gloves, and lab aprons.
- Minimize the amounts available in the lab (100 mL or less).
- Become familiar with first-aid measures for each chemical used.
- Become familiar with incompatibility issues for each chemical used.
- Emphasize how essential lab and storeroom cleanliness and personal hygiene are when dealing with hazardous materials.
- Keep hazardous chemicals in approved containers that are kept closed and stored away from sunlight and rapid temperature changes.
- Store chemicals of each hazard class away from those in other hazard classes.
- Keep a designated and locked storage cabinet for each hazard class.

FLAMMABLE

Prevention and Control Measures

- Store away from oxidizers and reactives.
- Keep containers closed when not in use.

- The ignition source is the easiest of the three components to remove. Check for the presence of lighted burners, sources of sparks (including static charge, friction, and electrical equipment), and hot objects such as hot plates or incandescent bulbs.

- Ground (electrically) all bulk metal containers when dispensing flammable liquids.

- Flammable vapors are usually heavier than air and can travel considerable distances before being diluted below ignitable concentrations.

- Ensure that there are class B fire extinguishers present in the laboratory and store room.

- Students should be drilled in EXACTLY what they must do if their clothes or hair catches fire. Practice "drop and roll" techniques. Both a safety shower and fire blankets should be available. Inform students that the shower is the best way to put out a fire on polyester clothing.

- Conduct a fire inspection with members of the local fire department at least once a year. Practice fire drills regularly.

- Provide adequate ventilation.

- Prepare for spills by having absorbent, vapor-reducing materials (available commercially) close at hand. Plan to have enough absorbent material to handle the maximum volume of flammable substances on hand.

Additional Protective Equipment

- Nitrile or butyl rubber gloves
- Approved storage containers
- Face shield (recommended)
- Fire blanket
- Safety shower
- Fire extinguishers (Class B)

CORROSIVE

Prevention and Control Measures

- Always wear a face shield along with safety goggles when handling solutions of any corrosive material with concentrations greater than 1 mol/L.

- Have an eyewash station in close proximity.

- Wear the correct type of hand protection that will be impervious to the corrosive being handled. (Nitrile or butyl rubber gloves are generally recommended.)

- Provide adequate ventilation.

- Prepare for spills by having neutralizing reagents close at hand and in sufficient quantities for materials on hand.

Additional Protective Equipment

- Nitrile or butyl rubber gloves
- Safety shower
- Face shield
- Eyewash stations
- Sleeve gauntlets

POISONOUS (TOXIC)

Prevention and Control Measures

- Treat all chemicals as toxic until proven otherwise. Above all, emphasize barriers, cleanliness, and avoidance of contact when handling any chemical.
- Wear protective equipment over exposed skin areas and eyes.
- Handle all contaminated glass and metal carefully. Remember that any sharp object can be a vehicle for introducing a toxic substance.
- Provide good ventilation. Use a chemical fume hood if possible.
- Recognize symptoms of overexposure and typical routes of introduction for each chemical used during a lab activity.
- Remember that the skin is not a good barrier to many toxic chemicals.
- Post the phone number of the nearest poison control center ON your phone.

Additional Protective Equipment

- Container for sharp objects
- Chemical fume hood
- Particle face mask

Chemical Storage

Never store chemicals alphabetically, as that greatly increases the risk of a violent reaction. Take these additional precautions.

1. Always lock the storeroom and all cabinets when not in use.
2. Do not allow students in the storeroom or preparation areas.
3. Avoid storing chemicals on the floor of the storeroom.
4. Do not store chemicals above eye level or on the top shelf in the storeroom.
5. Be sure shelf assemblies are firmly secured to walls.
6. Provide antiroll lips for all shelves.
7. Use shelving constructed of wood. Metal cabinets and shelves are easily corroded.
8. Avoid metal adjustable shelf supports and clips. They can corrode, causing shelves to collapse.
9. Store acids in their own locking storage cabinet.
10. Store flammables in their own locking storage cabinet.
11. Store poisons in their own locking storage cabinet.
12. Store oxidizers by classification, preferably in their own locking storage cabinets.

Additional Resources

Your school district may have more information on safety issues. Some districts have a safety officer responsible for safety throughout their schools. Other possible sources for information include your state education agency, teachers' and science teachers' associations, and local colleges or universities.

The **American Chemical Society Health and Safety Service** will refer inquiries about health and safety to appropriate resources.

American Chemical Society (ACS)
1155 Sixteenth Street, N.W.
Washington, D.C. 20036
(202) 872-4511

Hazardous Materials Information Exchange (HMIX)

Sponsored by the Federal Emergency Management Agency and the United States Department of Transportation, HMIX serves as a reliable on-line database. It can be accessed through an electronic bulletin board, and it provides information regarding instructional material and literature listings, hazardous materials, emergency procedures, and applicable laws and regulations.

HMIX can be accessed by a personal computer with a modem. Dial (312) 972-3275. The bulletin board is available 24 hours a day, seven days a week. The service is available free of charge. You pay only for the telephone call.

Safety Reference Works

Gessner, G. H., ed. *Hawley's Condensed Chemical Dictionary* (11th ed.). New York: Van Nostrand Reinhold, 1987.

A Guide to Information Sources Related to the Safety and Management of Laboratory Wastes from Secondary Schools. New York State Environmental Facilities Corp., 1985.

Lefevre, M. J. *The First Aid Manual for Chemical Accidents.* Stroudsberg, Pa: Dowdwen, 1989.

Pipitone, D., ed. *Safe Storage of Laboratory Chemicals.* New York: John Wiley, 1984.

Prudent Practices for Disposal of Chemicals from Laboratories. Washington, D.C.: National Academy Press, 1983.

Prudent Practices for Handling Hazardous Chemicals in Laboratories. Washington, D.C.: National Academy Press, 1981.

Strauss, H., and M. Kaufman, eds. *Handbook for Chemical Technicians.* New York: McGraw-Hill, 1981.

WARD'S MSDS Database and User's Guide. WARD'S CD-ROM, Catalog Number 74T5070.

Release and Disposal of Organisms

The ultimate responsibility falls on the teacher to ensure that each organism brought into the classroom receives adequate care during its stay, release, and final disposition.

The following information is provided to help you make informed decisions regarding the organisms used in the *Laboratory Techniques and Experimental Design* labs.

General Guidelines

- Aquatic Plants—Certain aquatic plants (*Elodea,* among others) should not be released or introduced into local habitats. *Elodea* is regulated as a pest organism in Canada and in parts of the northern United States.

- Insects—Permits are required to possess or release certain insects (cockroaches and termites). Check with your local office of the Animal and Plant Health Inspection Service (APHIS), U.S. Department of Agriculture, or contact WARD'S.

- Microbes—Bacteria, fungi, yeast, and growth media or materials that have come into contact with these organisms should not be discarded without decontamination (sterilization). See: "Safety With Microbes" in the *Laboratory Safety* section.

  Aquasprillum serpens—Destroy by autoclaving or disinfecting with household bleach (5% sodium hypochlorite).

  Bacillus cereus—Destroy by autoclaving or disinfecting with household bleach (5% sodium hypochlorite).

  Bacillus megaterium—Destroy by autoclaving or disinfecting with household bleach (5% sodium hypochlorite).

  Micrococcus luteus—Destroy by autoclaving or disinfecting with household bleach (5% sodium hypochlorite).

  Penicillium notatum—Destroy by autoclaving or disinfecting with household bleach (5% sodium hypochlorite).

  Staphylococcus epidermidis —Destroy by autoclaving or disinfecting with household bleach (5% sodium hypochlorite).

  Saccharomyces (dry; UV-sensitive yeast)—Destroy by autoclaving or disinfecting with household bleach (5% sodium hypochlorite).

- Microinvertebrates/Protists—These may be freely released in aquatic environments. Do not release nematodes unless specifically directed to do so.

Amoeba	*Paramecium*
Euglena	*Volvox*

- Macroinvertebrates/Vertebrates—Exotic, or nonindigenous forms, should not be released into native habitats. In many cases, these organisms may not survive climatic conditions or may interfere with native fauna and flora. Contact your local APHIS office or WARD'S for specific information about whether a particular organism would be considered nonindigenous to *your* area. In some cases, you may be able to find a home for an animal at a local pet shop.

   Owl pellets—Assure that these articles have been steam sterilized and are free from contagion prior to use. Dispose as inert solid waste.

• Plants—Locally cultivated native plants may be introduced by replanting. Ornamental plants that are not cultivated locally should not be introduced into a native habitat. Some states (including California) have strict rules regarding the procurement or introduction of nonindigenous plants because of the possible presence of root-damaging nematodes. Usually, plants shipped to these states must pass inspection.

   Moss clump—May be freely transplanted.
   Corn seedlings—May be freely transplanted.
   Rapid Radish plants—Discard.
   Tobacco plants infected with TMV—Discard.

Chemical Handling and Disposal

This information is furnished without warranty of any kind. Teachers should use it only as a supplement to other information they have and should make independent determinations of its suitability and completeness as it relates to their own district guidelines.

NONHAZARDOUS CHEMICALS

The following materials are considered nonhazardous; they do not meet the published criteria of established hazard characteristics or are not specifically regulated as hazardous substances.

- **STORAGE:** GREEN (general storage)
- **DISPOSAL:** LOW HAZARD for laboratory handling. Avoid creating dust when working with these materials. They may be disposed of as an inert solid or as liquid waste.
  - Carborundum powder—avoid dusting conditions
  - Table salt
  - Sugar

HAZARDOUS CHEMICALS

Before using or disposing of any of the materials listed below, familiarize yourself with the safety and handling procedures and storage information listed under "Reagents and Storage" in the *Laboratory Safety* section. Also refer to individual reagent labels and Material Safety Data Sheets for further information about hazards and precautions.

Container Labeling

Ensure that each container used by students in the laboratory is properly labeled with the following information:

- the name of the material and its concentration (if a solution)
- the names of individual components and their respective concentrations (if a mixture)
- the appropriate SIGNAL WORD
- a declarative statement of potential hazards
- immediate first-aid measures

Example:

> Lugol's Iodine Solution
> **WARNING: Poison if ingested. Irritant.**
> Do not ingest. Avoid skin and eye contact.
> Flush spills and splashes with water for 15 minutes; rinse mouth with water.
> Call your teacher immediately.

You should be aware of local, state, and federal regulations governing the disposal of hazardous materials. Contact a licensed Treatment, Storage, and Disposal (TSD) facility for disposal of large quantities of hazardous chemicals. Disposal protocols outlined below are ONLY for the substances (and quantities) specified.

Unless your school's drains are connected to a sanitary sewer system, no chemicals should ever be flushed down the drain. Never pour chemicals or reagents down the drain if you have a septic system. Even if you are connected to a sanitary sewer, do not pour any chemical down the drain unless you are certain it is safe and permitted.

The chemicals and reagents are classified by a color code to indicate hazard level. Remember to store chemicals of different hazard classes away from each other. YELLOW: Reactives, RED: Flammables, WHITE: Corrosives, BLUE: Toxics (poisons), GREEN: Low hazard for laboratory use.

Disposal Method A—Inert Solid Wastes (Low Hazard)

Items identified by *DISPOSAL: METHOD A* can be considered to be a LOW HAZARD for laboratory handling and disposal. Avoid creating or breathing dust from these materials. If necessary, seal these materials in a bag or other suitable container. These materials generally have a GREEN storage code (general storage). It is recommended that you check local, state, and federal regulations to be certain that these articles may be disposed of in a sanitary landfill.

Disposal Method B—Small Quantities of Liquid Wastes (Low Hazard)

Items identified by *DISPOSAL: METHOD B* can be considered to be a LOW HAZARD for laboratory handling and disposal. Wear the following PPE (personal protective equipment) when handling or disposing of these items: chemical safety goggles, apron, and polyethylene or nitrile gloves. Work in an area near an eyewash station. Use the following method to dispose of volumes of no more than 250 mL (unless specifically stated). Dilute the volume of waste material with 20 times as much tap water. Test pH with litmus or other indicator; if necessary, adjust pH to neutrality by adding small amounts of 1 M acid, base, or other reagent as required. (Exceptions are stated in the list below.) Place a beaker containing the diluted mixture in the sink, and run water to overflowing for 10 minutes, flushing to a known sanitary sewer. These materials generally have a GREEN storage code (general storage). It is recommended that you check local, state, and federal regulations to be certain that these articles may be disposed of in a sanitary landfill.

Aceto-orcein biological stain

- **CAUTION: Strong irritant/Poison.** Avoid skin and eye contact; can stain skin and clothing. In case of contact, flush affected areas with water for 15 minutes, including under the eyelids; rinse mouth with water. Avoid inhalation of vapors. Get immediate medical attention. (Contains 45% glacial acetic acid.)
- **PPE:** Chemical safety goggles, nitrile or polyethylene gloves, apron
- **STORAGE:** WHITE (corrosive liquid)
- **DISPOSAL:** Wear PPE: Chemical safety goggles, face shield, nitrile gloves, apron. Neutralize only small amounts (less than 250 mL) at any one time. Slowly add 1 M sodium bicarbonate solution to the acid until neutrality is confirmed by a litmus (pH) test. Place a beaker of the neutralized solution in a sink, and run water to overflowing for 10 minutes, flushing to a sanitary sewer.

L-Ascorbic acid (Vitamin C) *NOT FOR STUDENT USE*

• **STORAGE:** GREEN (general storage)

L-Ascorbic acid (Vitamin C) solutions (0.1%, 0.05%)

• **PREPARATION:** Avoid dusting conditions.

  0.05% solution—Carefully dissolve 0.05 g of ascorbic acid in 100 mL of distilled water.
  0.1% solution—Carefully dissolve 0.1 g of ascorbic acid in 100 mL of distilled water.
  Prepare before use where possible.
• **STORAGE:** GREEN (general storage)
• **DISPOSAL:** METHOD B

Benedict's solution

• **CAUTION: Irritant/Poison.** Avoid skin and eye contact; do not ingest. Prolonged or repeated contact may cause skin irritation. May be harmful if swallowed. If contact occurs, flush affected areas for 15 minutes, including under the eyelids; rinse mouth with water.
• **PPE:** Chemical safety goggles, nitrile or polyethylene gloves, apron
• **PREPARATION:** Wear PPE. Heat to dissolve 173 g of sodium citrate and 100 g of anhydrous sodium carbonate in 800 mL distilled water. Filter and dilute to 850 mL. Dissolve 17.3 g of crystalline copper sulfate, $CuSO_4$, in 100 mL of distilled water. Pour the $CuSO_4$ solution into the carbonate citrate solution, stir, and add enough water to make 1 L. Label as follows: *CAUTION—Irritant*
• **STORAGE:** BLUE (poison)
• **DISPOSAL:** METHOD B

WARD'S Blood, human—simulated

• **CAUTION:** Mild irritant. Avoid skin and eye contact; can stain skin and clothing. In case of contact, flush affected areas with water for 15 minutes, including under the eyelids; rinse mouth with water.
• **STORAGE:** GREEN (general storage)
• **DISPOSAL:** METHOD B

Bleach (5% sodium hypochlorite solution) *NOT FOR STUDENT USE*

• **CAUTION: Reactive material; strong oxidant.** Strong irritant. Avoid skin and eye contact; avoid vapor inhalation. Vapors are irritating to the upper respiratory tract; prolonged inhalation may cause edema. In case of contact, flush affected areas with water for 15 minutes, including under the eyelids; get medical attention if redness or irritation persists. If ingested, give one or two glasses of water if conscious; contact a physician.
• **PPE:** Chemical safety goggles, nitrile gloves, apron. Working under a chemical fume hood is recommended.
• **STORAGE:** YELLOW (reactive material) Do not mix with acids or other oxidizers.
• **DISPOSAL:** METHOD B

Biuret solution

• **CAUTION: Strong irritant.** Avoid skin and eye contact. In case of contact, flush affected areas with water for 15 minutes, including under the eyelids; rinse mouth with water. Get immediate medical attention if irritation symptoms following contact persist.
• **PPE:** Chemical safety goggles, nitrile or polyethylene gloves, apron
• **PREPARATION:** Wear PPE: dissolve 3 g of copper sulfate and 12 g of potassium sodium tartrate in 1 L of distilled water. Slowly add 600 mL of 10% sodium hydroxide while constantly stirring. Label as follows: *CAUTION—Strong Irritant*
• **DISPOSAL:** Wear PPE. Place 250 mL in a 1 L beaker, and dilute with water to 750 mL. Place the beaker in a sink, and run water to overflowing for 10 minutes.

Chemical Handling and Disposal

Caffeine solution (0.5%)
- **CAUTION:** Irritant. Avoid skin and eye contact; do not ingest. If contact occurs, flush affected areas for 15 minutes; rinse mouth with water.
- **PPE:** Chemical safety goggles, nitrile or polyethylene gloves, apron
- **PREPARATION:** Wear PPE. Avoid dusting conditions. Carefully dissolve 0.5 g of caffeine in 100 mL of distilled water. Prepare before use where possible.
- **STORAGE:** GREEN (general storage) Extend shelf life by storing in amber bottles under refrigeration.
- **DISPOSAL:** METHOD B

Calcium hydroxide NOT FOR STUDENT USE
- **STORAGE:** WHITE (corrosive solid) Keep container tightly closed.

Catalase—concentrate (hydrogen peroxide oxidoreductance)
- **CAUTION:** Irritant. Avoid contact with skin and eyes. Do not inhale mist; potential strong allergen. Do not ingest.
- **PPE:** Chemical safety goggles, nitrile or polyethylene gloves, apron
- **STORAGE:** GREEN (general storage) Store in amber bottle; keep away from light. Refrigeration recommended to increase shelf life.
- **DISPOSAL:** METHOD B

WARD'S Chlorophyll extract—simulated
- **CAUTION:** Mild Irritant. Avoid skin and eye contact; can stain skin and clothing. In case of contact, flush affected areas with water for 15 minutes, including under the eyelids; rinse mouth with water.
- **STORAGE:** GREEN (general storage)
- **DISPOSAL:** METHOD B

WARD'S Chromatography solvent
- **STORAGE:** GREEN (general storage)
- **DISPOSAL:** METHOD B

Crystal violet stain (1.6%)
- **CAUTION:** Irritant. Avoid skin and eye contact; can stain skin and clothing. In case of contact, flush affected areas with water for 15 minutes, including under the eyelids; rinse mouth with water.
- **PREPARATION:** Avoid dusting conditions. Dissolve 1.6 g of crystal violet stain powder into 10 mL of denatured ethyl alcohol; to this add 90 mL of distilled water. Label as follows: *CAUTION—Irritant.*
- **PPE:** Chemical safety goggles, polyethylene gloves, apron
- **DISPOSAL:** METHOD B

WARD'S Detain™ (nontoxic)
- **STORAGE:** GREEN (general storage)
- **DISPOSAL:** METHOD B

Disinfectant solution
- **CAUTION:** Irritant. Avoid contact with skin and eyes. Do not inhale mist; potential strong allergen. Do not ingest.
- **PPE:** Chemical safety goggles, nitrile or polyethylene gloves, apron
- **PREPARATION:** Add 100 mL of household bleach solution to 900 mL of distilled water.
- **STORAGE:** GREEN (general storage) Store in amber bottles; keep away from light.
- **DISPOSAL:** METHOD B

Enzyme concentrate (catalase)

- **CAUTION:** Irritant. Avoid contact with skin and eyes. Do not inhale mist; potential strong allergen. Do not ingest.
- **PPE:** Chemical safety goggles, nitrile or polyethylene gloves, apron
- **PREPARATION:** Wear PPE. Carefully add 1 mL of catalase concentrate to 700 mL of distilled water. Dispense to an amber dropping bottle. Prepare this solution just before use.
- **STORAGE:** GREEN (general storage) Store in amber bottles; keep away from light. Refrigeration required to increase shelf life.
- **DISPOSAL:** METHOD B

Ethyl alcohol (70%)

- **WARNING: Flammable liquid.** Avoid open flames, excessive heat, sparks, and other potential ignition sources; do not ingest; avoid eye contact; avoid prolonged skin contact. In case of contact, flush affected areas with water for 15 minutes; including under the eyelids; rinse mouth with water. Get prompt medical attention.
- **PPE:** Chemical safety goggles, nitrile or polyethylene gloves, apron
- **STORAGE:** RED (flammable liquid)
- **DISPOSAL:** Wear PPE. Dilute small volumes (less than 250 mL) in a ratio of 1 part solution to 20 parts water. Place a beaker of the diluted solution in the sink, and run water to overflowing, flushing to a sanitary sewer.

Ethyl alcohol (95%) NOT FOR STUDENT USE

- **WARNING: Flammable liquid.** Avoid open flames, excessive heat, sparks, and other potential ignition sources; do not ingest; avoid eye contact; avoid prolonged skin contact. In case of contact, flush affected areas with water for 15 minutes; including under the eyelids; rinse mouth with water. Get prompt medical attention.
- **PPE:** Chemical safety goggles, nitrile or polyethylene gloves, apron
- **STORAGE:** RED (flammable liquid)
- **DISPOSAL:** Wear PPE. Dilute small volumes (less than 250 mL) in a ratio of 1 part solution to 20 parts water. Place a beaker of the diluted solution in a sink, and run water to overflowing, flushing to a sanitary sewer.

Fertilizers, plant

- **PPE:** Chemical safety goggles, nitrile or polyethylene gloves, apron
- **STORAGE:** GREEN (general storage)
- **DISPOSAL:** METHOD B

Fructose solution (20%)

- **PREPARATION:** Avoid dusting conditions. Carefully dissolve 100 g of fructose in 400 mL of distilled water; add distilled water to make a final volume of 500 mL.
- **STORAGE:** GREEN (general storage)
- **DISPOSAL:** METHOD B

Giemsa stain

- **CAUTION: Combustible liquid; irritant; poison.** Avoid eye and skin contact; do not ingest. In case of contact, flush affected areas with water for 15 minutes, including under the eyelids; rinse mouth with water. Get prompt medical attention.
- **PPE:** Chemical safety goggles, polyethylene gloves, apron
- **STORAGE:** RED (combustible liquid) Avoid contact with strong oxidizers and halogenated solvents.
- **DISPOSAL:** Wear PPE. Dilute small volumes (less than 250 mL) in a ratio of 1 part solution to 20 parts water. Place a beaker of the diluted solution in a sink, and run water to overflowing, flushing to a sanitary sewer.

Gram's iodine stain

- **CAUTION: Irritant; poison.** Avoid eye and skin contact; do not ingest. In case of contact, flush affected areas with water for 15 minutes, including under the eyelids; rinse mouth with water. Get prompt medical attention.
- **PPE:** Chemical safety goggles, polyethylene gloves, apron
- **PREPARATION:** Wear PPE. Dissolve 1 g of iodine and 2 g of potassium iodide in 300 mL of distilled water. Label as follows: *CAUTION—Irritant /Poison.*
- **STORAGE:** BLUE (poison) Store in amber bottles in a cool environment, isolated from all corrosives, acids, flammables, and combustibles.
- **DISPOSAL:** Wear PPE: Chemical safety goggles, face shield, nitrile gloves, apron. Add 0.1 M sodium thiosulfate solution slowly to a small quantity (less than 50 mL) of Gram's stain, and mix well until all iodine has reacted and the solution is colorless, with no solids visible. (This will require several liters of sodium thiosulfate solution in most cases.) If necessary, neutralize with 1 M sodium bicarbonate solution. Ensure neutrality with a litmus (pH) test. Place a beaker of the neutralized and decolorized solution in a sink, and run water to overflowing for 10 minutes, flushing to a sanitary sewer.

Hydrochloric acid—concentrate

- **STORAGE:** WHITE (corrosive liquid)

Hydrochloric acid (1 M)

- **CAUTION:** Irritant. Avoid skin and eye contact; do not ingest. If contact occurs, flush affected areas for 15 minutes; rinse mouth with water.
- **PPE:** Chemical safety goggles, nitrile or polyethylene gloves, apron
- **PREPARATION:** Wear PPE: Chemical goggles, face shield, nitrile gloves, apron; work in chemical fume hood; have an eyewash station within a 15-second walk. *Slowly* add 82 mL of concentrated HCl to 500 mL of distilled water in a volumetric flask, and fill to the 1000 mL mark.
- **STORAGE:** GREEN (general storage)
- **DISPOSAL:** Wear PPE: Chemical safety goggles, face shield, nitrile gloves, apron. Neutralize only small amounts (less than 250 mL) at any one time. Slowly add 1 M sodium bicarbonate solution to the acid until neutrality is confirmed by a litmus (pH) test. Place a beaker of the neutralized solution in a sink, and run water to overflowing for 10 minutes, flushing to a sanitary sewer.

Hydrogen peroxide solutions (3%, 0.03%)

- **CAUTION: Reactive;** irritant. Avoid mixing this chemical with any other chemical; avoid skin and eye contact. In case of contact, flush affected areas with water for 15 minutes, including under the eyelids; rinse mouth with water.
- **PPE:** Chemical safety goggles, nitrile or polyethylene gloves, apron
- **PREPARATION:** 0.03% solution—Wear PPE. Carefully add 33 mL of hydrogen peroxide solution (3%) to 1 L of distilled water. Prepare this solution just before use.
- **STORAGE:** YELLOW (reactive substance) Store in amber bottles; keep away from light.
- **DISPOSAL:** 3% solution—Wear PPE. Place a small amount (less than 50 mL) in a beaker in a sink, and run water to overflowing for 10 minutes, flushing to a sanitary sewer. 0.03% solution— METHOD B

Indophenol, sodium salt (2,6 Dichloroindophenol, sodium salt) NOT FOR STUDENT USE

- **STORAGE:** GREEN (general storage) Avoid oxidizers

Indophenol solution (0.1%)

- **CAUTION:** Irritant. Avoid skin and eye contact; do not ingest. If contact occurs, flush affected areas for 15 minutes; rinse mouth with water.
- **PPE:** Chemical safety goggles, nitrile or polyethylene gloves, apron
- **PREPARATION:** Wear PPE. Avoid dusting conditions. Carefully dissolve 0.1 g of 2,6 Dichloroindophenol, sodium salt (Indophenol, sodium salt) in 100 mL of distilled water. Prepare before use when possible.
- **STORAGE:** GREEN (general storage) Extend shelf life by storing in amber bottles under refrigeration.
- **DISPOSAL:** METHOD B

Iodine *NOT FOR STUDENT USE*

- **STORAGE:** BLUE (poison). Store in a cool, dry location away from combustibles, organics, or other oxidizers.

Lactose solution (20%)

- **PREPARATION:** Avoid dusting conditions. Carefully dissolve 100 g of lactose in 400 mL of distilled water; add distilled water to a final volume of 500 mL.
- **STORAGE:** GREEN (general storage)
- **DISPOSAL:** METHOD B

Limewater (0.14% calcium hydroxide solution)

- **CAUTION:** Irritant. Avoid skin and eye contact; do not ingest. If contact occurs, flush affected areas for 15 minutes; rinse mouth with water.
- **PPE:** Chemical safety goggles, nitrile or polyethylene gloves, apron
- **PREPARATION:** Wear PPE. Avoid dusts. Carefully dissolve 140 mg of calcium hydroxide in 100 mL of distilled water. Prepare prior to use. Label as follows: *CAUTION—Irritant.*
- **STORAGE:** GREEN (general storage) Keep container tightly closed; absorbs carbon dioxide.
- **DISPOSAL:** METHOD B

Lugol's iodine solution

- **CAUTION:** Poison; irritant. Avoid eye and skin contact; do not ingest. This material will stain skin and clothing. In case of contact: *Ingestion:* If swallowed, give one or two glasses of water to drink if conscious. Induce vomiting. Follow by giving starch solution, milk, egg white, or gruel. Contact a physician immediately. *Eye Contact:* Flush eyes, including under the eyelids, thoroughly under running water for 15 minutes. Get immediate medical attention. *Inhalation:* Remove to fresh air; get medical attention if overexposure symptoms persist. *Skin:* Flush thoroughly with water. Contact a physician if irritation develops.
- **PPE:** Chemical safety goggles, polyethylene gloves, apron
- **PREPARATION:** Additional protection for preparation and disposal: face shield, nitrile rubber gloves, fume hood, eyewash station in close proximity to dispensing activities. See remarks under Iodine before preparing. Dissolve 10 g of potassium iodide in 100 mL distilled water. To this solution add 5 g of iodine crystals. Dilute fivefold with distilled water. Label as follows: *WARNING—Poison if ingested; Irritant*
- **STORAGE:** BLUE (poison) Store in amber bottles, away from direct sunlight. Avoid oxidizing materials.
- **DISPOSAL:** Wear PPE. Place 250 mL in a 1 L beaker. Slowly add 0.1 M sodium thiosulfate, and mix until the solution is decolorized (fully reduced). Place a beaker of the decolorized solution in a sink, and run water to overflowing for 10 minutes, flushing to a sanitary sewer.

WARD'S Mounting medium (Piccolyte II)

- **CAUTION: Combustible liquid;** irritant. Avoid skin and eye contact, and open flames or ignition sources; do not ingest. If contact occurs, flush affected areas for 15 minutes; rinse mouth with water.

- **PPE:** Chemical safety goggles, polyethylene gloves, apron
- **STORAGE:** RED (combustible liquid)
- **DISPOSAL:** Open container to air inside a fume hood and evaporate mixture to dryness; dispose of as an insert solid waste. Volatiles include food oil distillates.

Nitrate pollutant (0.5% urea nitrogen)

- **CAUTION:** Mild irritant. Avoid skin and eye contact; do not ingest. If contact occurs, flush affected areas for 15 minutes; rinse mouth with water.
- **PPE:** Chemical safety goggles, apron
- **STORAGE:** GREEN (general storage)
- **DISPOSAL:** METHOD B

Nutrient agar (plates)

- **PREPARATION:** Dissolve 23 g of nutrient agar dehydrated microbiological media in 1 L of distilled water. Bring the mixture to a boil, and dispense into bottles or tubes. Autoclave at 121°C and 15–18 psi for 15–20 minutes. Allow to cool. To dispense, melt in a microwave in 30-second heating cycles (make sure caps are open to allow gas release) and dispense 15–20 mL into sterile petri dishes. Allow to cool. Refrigerate upside down to avoid condensation contamination.
- **STORAGE:** GREEN (general storage) Store under refrigeration (6°C) until needed.
- **DISPOSAL:** LOW HAZARD for laboratory handling when sterile. Autoclave or steam-sterilize (121°C and 15 psi for 15–20 minutes) contaminated microbial material. The plates may also be chemically sterilized by applying a thin layer of household bleach (5% sodium hypochlorite) to the surface of a plate, waiting 20 minutes, taping plates, and discarding as inert solid waste.

Nutrient broth (prepared, sterile)

- **STORAGE:** GREEN (general storage); store under refrigeration (6°C) until needed.
- **DISPOSAL:** LOW HAZARD for laboratory handling when sterile. Autoclave or steam-sterilize (121°C and 15 psi for 15–20 minutes) contaminated material. May also be chemically sterilized by applying a thin layer of household bleach (5% sodium hypochlorite) to the surface of a plate, waiting 20 minutes, taping plates, and discarding as inert solid waste.

Onion root tip storage medium (83% isopropyl alcohol)

- **WARNING: Flammable liquid.** Avoid open flames, excessive heat, sparks, and other potential ignition sources; do not ingest; avoid eye contact; avoid prolonged skin contact. In case of contact, flush affected areas with water for 15 minutes; including under the eyelids; rinse mouth with water. Get prompt medical attention.
- **PPE:** Chemical safety goggles, nitrile or polyethylene gloves; apron
- **STORAGE:** RED (flammable liquid)
- **DISPOSAL:** Wear PPE. Dilute small volumes (less than 50 mL) in a ratio of 1 part solution to 20 parts water. Place a beaker of the diluted solution in a sink, and run water to overflowing, flushing to a sanitary sewer.

Phosphate buffers

- **PPE:** Chemical safety goggles, apron
- **PREPARATION:** Wear PPE. Avoid creating dusts.
- **STORAGE:** GREEN (general storage)
- **DISPOSAL:** METHOD B

Phosphate pollutant (0.5% monosodium phosphate)

- **CAUTION:** Mild irritant. Avoid skin and eye contact; do not ingest. If contact occurs, flush affected areas for 15 minutes; rinse mouth with water.
- **PPE:** Chemical safety goggles, apron

- **STORAGE:** GREEN (general storage)
- **DISPOSAL:** METHOD B

Potassium iodide *NOT FOR STUDENT USE*

- **STORAGE:** BLUE (poison) Store in a cool, dry location away from combustibles, organics, or other oxidizers.

Potassium phosphate solution (0.1 M)

- **PPE:** Chemical safety goggles, polyethylene gloves, apron
- **PREPARATION:** Wear PPE. Avoid creating dusts. Dissolve 17.4 g of potassium hydrogen phosphate in 500 mL of distilled water in a volumetric flask, and fill to the 1 L mark.
- **STORAGE:** GREEN (general storage)
- **DISPOSAL:** METHOD B

- **WARNING:** Flammable liquid; irritant. Avoid skin and eye contact, and open flames or ignition sources; do not ingest. If contact occurs, flush affected areas for 15 minutes; rinse mouth with water.
- **PPE:** Chemical safety goggles, polyethylene gloves, apron
- **PREPARATION:** Wear PPE: chemical safety goggles, face shield, nitrile gloves, apron. Work in a chemical fume hood; have an eyewash station within a 15-second walking distance. Add 25 mL of glacial acetic acid to 75 mL of 95% ethanol.
- **STORAGE:** RED (flammable liquid)
- **DISPOSAL:** Wear PPE. Dilute small volumes (less than 250 mL) in a ratio of 1 part solution to 20 parts water. Place a beaker of the diluted solution in a sink, and run water to overflowing, flushing to a sanitary sewer.

Root tip fixative—acetic acid/alcohol (Carnoy's fixative)

- **WARNING: Flammable liquid;** irritant. Avoid skin and eye contact, and open flames or ignition sources; do not ingest. If contact occurs, flush affected areas for 15 minutes; rinse mouth with water.
- **PPE:** Chemical safety goggles, polyethylene gloves, apron
- **PREPARATION:** Wear PPE: chemical safety goggles, face shield, nitrile gloves, apron. Work in a chemical fume hood; have an eyewash station within a 15-second walking distance. Add 25 mL of glacial acetic acid to 75 mL of 95% ethanol.
- **STORAGE:** RED (flammable liquid)
- **DISPOSAL:** Wear PPE. Dilute small volumes (less than 250 mL) in a ratio of 1 part solution to 20 parts water. Place a beaker of the diluted solution in a sink, and run water to overflowing, flushing to a sanitary sewer.

Safranin O powder *NOT FOR STUDENT USE*

- **STORAGE:** GREEN (general storage)

Safranin stain solution—aqueous (0.1%)

- **CAUTION:** Irritant. Avoid skin and eye contact; can stain skin and clothing. In case of contact, flush affected areas with water for 15 minutes, including under the eyelids; rinse mouth with water.
- **PREPARATION:** Avoid dusting conditions, dissolve 0.1 g of Safranin O powder in 100 mL of distilled water. Label as follows: *CAUTION—Irritant.*
- **PPE:** Chemical safety goggles, polyethylene gloves, apron
- **DISPOSAL:** METHOD B

Chemical Handling and Disposal

WARD'S Sera, animal—simulated

- **STORAGE:** GREEN (general storage)
- **DISPOSAL:** METHOD B

Sodium bicarbonate NOT FOR STUDENT USE

- **STORAGE:** GREEN (general storage)

Sodium bicarbonate (1 M)—for neutralizing acids NOT FOR STUDENT USE

- **PPE:** Chemical safety goggles, polyethylene gloves, apron
- **PREPARATION:** Wear PPE. Avoid creating dusts. Dissolve 84 g in 500 mL of distilled water in a volumetric flask, and fill to the 1 L mark.
- **USE & DISPOSAL:** Wear PPE: chemical safety goggles, face shield, nitrile gloves, apron. Add this solution slowly to acidic solutions until neutrality is confirmed by a litmus (pH) test. Never neutralize an acid solution more concentrated than 1 M. Place a beaker of the neutralized solution in a sink, and run water to overflowing for 10 minutes, flushing to a sanitary sewer.

Sodium chloride solution (0.15 M)

- **PPE:** Chemical safety goggles, polyethylene gloves, apron
- **PREPARATION:** Wear PPE. Avoid creating dusts. Dissolve 8.7 g in 500 mL of distilled water in a volumetric flask, and fill to the 1 L mark.
- **STORAGE:** GREEN (general storage)
- **DISPOSAL:** METHOD B

Sodium phosphate, monobasic (1.8%)

- **PPE:** Chemical safety goggles, apron
- **PREPARATION:** Wear PPE. Avoid creating dusts. Add 1.8 g of sodium dihydrogen phosphate to 100 mL of distilled water.
- **STORAGE:** GREEN (general storage)
- **DISPOSAL:** METHOD B

Sodium phosphate, dibasic (0.1 M)

- **PPE:** Chemical safety goggles, apron
- **PREPARATION:** Wear PPE. Avoid creating dusts. Dissolve 17.4 g of disodium hydrogen phosphate in 500 mL of distilled water in a volumetric flask. Then add distilled water to the 1000 mL mark.
- **STORAGE:** GREEN (general storage)
- **DISPOSAL:** METHOD B

Sodium thiosulfate pentahydrate (hypo) solution (0.1 M)—for neutralizing Lugol's

- **PREPARATION:** Wear PPE: goggles, apron, nitrile gloves. Avoid dusting conditions; avoid breathing dust. Dissolve 25 g of crystals in 500 mL of distilled water in a volumetric flask. Then add distilled water to the 1000 mL mark. Label as follows: *CAUTION—Irritant.*
- **STORAGE:** GREEN (general storage)
- **DISPOSAL:** Wear PPE: goggles, apron, nitrile gloves. Flush to a sanitary sewer with a copious amount of running water for 10 minutes.

Sucrose solution (20%)

- **PREPARATION:** Avoid dusting conditions. Carefully dissolve 100 g of sucrose in 400 mL of distilled water, and add distilled water to a final volume of 500 mL.
- **STORAGE:** GREEN (general storage)
- **DISPOSAL:** METHOD B

Sudan III stain (0.5% solution)

- **WARNING:** Flammable liquid. Avoid open flames, excessive heat, sparks, and other potential ignition sources; do not ingest; avoid eye contact; avoid prolonged skin contact. In case of contact, flush affected areas with water for 15 minutes; including under the eyelids; rinse mouth with water. Get prompt medical attention.
- **PREPARATION:** Wear PPE. Avoid creating dusts. Dissolve 0.5 g of Sudan III powdered stain in 100 mL of 99% denatured ethyl alcohol. Dispense to amber bottle; protect from light.
- **PPE:** Chemical safety goggles, nitrile or polyethylene gloves, apron
- **STORAGE:** RED (flammable liquid) Avoid strong oxidizers.
- **DISPOSAL:** Wear PPE. Dilute small volumes (less than 250 mL) in a ratio of 1 part solution to 20 parts water. Place a beaker of the diluted solution in a sink, and run water to overflowing, flushing to a sanitary sewer.

Trichrome stain

- **CAUTION:** Irritant. Avoid skin and eye contact; can stain skin and clothing. In case of contact, flush affected areas with water for 15 minutes, including under the eyelids; rinse mouth with water.
- **PPE:** Chemical safety goggles, polyethylene gloves, apron
- **DISPOSAL:** Wear PPE: Chemical safety goggles, face shield, nitrile gloves, apron. Neutralize only small amounts (less than 250 mL) at any one time. Slowly add 1 M sodium bicarbonate solution to the acid until neutrality is confirmed by a litmus (pH) test. Place a beaker of the neutralized solution in a sink, and run water to overflowing for 10 minutes, flushing to a sanitary sewer.

Trypsin solution (2.5%)

- **PREPARATION:** Add 5 mL of distilled water to a vial of powdered Bacto-trypsin. Mix well.
- **STORAGE:** GREEN (general storage) Store hydrated solution in amber bottles; store powder and solution under refrigeration.
- **DISPOSAL:** METHOD B

Tryptic soy agar (plates)

- **PREPARATION:** Dissolve 40 g of tryptic soy agar dehydrated microbiological media in 1 L of distilled water. Bring mixture to a boil, and dispense into bottles or tubes. Autoclave at 121°C and 15–18 psi for 15–20 minutes. Allow to cool. To dispense, melt in a microwave in 30-second heating cycles (make sure caps are open to allow gas release) and dispense 15–20 mL into sterile petri dishes. Allow to cool; refrigerate upside down to avoid condensation contamination.
- **STORAGE:** GREEN (general storage); store under refrigeration (6°C) until needed.
- **DISPOSAL:** LOW HAZARD for laboratory handling when sterile. Autoclave or steam-sterilize (121°C and 15 psi for 15–20 minutes) contaminated material. The plates may also be chemically sterilized by applying a thin layer of household bleach (5% sodium hypochlorite), waiting 20 minutes, taping the plates, and discarding them as inert solid waste.

WARD'S Urine samples—simulated

- **CAUTION:** Mild irritant. Avoid skin and eye contact; can stain skin and clothing. In case of contact, flush affected areas with water for 15 minutes, including under the eyelids; rinse mouth with water.
- **STORAGE:** GREEN (general storage)
- **DISPOSAL:** METHOD B

Master Materials List, by Category

This materials cross-reference guide was prepared by WARD'S Natural Science, the preferred science supplier for the Holt BioSources Lab Program published by Holt, Rinehart and Winston.

Materials are grouped into five categories: Biological Supplies, Chemicals and Media, Laboratory Equipment, Kits, and Miscellaneous. Each entry is listed alphabetically, followed by package size and the WARD'S catalog number. The second column gives the number of the lab in which the material is used. A list of the materials needed for each lab follows.

WARD'S also has available a convenient and easy computer software-ordering system specifically designed for use with the Holt BioSources Lab Program. The software-ordering system lists all required and supplemental materials needed for every lab. Click on the products you need, and the software automatically creates your shopping list, keeping track of the materials you ordered and their costs. The software-ordering system is available for both Macintosh and IBM-compatible computers. Call WARD'S at 1-800-962-2660 for your free copy.

To order, or for questions concerning the use of WARD'S materials, call toll-free 1-800-962-2660 or fax WARD'S at 1-800-635-8439.

WARD'S™ WARD'S Natural Science Establishment, Inc.
5100 W. Henrietta Road
P.O. Box 92912
Rochester, NY 14692-9012
Internet address: http://www.wardsci.com
E-mail address: customer_service@wardsci.com

Biological Supplies

Item	Lab
Albino corn seed, pkt of 100 (86 T 8085)	**C12**
Animal hair (local)	**C2**
Aquaspirillum serpens, culture (85 T 1177)	**C32**
Bacillus megaterium, culture (85 T 1154)	**C32, C33**
Cells, fixed for karyotyping, vial (36 T 6027)	**C13**
Chlorella, culture (86 T 0126)	**C27**
Dicot stem CS, vial (63 T 1062)	**C35**
Dirt, unsterilized (local)	**C27**
Drosophila, wild type, culture (87 T 6550)	**C39**
Elodea, living, pkg of 10 (86 T 7500)	**C1**
Euglena sp., culture (87 T 0100)	**C3**
Flower petals (local)	**C2**
Grass (local)	**C2**
Hair, curly black (local)	**C2**
Hair, straight blonde (local)	**C2**
Leaves, crushed (local)	**C2**
Micrococcus luteus (85 T 1915)	**C31**
Mixed pond protozoa, culture (87 T 1510)	**C29**

Item	Lab
Monocot & Dicot Roots (CS), slide, each (91 T 9910)	**C35**
Monocot & Dicot Stems (CS), slide, each (91 T 9914)	**C35**
Monocot stem CS, vial (63 T 1063)	**C35**
Moss *(Polytrichum),* living, portion (86 T 4360)	**C1**
Onion root tips, vial (63 T 1212)	**C9, C10**
Owl pellets, pkg of 10 (69 T 3392)	**C22**
Owl pellets, NW, pkg of 60 (69 T 3393)	**C23**
Owl pellets, SE, pkg of 60 (69 T 3403)	**C23**
Paramecium caudatum, culture (87 T 1310)	**C3**
Penicillium notatum, culture (85 T 4700)	**C34**
Plant Mitosis LS (QS), slide, each (91 T 7042)	**C1**
Plant Mitosis-Metaphase (QS) FS&FG, each (93 T 1922)	**C10**
Rapid Radish seeds, pkg of 50 (86 T 8020)	**C37, C38**
Saccharomyces, UV-sensitive, culture (85 T 5001)	**C45**

Sweet corn seed, untreated, 2 oz pkt
(86 T 8080) **C12**
Tomato seeds, Rutgers, pkt of 100 (86 T 8340) . . **C30**
Yeast, viable, 10 g pkt (88 T 0929) **C7, C8**

Chemicals and Media

Item	Lab
Aceto-orcein, biological stain, 2% sol., 30 mL btl (38 T 9050) **C9, C10**	
Agar, sabouraud-dextrose, pkg of 6 (88 T 0920) **C45**	
Albumin, egg powder, LABgr, 100 g btl (39 T 0197) **C46**	
Alcohol pads, sterile, pkg of 200 (36 T 5519) **C13, C35**	
Benedict's solution, qualitative, 120 mL btl (37 T 0697) **C46**	
Biuret reagent, urea protein test, 120 mL btl (37 T 0790) **C46, C47**	
Bleach, chlorine, 1 pt btl (37 T 5554). . . . **C31, C32, C33, C34, C45**	
Caffeine, 0.5% solution (39 T 0866). **C10**	
Catalase, 5 mL btl (38 T 2371). **C5, C6**	
Corn starch, powder, LABgr, 500 g btl (39 T 3271) **C34**	
Cupric sulfate, 20% aqueous sol., 500 mL btl (37 T 2242). **C37, C38**	
D (+) fructose (levulose), 100 g btl (39 T 1356) . . . **C8**	
D (+) lactose hydrate, 500 g btl (39 T 2076). **C8**	
Fertilizer, nitrogen, pkt (20 T 6043) **C38**	
Fertilizer, phosphorus, pkt (20 T 6041) **C38**	
Fertilizer, potash, pkt (20 T 6042). **C38**	
Giemsa stain, 2.0 g btl (326 T 8522) **C13**	
Glucose solution, 15%, 250 mL btl (37 T 9000) **C46**	
Glucose test paper, pkg of 50 (14 T 4107) **C47**	
Hydrochloric acid, 1 M, volumetric, 150 mL btl (37 T 8605) **C9, C10**	
Hydrogen peroxide, 3%, LABgr., 473 mL btl (37 T 8450). **C5, C6**	
Indophenol solution, 0.1%, 500 mL btl (37 T 9542) **C4**	
Lake water (local). **C28**	
Limewater solution, LABgr., 500 mL btl (37 T 2630). **C7, C8**	
Lugol's iodine solution, 500 mL btl (39 T 1685) **C46, C47**	
Phosphate buffer, pkt (326 T 8521) **C13**	
Piccolyte II, 120 mL btl (37 T 9530) **C32, C35**	
Plant fertilizer, 2.75 g pkg (20 T 6040) **C37, C38**	

Item	Lab
Potassium phosphate, dibasic, 500 g btl (37 T 4141) **C30**	
Protist-slowing agent: Detain, 0.5 oz btl (37 T 7950) **C3, C29**	
Rapid soil test kit, each (20 T 7858) **C25, C26**	
Salt, iodized, 26 oz pkg (37 T 5482) **C2**	
Silicon carbide abrasive, 400 grit, lb (28 T 2540) **C30**	
Sodium chloride, fine white granule, 500 g btl (37 T 5487) **C13**	
Sodium phosphate, monobasic, Reagent, 100 g btl (37 T 5655) **C28**	
Starch solution, 1%, 250 mL btl (37 T 9001) . . . **C46**	
Sucrose, granular, SCIgr, 500 g btl (39 T 3182). **C7, C8**	
Sudan III (solvent red 23), powder, 10 g btl (38 T 8787) **C46, C47**	
Sugar, granular, 5 lb pkg (39 T 3180) **C2**	
Trichome stain, modified, each (38 T 7000). . . . **C35**	
Trypsin, 5 mL btl (326 T 8524) **C13**	
Tryptic soy agar base plates, pkg of 6 (88 T 0925) **C31**	
Uric Acid (WM) slide (94 T 0418) **C48**	
Vitamin C (ascorbic acid) pkt, L (37 T 9558). . . . **C4**	
Water (local) . . **C1, C7, C12, C37, C38, C46, C48**	
Water, distilled, 1 gal btl (88 T 7005). . . **C5, C8, C9, C18, C32, C33, C43, C45, C47**	

Laboratory Equipment

Item	Lab
Apron, disposable polyethylene, box of 100, (15 T 1050) **C3, C4, C5, C6, C7, C8, C9, C10, C13, C17, C18, C22, C23, C25, C26, C27, C28, C29, C30, C31, C32, C33, C34, C35, C36, C37, C38, C42, C43, C44, C45, C46, C47, C48**	
Balance, triple beam, each (15 T 6057) . . . **C25, C26**	
Beaker, low-form 100 mL Griffin, each (17 T 4020). **C2, C30, C33**	
Beaker, low-form 1000 mL Griffin, each (17 T 4080) **C7**	
Beaker, low-form 250 mL Griffin, each (17 T 4040) **C32**	
Beaker, low-form 50 mL Griffin, each (17 T 4010). **C6, C18, C28**	
Beaker, low-form 600 mL Griffin, each (17 T 4060) **C10, C17, C46, C48**	
Bunsen burner, standard natural gas, each (15 T 0612) **C31, C32, C33**	

Item	Lab
Clamp, Stoddard test-tube, pkg of 6 (15 T 0841)	**C46, C48**
Coverslips, 22 mm plastic, box of 100 (14 T 3555)	**C1, C2, C3, C9, C10, C29, C32, C33, C35**
Dissection pan set, economy, set (18 T 3665)	**C22, C23**
Dual-wave ultraviolet lamp, each (29 T 3000)	**C45**
Erlenmeyer flask, economy 250 mL, each (17 T 2803)	**C7**
Erlenmeyer flask, economy 500 mL, each (17 T 2804)	**C8**
Filter paper, medium grade, 11.0 cm, box of 100 (15 T 2817)	**C25, C26**
Filter paper, medium grade, 9.0 cm, box (15 T 2815)	**C5, C6**
Flask, 500 mL Pyrex Florence boiling, each (17 T 3030)	**C39**
Forceps, dissecting, medium, each (14 T 1001)	**C1, C6, C9, C10, C22, C35, C37**
Forceps, Kirkbridge slide, each (15 T 1640)	**C32, C33**
Forceps, student dissecting, broad, each (14 T 0512)	**C5, C23**
Gas lighter, flat file, pkg of 10 (15 T 0683)	**C31, C32, C33**
Gloves, disposable, medium, box of 100, (15 T 1071)	**C6, C13, C17, C18, C22, C23, C25, C26, C30, C31, C32, C33, C34, C35, C36, C37, C42, C43, C44, C45, C46, C47, C48**
Gloves, heat fefier kelnit cotton, pair (15 T 1095)	**C6, C25, C26**
Graduated cylinder, 10 mL PP, each (18 T 1705)	**C5, C6, C36, C47**
Graduated cylinder, 25 mL PP, each (18 T 1710)	**C28**
Hot plate, 700 W single burner, each (15 T 7999)	**C6, C10, C46, C 48**
Incubator, lab, each (15 T 0060)	**C31, C45**
Inoculating loop, disposable, pkg of 25 (14 T 0954)	**C33, C45**
Inoculating loop, nichrome, each (14 T 0957)	**C31, C32**
Jar caps, white metal, each (17 T 2133)	**C34**
Jar, staining, pkg of 10 (18 T 2102)	**C13**
Jar, wide mouth glass, each (17 T 2021)	**C34**
Lamp, clamp, with reflector, each (36 T 4168)	**C28, C39**

Item	Lab
Lens tissue, box of 50 (15 T 8250)	**C1, C32**
Magnifier, dual 3× & 6×, each (24 T 1112)	**C34**
Marker, black lab, each (15 T 3083)	**C25, C26**
Medicine cup, polypropylene each (250 T 5454)	**C5, C6**
Micro spoon, 9 in. stainless steel, each (15 T 4330)	**C47**
Microscope slide, qual. precleaned, pkg of 72 (14 T 3500)	**C1, C2, C9, C10, C13, C32, C33, C35**
Microscope slide, ruled, 80 squares, pkg of 15 (14 T 3120)	**C28**
Microscope slide, single concavity, pkg of 12 (14 T 3510)	**C3, C29**
Microscope, the WARD'S Scope, each (24 T 2310)	**C1, C2, C3, C9, C10, C13, C27, C28, C29, C32, C33, C35, C42, C43, C48**
Mortar, porcelain size 0, 50 mL, each (15 T 3334)	**C30**
Non-nutrient agar plates, pkg of 6 (88 T 0926)	**C31**
Oven, gravity convection laboratory, each (15 T 0061)	**C25, C26**
Pen, black wax marker for glass, each (15 T 1155)	**C2, C5, C7, C8, C10, C28, C30, C31, C32, C34, C37, C42, C43, C45, C46, C47, C48**
Pestle, porcelain size 0, each (15 T 3335)	**C30**
Petri culture dishes, pyrex 60 × 15 mm, each (17 T 0730)	**C9**
Pipet bulbs, pkg of 10 (15 T 0511)	**C13, C31, C48**
Pipet, 3 in. glass dropping, pkg of 12 (17 T 0230)	**C1, C2, C3, C4, C28, C29, C32, C33, C38, C48**
Pipet, 6 in. nonsterile, pkg of 500 (18 T 2971)	**C10, C37, C46, C48**
Pipet, serological, glass disposable, pkg of 100 (17 T 4853)	**C31**
Pipets, Pasteur 5 3/4 in., pkg of 250 (17 T 1145)	**C13**
Safety goggles, SG34 regular, each (15 T 3046)	**C3, C4, C5, C6, C7, C8, C9, C10, C13, C17, C18, C25, C26, C27, C28, C29, C30, C31, C32, C33, C34, C35, C38, C36, C42, C43, C44, C45, C46, C47, C48**
Scissors, dissecting, student economy, each (14 T 0525)	**C1, C11, C14, C15, C36**
Scoop, laboratory, pkg of 12 (15 T 4339)	**C47**
Slide staining dish, slotted, each (17 T 0205)	**C32, C33**

Item	Lab
Specimen container w/lid, pkg of 25 (18 T 1450)	**C45**
Stereomicroscope, wide-field, each (24 T 4601)	**C1**
Sterile graduated pipet, 6 in., pkg of 500 (18 T 2972)	**C45**
Stirring rods, 6 in. glass, 150 × 5 mm, pkg of 10 (17 T 6005)	**C5, C7, C47, C48**
Stopper, 2-hole rubber, size 8, lb (15 T 8518)	**C7, C8**
Stopwatch, digital, each (15 T 0512)	**C3, C5 C6, C8, C9, C13, C17, C32, C33, C35, C36, C41, C45, C48**
Swab applicator, pkg of 100 (14 T 5502)	**C30, C31, C34, C45**
Teasing needle, wood handle, each (14 T 0650)	**C1, C22, C23**
Test-tube rack, 6-well LDPE, each (18 T 4231)	**C10, C31, C46, C47, C48**
Test-tube rack, 80 tubes, each (18 T 4350)	**C4**
Test tube with cap, 16 × 150 mm Pyrex, each (17 T 1340)	**C31, C45**
Test tube with rim, 13 × 100 mm Pyrex, each (17 T 0610)	**C4**
Test tube with rim, 15 × 125 mm Pyrex, each (17 T 0620)	**C10, C46, C47, C48**
Thermometer, Lab −20 to 110°C, each (15 T 1416)	**C6, C7, C8, C25, C26**
Thermometer, pocket dial, each (15 T 0832)	**C25**
Tubing, 7 mm rigid plastic, lb (18 T 7700)	**C7, C8**
Tubing, latex, 1/4 × 1/16 in., 10-ft roll (15 T 1134)	**C7, C8**
Vacuum bottle, 500 mL, each (17 T 3805)	**C7, C8**
Vial, opticlear w/cap 27 × 55 mm, each, box of 12 (17 T 0143)	**C5, C35**
Washing bottle, fine stream, 150 mL, each, pkg of 6 (18 T 4140)	**C31, C32, C33, C34**
Wind meter, handheld w/case, each (23 T 1341)	**C25, C26**
Wood splints, 5 1/2 in., pkg of 500 (14 T 0105)	**C9, C35**

Miscellaneous

Item	Lab
Aluminum foil, 12 in. wide, roll of 25-ft (15 T 1009)	**C39**
Audubon *Field Guide to North American Birds, Eastern Region*, each (32 T 2112)	**C24, C26, C40**
Audubon *Field Guide to North American Birds, Western Region*, each (32 T 2113)	**C24, C26, C40**
Audubon *Field Guide to North American Mammals*, each (32 T 2114)	**C24, C26**
Audubon *Field Guide to North American Mushrooms*, each (32 T 2118)	**C24**
Audubon *Field Guide to North American Reptiles and Amphibians*, each (32 T 2123)	**C24, C26**
Audubon *Field Guide to North American Trees, Eastern Region*, each (32 T 2121)	**C24, C26**
Audubon *Field Guide to North American Trees, Western Region*, each (32 T 2122)	**C24, C26**
Audubon *Field Guide to North American Wildflowers, Eastern Region*, each (32 T 2119)	**C24, C26**
Audubon *Field Guide to North American Wildflowers, Western Region*, each (32 T 2120)	**C24, C26**
Bags, resealable zipper, 6 × 9 in., pkg of 10 (18 T 6922)	**C2, C25, C26**
Banana slices (local)	**C39**
Binoculars, wide-angle insta-focus, each (25 T 4523)	**C40**
Bird identification form (local)	**C40**
Bottle, spray mist dispenser, 475 mL, each (18 T 2260)	**C34**
Brush, camel hair #1, each (15 T 2111)	**C37**
Burrito filling (local)	**C47**
Calculator, slimline TI-1100+, each (27 T 3055)	**C24**
Carrot juice (local)	**C4**
Clipboard, 9 × 12 1/2 in., each (15 T 9852)	**C24**
Container, plastic, (local)	**C37**
Corn meal (local)	**C2**
Flags, colored (local)	**C26**
Flower pot, plastic 3 in., pkg of 10 (20 T 2130)	**C30**
Hammer, 16 oz claw, wood handle, each (12 T 0110)	**C24**
Hand exerciser (local)	**C41**
Human Karyotype Form, pad of 30 (33 T 1045)	**C14, C15**
Ice (local)	**C6, C13, C39**
Index cards, 5 × 8 in. lined, pkg of 5 (15 T 9820)	**C40**
Karyotyping set (33 T 1041)	**C15**
Labels, polypaper 1 1/2 × 3 in., pkg of 120 (15 T 1833)	**C37, C38**
Lemon juice (local)	**C4**

Item	Lab
Light bulb, clear 150 W, 120 V, each (36 T 4173)	**C28, C39**
Metal can, with ends cut (local)	**C25, C26**
Meter stick, maple, each (15 T 4065)	**C24, C25, C26**
Newspaper (local)	**C1**
Normal Male Chromosome Spread, pad (33 T 1046)	**C14**
Orange juice, boiled (local)	**C4**
Orange juice, canned (local)	**C4**
Orange juice, fresh squeezed (local)	**C4**
Orange juice, frozen (local)	**C4**
Orange juice, squeezed and exposed (local)	**C4**
Orange peels (local)	**C34**
Pan, plate, or piece of plastic (local)	**C38**
Paper clips, pkg of 1000 (15 T 9815)	**C11**
Paper punch, one-hole, each (15 T 9810)	**C5, C6**
Paper strip, 2 × 6 cm, dark blue (local)	**C11**
Paper strip, 2 × 6 cm, dark green (local)	**C11**
Paper strip, 2 × 6 cm, light blue (local)	**C11**
Paper strip, 2 × 6 cm, light gree (local)	**C11**
Paper towel, 100-sheet 2-ply roll, each (15 T 9844)	**C1, C5, C6, C9, C31, C32, C33, C34, C45**
Paper, assorted construction, pkg of 50 (15 T 9841)	**C24**
Paper, white (local)	**C19, C22, C23, C24, C40, C45**
Pencils, Ticonderoga No. 2, box of 12 (15 T 9816)	**C13, C19, C24, C36, C40**
Pineapple juice (local)	**C4**
Plant medium, Rapid Radish, ft (20 T 8640)	**C37, C38**
Plant stand with fluorescent light, each (20 T 5100)	**C27, C37, C38**
Plant tray, nesting 13 × 15 × 3.5 in., each (20 T 3210)	**C37, C38**
Potato flakes, russet, each (14 T 0010)	**C34**
Raisins (local)	**C34**
Razor blades, single-edge, box of 100 (14 T 4172)	**C10**
Refrigerator, explosionproof, each (15 T 8810)	**C31**
Rubber bands, assorted, 0.25 lb pkg (15 T 9824)	**C25, C26**
Ruler, 6 in. white vinylite, each (14 T 0810)	**C5, C6, C11, C14, C15, C22, C36, C37, C38**
Sand, fine white, 32 oz pkg (45 T 1983)	**C2**

Item	Lab
Seed starter tray, self-watering, each (20 T 2100)	**C12**
Soap, liquid antibacterial, each (18 T 1533)	**C31**
Soil, garden potting, 8 lb bag (20 T 8306)	**C2, C12, C30**
Stake/twist tie set, Rapid Radish, pkg of 40 (20 T 2110)	**C37, C38**
Stakes (local)	**C25, C26**
Stakes, wooden (local)	**C24**
String, 0.5 lb pkg (15 T 9863)	**C11, C24, C25, C26**
Sunscreen (local)	**C45**
Tape, masking 3/4 in. × 60 yd roll, pkg of 3 (15 T 9828)	**C31**
Tape, transparent w/dispenser, each (15 T 1959)	**C11, C14, C15, C22, C31, C37, C39, C45**
Thread, green and red (local)	**C2**
Thread, nylon, 196 yd/spool, pkg of 3 (15 T 9837)	**C1**
Tobacco from cigarettes (local)	**C30**
Tomato juice (local)	**C4**
Toothpicks, round, box of 800 (15 T 9840)	**C3, C18, C43**
Tray, Rapid Radish self-watering, each (20 T 2101)	**C37, C38**
Trowel, metal, each (20 T 7015)	**C25, C26**
Vegetable oil, 16 oz btl (37 T 9540)	**C46**
Watch or clock (local)	**C18**

Kits

Item	Lab
Biochemical Evidence for Evolution (36 T 1222)	**C18**
Blood Typing Lab (36 T 0252)	**C42**
Blood Typing Whodunnit Lab (36 T 0021)	**C43**
Blood Typing—Pregnancy and Hemolytic Disease (36 T 0251)	**C44**
Chromatography of Plant Pigments (36 T 0062)	**C36**
Detergent and Fertilizer as Pollutants (36 T 1221)	**C27, C28**
DNA Whodunnit Kit (36 T 1610)	**C16**
Enzyme Catalysis Lab Activity (36 T 6013)	**C5, C6**
Evolution and Blood Serum (36 T 0250)	**C17**
Forensic Mystery at 323 Maple (36 T 6008)	**C2**
Gram Stain Kit (38 T 9986)	**C32, C33**
Introduction to Owl Pellets (36 T 5487)	**C22**

Item	Lab
Investigation of NW vs. SE Owl Pellets (36 T 5488)	**C23**
Investigating Plant Cells Study (36 T 1213)	**C35**
Karyotyping Human Chromosomes (36 T 6009)	**C13**
Microscope Slide Techniques (36 T 6057)	**C1**
Mineral Nutrition of Plants Lab (36 T 8003) . . .	**C38**
Plant and Animal Mitosis (36 T 1212)	**C9**
Plant Growth and Life Cycle Lab (36 T 8000)	**C37**
Protist Structure Study Kit (87 T 1570)	**C3**
Simulated Urinalysis Lab (36 T 6012)	**C48**

Supplemental Materials

Item	Lab
A New Look at Algae, VHS video, each (193 T 2015)	**C27, C28**
ATP Muscle Kit, each (36 T 5417).	**C41**
Antibody-Antigen Reactions Lab, kit (36 T 6011)	**C42, C43, C44**
Atlas of Urinary Sediment, book, each (32 T 0808)	**C48**
Bacteria (2nd Edition), VHS video, each (193 T 0267)	**C31**
Bacteriology Introductory Set, live, set (85 T 3900)	**C31**
Beyond the Crime Lab, hardcover, each (32 T 1779)	**C2**
Bird Studies, chart, each (33 T 0566)	**C40**
Birds, VHS video, each (193 T 0432)	**C40**
Blood Cell, 18 × 24 in. poster, each (33 T 4015)	**C42, C43, C44**
Blood Types Slide Set, set of 6 (95 T 2641)	**C42, C43, C44**
Blood Typing Whodunnit Lab, kit (36 T 0021)	**C16, C42**
Chlorophyll Chromatogram Kit, each (38 T 6822)	**C36**
Chromosomal Phenomena Meiosis Set, each (81 T 4501)	**C11**
Corn Genetics Set, each (86 T 8889).	**C12**
Corn, Riker mount, each (20 T 8005).	**C12**
Counting and Identifying Polluted Water Algae, kit, each (36 T 5413)	**C27, C28**
DNA Fingerprinting Electrophoresis, set (36 T 5167)	**C16**
DNA Fingerprinting, hardcover, each (32 T 0258)	**C16**
Environment, VHS video, each (193 T 0119)	**C24, C25, C26**

Item	Lab
Environmental Science Activities, each (32 T 4080)	**C24, C25, C26**
Enzyme Action Kit, each (85 T 3984)	**C5**
Evolution CD-ROM, Mac/IBM, each (74 T 0518)	**C17, C18**
Evolution, each (32 T 0031)	**C17, C18**
Eyewitness Bird, VHS video, each (193 T 4031)	**C40**
Five Kingdom Classification Kit, each (32 T 2209)	**C29**
Field Guide to Microlife, each (32 T 8063).	**C3,C27,C28,C29**
Food Chain, VHS video, each (386 T 0426)	**C47**
Food Chain/Nutrient Cycle, laserdisc, each (196 T 0417)	**C46, C47**
Food Chemistry Module Kit, each (36 T 2041)	**C46, C47**
Food Composition & Calories, chart, each (33 T 0586)	**C46, C47**
Fungi Video Investigation, kit (87 T 3503).	**C34**
Fungi, VHS video, each (193 T 1200)	**C34**
Gram Stain Concept-Study Kit, each (85 T 3983)	**C32, C33**
Grow a Plant Kit, each (36 T 6016).	**C37, C38**
Guide to Gardening Class, each (32 T 0119)	**C37, C38**
How Things Are Classified, VHS video, each (193 T 5550)	**C29**
How to Use Light Microscope, each (32 T 1216)	**C1**
Human Body: Urinary System, IBM, each (74 T 2073)	**C48**
Human Karyotypes Slide Set, set of 3 (95 T 2656)	**C13, C14, C15**
Introduction to Chromatography, kit (36 T 1211)	**C36**
Karyotyping/Human Genetics Model, set (36 T 6023).	**C13, C14, C15**
Learning About the Environment CD-ROM, IBM, each (74 T 0507)	**C24, C25, C26**
Learning About the Environment CD-ROM, Mac, each (74 T 0506).	**C24, C25, C26**
Making a Water Drop Microscope, kit (36 T 0700)	**C1**
Meiosis Model Activity Set, each (82 T 1229)	**C11**
Meiosis, 18 × 24 in. wall chart, each (33 T 4018)	**C11**
Meiosis, VHS video, each (193 T 0278).	**C11**

Item	Lab
Meiosis and Mitosis, laserdisc, each (196 T 0275)	**C9, C10, C11**
Mendelian Genetics, Dihybrid, mount, each (67 T 1161)	**C12**
Mendelian Genetics, Hybrid, mount, each (67 T 1160)	**C12**
Mitosis, VHS video, each (193 T 0277)	**C9, C10**
Nutrient Cycle, VHS video, each (193 T 0427)	**C46, C47**
Organic Evolution, VHS video, set of 6 (193 T 0313)	**C19**
Owl By Day and Night, softcover, each (32 T 0443)	**C22, C23**
Owl Pellet, display mount, each (67 T 1220)	**C22, C23**
Owl Pellet Multimedia Lab, kit (36 T 5494)	**C22, C23**
Owl Pellet Resource Manual, each (32 T 0800)	**C22**
Owl Pellets, VHS video, each (193 T 2150)	**C22, C23**
Passage of Food Through the Digestive Tract, VHS video, each (193 T 2094)	**C46, C47**
Patterns of Evolution, VHS video, each (193 T 0441)	**C17, C18**
Plant Growth-Development, chart, set of 3 (33 T 0601	**C37, C38**
Plant Mitosis, chart, each (33 T 0567)	**C9, C10**
Population Genetics and Evolution Lab, each (36 T 1510)	**C19**
Protist Culture Set, Introductory, set of 3 (87 T 1530)	**C29**
Protist Kingdom, VHS video, each (193 T 6437)	**C3, C29**
Protists of Puddles and Ponds, poster, each (33 T 0133)	**C3**

Item	Lab
Protists: Form, Function, VHS video, each (193 T 0289)	**C3**
Rapid Radish Manual, each (32 T 0813)	**C37, C38**
Raptors, Birds of Prey CD-ROM, each (74 T 5071)	**C40**
Reptiles & Amphibians Care, each (32 T 0731)	**C20, C21**
Stems, chart, each (33 T 0576)	**C35**
Surveying & Mapping, each (32 T 1765)	**C24, C25, C26**
Tobacco Mosaic Virus (SECT) QS, slide, each (90 T 7580)	**C30**
Two Views: Water Pollution, VHS video, each (193 T 0323)	**C27, C28**
Urinary System Model Activity Set, each (82 T 1234)	**C48**
Urine Sediment WM, slide, each (94 T 0258)	**C48**
Urine Sediment—Unstained (WM), slide, each (94 T 0252)	**C48**
Using A Compound Microscope, VHS video, each (193 T 2203)	**C1**
Using Yeast to Teach Genetics, each (32 T 1003)	**C6**
WARD'S Effects of Sunscreen on DNA Replication, kit (85 T 3523)	**C45**
Water Pollution, chart, each (33 T 2444)	**C27, C28**
Water Quality Assessment Kit, each (86 T 3056)	**C27, C28**
What Influences Enzyme Activity?, kit (36 T 1216)	**C5, C6**

Master Materials List, by Lab

C1 Using a Microscope

Required Materials	Quantity Needed
Coverslips, 22 mm Plastic	2 per team
Elodea, living	1 leaf per team
Forceps, dissecting, medium	1 per team
Lens tissue	1 sheet per team
Microscope, the WARD'S Scope	1 per team
Microscope slide, qual. precleaned	2 per team
Moss *(Polytrichum)*, living	1 per class
Newspaper	1 per class
Paper towel, 100-sheet 2-ply roll	1 sheet per team
Pipet, 3 in. glass dropping	1 per team
Plant Mitosis (LS) QS, slide	1 per team
Scissors, student econ. dissecting	1 per team
Stereomicroscope, wide-field	1 per team
Teasing needle, wooden handle	1 per team
Thread, nylon, 196 yd/spool	4 cm per team
Water	1 mL per team

Alternative Activity

Microscope Slide Techniques	1 per 30 students

C2 Using a Microscope—Criminal Investigation

Required Materials	Quantity Needed
Animal hair	10 strands per team
Bags, resealable zipper, 6 × 9 in.	4 per team
Beaker, low-form 100 mL Griffin	1 per team
Corn meal	1 oz per team
Coverslips, 22 mm plastic	4 per team
Flower petals	20 per team
Grass	1 oz per team
Hair, curly black	10 strands per team
Hair, straight blonde	10 strands per team
Leaves, crushed	10 per team
Microscope slide, qual. precleaned	4 per team
Microscope, the WARD'S Scope	1 per team
Pen, black wax marker for glass	1 per class
Pipet, 3 in. glass dropping	1 per team
Salt, iodized	1 per class
Sand, fine white	1 per class
Soil, garden potting	1 per class
Sugar, granular	1 per class
Thread, green and red	10 strands per team

Alternative Activity

Forensic Mystery at 323 Maple	1 per 30 students

C3 Using a Microscope—Slowing Protozoans

Required Materials	Quantity Needed
Apron, disposable polyethylene	1 per student
Coverslips, 22 mm plastic	10 per team
Euglena sp., culture	11 drops per team
Microscope, the WARD'S Scope	1 per team
Microscope slide, single concavity	10 per team
Paramecium caudatum, culture	11 drops per team
Pipet, 3 in. glass dropping	3 per team
Protist-slowing agent: Detain	8 drops per team
Safety goggles, SG34 regular	1 per student
Stopwatch, digital	1 per team
Toothpicks, round	10 per team

Alternative Activity

Protist Structure Study Kit	1 per class

C4 Testing for Vitamin C

Required Materials	Quantity Needed
Apron, disposable polyethylene	1 per student
Carrot juice	10 mL per team
Indophenol solution, 0.1%	1 per class
Lemon juice	10 mL per team
Orange juice, boiled	10 mL per team
Orange juice, canned	10 mL per team
Orange juice, fresh squeezed	10 mL per team
Orange juice, frozen	10 mL per team
Orange juice, squeezed and exposed	10 mL per team
Pineapple juice	10 mL per team
Pipet, 3 in. glass dropping	4 per team
Safety goggles, SG34 regular	1 per student
Test-tube rack, 80 tubes	1 per team
Test tube with rim, 13 × 100 mm Pyrex	11 per team
Tomato juice	10 mL per team
Vitamin C (ascorbic acid)	1 pkt per class

C5 Observing the Effect of Concentration on Enzyme Activity

Required Materials	Quantity Needed
Apron, disposable polyethylene	1 per student
Catalase	30 mL per team
Filter paper, med. grade, 9.0 cm	1 sheet per team
Forceps, student dissecting, broad	1 per team
Graduated cylinder, 10 mL PP	1 per team
Hydrogen peroxide, 3%, LABgr	1 per class
Medicine cup, polypropylene	6 per team
Paper punch, one-hole	1 per class
Paper towel, 100-sheet 2-ply roll	2 sheets per team
Pen, black wax marker for glass	1 per team
Ruler, 6 in. white vinylite	1 per team
Safety goggles, SG34 regular	1 per student
Stirring rods, 6 in. glass, 150 × 5 mm	1 per team
Stopwatch, digital	1 per team
Vial, opticlear w/cap 27 × 55 mm	1 per team

Water, distilled. . . . . 50 mL per team
Alternative Activity
Enzyme Catalysis Lab Activity . . . 1 per 30 students

C6 Observing the Effect of Temperature on Enzyme Activity

Required Materials	Quantity Needed
Apron, disposable polyethylene.	1 per student
Beaker, low-form 50 mL Griffin	1 per team
Catalase.	10 mL per team
Filter paper, med. grade, 9.0 cm. . .	1 sheet per team
Forceps, dissecting, medium	1 per team
Gloves, disposable, medium.	2 per student
Gloves, heat defier kelnit cotton	1 per team
Graduated cylinder, 10 mL PP.	1 per team
Hot plate, 700 W single-burner.	1 per team
Hydrogen peroxide, 3%, LABgr. . . .	50 mL per team
Ice	1 bucket per class
Medicine cup, polypropylene.	1 per team
Paper punch, one hole	1 per team
Paper towel, 100-sheet 2-ply roll. .	4 sheets per team
Ruler, 6 in. white vinylite	1 per team
Safety goggles, SG34 regular	1 per student
Stopwatch, digital	1 per team
Thermometer, Lab −20 to 110°C.	1 per team

Alternative Activity
Enzyme Catalysis Lab Activity . . . 1 per 30 students

C7 Measuring the Release of Energy from Sucrose

Required Materials	Quantity Needed
Apron, disposable polyethylene.	1 per student
Beaker, 1000 mL low-form Griffin	1 per team
Erlenmeyer flask, 250 mL economy.	2 per team
Limewater Solution, LABgr.	200 mL per team
Pen, black wax marker for glass.	1 per team
Safety goggles, SG34 regular	1 per student
Stirring rods, 6 in. glass, 150 × 5 mm. . .	1 per team
Stopper, 2-hole rubber, size 8.	2 per team
Sucrose, granular, SCIgr	150 g per team
Thermometer, lab −20 to 110°C	2 per team
Tubing, latex, 1/4 × 1/16 in.	80 cm per team
Tubing, rigid plastic, 7 mm	20 cm per team
Vacuum bottle, 500 mL	2 per team
Water	800 mL per team
Yeast, Viable	1 per team

C8 Measuring the Release of Energy—Best Food for Yeast

Required Materials	Quantity Needed
Apron, disposable polyethylene.	1 per student
D(+) fructose (levulose).	100 g per team
D(+) lactose hydrate.	100 g per team
Erlenmeyer flask, 500 mL economy.	4 per team
Limewater solution, LABgr..	80 mL per team
Pen, black wax marker for glass.	1 per team
Safety goggles, SG34 regular	1 per student
Stopper, 2-hole rubber, size 8.	4 per team
Stopwatch, digital	1 per team
Sucrose, granular, SCIgr	100 g per team
Thermometer, lab −20 to 110°C	4 per team
Tubing, latex, 1/4 × 1/16 in.	160 cm per team
Tubing, rigid plastic, 7 mm	2 pieces per team
Vacuum bottle, 500 mL	4 per team
Water, distilled	400 mL per team
Yeast, Viable	2 pkg per team

C9 Preparing a Root Tip Squash

Required Materials	Quantity Needed
Aceto-orcein, bio. stain, 2% sol.	1 mL per team
Apron, disposable polyethylene.	1 per student
Coverslips, 22 mm plastic	3 per team
Forceps, dissecting, medium	1 per team
Hydrochloric acid, 1 M	6 mL per team
Microscope slide, qual. precleaned	3 per team
Microscope, the WARD'S Scope	1 per team
Onion root tips.	1 vial per class
Paper towel, 100-sheet 2-ply roll. .	3 sheets per team
Petri culture dish, 60 × 15 mm Pyrex. . .	1 per team
Safety goggles, SG34 regular	1 per student
Stopwatch, digital	1 per team
Water, distilled	10 mL per team
Wood splints, 5 1/2 in..	1 per team

Alternative Activity
Plant and Animal Mitosis . . . . 1 per 30 students

C10 Preparing a Root Tip Squash—Stopping Mitosis

Required Materials	Quantity Needed
Aceto-orcein, bio. stain, 2% sol. . . .	10 mL per team
Apron, disposable polyethylene.	1 per student
Beaker, low-form 600 mL Griffin	1 per team
Caffeine, 0.5% solution	30 mL per class
Coverslips, 22 mm plastic	6 per team
Forceps, dissecting, medium	1 per team
Hot plate, 700 W single-burner.	1 per team
Hydrochloric acid 1 M.	5 mL per team

Microscope slide, qual. precleaned . . . . . 6 per team
Microscope, the WARD'S Scope . . . . 1 per team
Onion root tips. 6 per team
Pen, black wax marker for glass. . . . . . . 1 per team
Pipet, nonsterile 6 in. . . . . 1 per team
Plant Mitosis-Metaphase (QS) FS&FG,
    slide . . . . 1 per team
Razor blades, single-edge . . . . 1 per team
Safety goggles, SG34 regular . . . . 1 per student
Test-tube rack, 6-well LDPE. . . . . 1 per team
Test tube, 15 × 125 mm Pyrex . . . . 2 per team

C11 Modeling Meiosis

Required Materials	Quantity Needed
Paper clips.	8 per team
Paper strip 2 × 6 cm, dark blue.	1 per team
Paper strip 2 × 6 cm, dark green.	1 per team
Paper strip 2 × 6 cm, light gree.	1 per team
Paper strip 2 × 6 cm, light blue.	1 per team
Ruler, 6 in. white vinylite	1 per team
Scissors, student econ. dissecting.	1 per team
String.	3.5 m per team
Tape, transparent w/dispenser	1 per team

C12 Analyzing Corn Genetics

Required Materials	Quantity Needed
Albino Corn Seed.	10 per team
Seed starter tray, self-watering.	1 per team
Soil, garden potting.	1 per class
Sweet corn seed, untreated.	20 per team
Water	1 L per team

C13 Preparing Tissue for Karyotyping

Required Materials	Quantity Needed
Alcohol pads, sterile	1 pad per team
Apron, disposable polyethylene.	1 per student
Cells, fixed for karyotyping	1 per team
Giemsa Stain	2 mL per team
Gloves, disposable, medium.	2 per student
Ice	1 bucket per team
Jar, staining	5 per team
Microscope slide, qual. precleaned	1 per team
Microscope, the WARD'S Scope	1 per team
Pencils, Ticonderoga #2	1 per team
Phosphate buffer.	1 per team
Pipet bulbs	1 per team
Pipets, 5 3/4 in. Pasteur	1 per team
Safety goggles, SG34 regular	1 per student
Sodium chloride, fine granular.	15 g per team
Stopwatch, digital	1 per team

Trypsin . . . . 2 mL per team
Alternative Activity
Karyotyping Human
    Chromosomes . . . . 1 per 2 groups

C14 Karyotyping

Required Materials	Quantity Needed
Human Karyotype Form	1 per team
Normal Male Chromosome Spread	1 per team
Ruler, 6 in. white vinylite	1 per team
Scissors, student econ. dissecting.	1 per team
Tape, transparent w/dispenser	1 per team

C15 Karyotyping—Genetic Disorders

Required Materials	Quantity Needed
Human Karyotype Form	1 per team
Karyotyping Set.	1 per class
Ruler, 6 in. white vinylite	1 per team
Scissors, student econ. dissecting.	1 per team
Tape, transparent w/dispenser	1 per team

C16 DNA Whodunit

Required Materials	Quantity Needed
DNA Whodunnit Kit.	1 per 14 teams

C17 Analyzing Blood Serum to Determine Evolutionary Relationships

Required Materials	Quantity Needed
Apron, disposable polyethylene.	1 per student
Beaker, 600 mL low-form Griffin	1 per team
Evolution and Blood Serum, kit	1 per team
Gloves, disposable, medium.	2 per student
Safety goggles, SG34 regular	1 per student
Stopwatch, digital	1 per team

C18 Analyzing Blood Serum— Evolution of Primates

Required Materials	Quantity Needed
Apron, disposable polyethylene	1 per student
Beaker, 50 mL low-form Griffin.	1 per team
Biochemical Evidence for Evolution	1 per class
Gloves, disposable, medium.	2 per student
Safety goggles, SG34 regular	1 per student
Toothpicks (additional).	300 per team
Watch or clock.	1 per team
Water, distilled.	20 mL per team

C19 Analyzing Amino-Acid Sequences to Determine Evolutionary Relationships

Required Materials	*Quantity Needed*
Paper, white	1 sheet per team
Pencils, Ticonderoga #2	1 per team

C20 Observing Animal Behavior

Required Materials	*Quantity Needed*

No materials needed.

C21 Observing Animal Behavior— Grant Application

Required Materials	*Quantity Needed*

No materials needed.

C22 Examining Owl Pellets

Required Materials	*Quantity Needed*
Apron, disposable polyethylene	1 per student
Dissection pan set, economy	1 per team
Forceps, dissecting, medium	1 per team
Gloves, disposable, medium	2 per student
Owl pellets	1 per team
Paper, white	1 sheet per team
Ruler, 6 in. white vinylite	1 per team
Tape, transparent w/dispenser	1 per team
Teasing needle, wood handle	1 per team

Alternative Activity

Introduction to Owl Pellets	1 per 15 teams

C23 Examining Owl Pellets—NW vs. SE

Required Materials	*Quantity Needed*
Apron, disposable polyethylene	1 per student
Forceps, dissecting, medium	1 per team
Dissection pan set, economy	1 per team
Gloves, disposable, medium	2 per student
Owl pellets, NW	1 per team
Owl pellets, SE	1 per team
Paper, white	2 sheets per team
Teasing needle, wood handle	1 per team

C24 Mapping Biotic Factors in the Environment

Required Materials	*Quantity Needed*
Audubon Field Guides	1 each per class
Calculator, slimline TI-1100+	1 per team
Clipboard, 9 × 12 1/2 in.	1 per team
Hammer, claw, wood handle 16 oz	1 per team
Meter stick, maple	1 per team
Paper, assorted construction	1 sheet per team
Paper, white	1 sheet per team
Pencils, Ticonderoga #2	1 per team
Protractor	1 per team
Stakes, wooden	4 per team
String	16 m per team

C25 Assessing Abiotic Factors in the Environment

Required Materials	*Quantity Needed*
Apron, disposable polyethylene	1 per student
Bags, resealable zipper, 6 × 9 in.	1 per team
Balance, triple beam	1 per team
Filter Paper, med. grade, 11.0 cm	4 sheets per team
Gloves, disposable, medium	2 per student
Gloves, heat defier kelnit cotton	1 per team
Gravity convection laboratory oven	1 per class
Marker, black lab	1 per team
Metal can, with ends cut out	1 per team
Meter stick, maple	1 per team
Rapid soil test kit	1 per team
Rubber bands, assorted	4 per team
Safety goggles, SG34 regular	1 per student
Stakes	9 per team
String	32 m per team
Thermometer, lab −20 to 110°C	1 per team
Thermometer, pocket dial	1 per team
Trowel, metal	1 per team
Wind meter, handheld w/case	1 per team

C26 Assessing and Mapping Factors in the Environment

Required Materials	*Quantity Needed*
Apron, disposable polyethylene	1 per student
Audubon Field Guides	1 each per class
Bags, resealable zipper, 6 × 9 in.	1 per team
Balance, triple beam	1 per team
Filter paper, med. grade, 11.0 cm	1 sheet per team
Flags, colored	4 per team
Gloves, disposable, medium	2 per student
Gloves, heat defier kelnit cotton	1 per team
Gravity convection laboratory oven	1 per team
Marker, black lab	1 per class
Metal can, with ends cut out	1 per team
Meter stick, maple	1 per team
Protractor	1 per team
Rapid soil test kit	1 per team
Rubber bands, assorted	1 per team
Safety goggles, SG34 regular	1 per student
Stakes	8 per team
String	40 m per team
Thermometer, Lab −20 to 110°C	1 per team

Trowel, metal. . . . . 1 per team
Wind meter, handheld w/case . . . . 1 per team

C27 Studying an Algal Bloom

Required Materials	*Quantity Needed*
Apron, disposable polyethylene.	1 per student
Chlorella, culture	1 per class
Detergent and Fertilizer as Pollutants, kit	1 per class
Dirt, unsterilized	5 g per team
Microscope, the WARD'S Scope	1 per team
Plant stand with fluorescent light.	1 per class
Safety goggles, SG34 regular	1 per student

C28 Studying an Algal Bloom— Phosphate Pollution

Required Materials	*Quantity Needed*
Apron, disposable polyethylene.	1 per student
Beaker, low-form 50mL Griffin	3 per team
Graduated cylinder, 25 mL PP.	1 per team
Lake water	75 mL per team
Lamp, clamp, with reflector	1 per team
Light bulb, clear 150 W, 120 V	1 per team
Microscope slide, ruled 80 squares	24 per team
Microscope, the WARD'S Scope	1 per team
Pen, black wax marker for glass.	1 per team
Pipet, 3 in. glass dropping	1 per team
Safety goggles, SG34 regular	1 per student
Sodium phosphate, monobasic	2 g per class

Alternative Activity

Detergent/Fertilizer as Pollutants . 1 per 30 students

C29 Classifying Mysterious Organisms

Required Materials	*Quantity Needed*
Apron, disposable polyethylene.	1 per student
Coverslips, 22 mm plastic	1 per team
Microscope slide, single concavity.	1 per team
Microscope, the WARD'S Scope	1 per team
Mixed Pond Protozoa, culture	1 per class
Pipet, 3 in. glass dropping	1 per team
Protist-slowing agent: Detain	3 drops per team
Safety goggles, SG34 regular	1 per student

C30 Screening for Resistance to Tobacco Mosaic Virus

Required Materials	*Quantity Needed*
Apron, disposable polyethylene.	1 per student
Beaker, low-form 100 mL Griffin	2 per team
Flower pot, plastic 3 in.	5 per team

Gloves, disposable, medium. . . . . 2 per team
Mortar, 50 mL porcelain, size 0 . . . . 1 per team
Pen, black wax marker for glass. . . . . 1 per team
Pestle, porcelain size 0 . . . . 1 per team
Potassium phosphate, dibasic . . . . 18 g per class
Safety goggles, SG34 regular . . . . 1 per student
Silicon carbide abrasive, 400 Grit . . . . . 1 g per team
Soil, garden potting . . . . 1 per team
Swab applicator. 5 per team
Tobacco from cigarettes . . . . 5 per team
Tomato seeds, Rutgers . . . . 5 per team

C31 Using Aseptic Technique

Required Materials	*Quantity Needed*
Apron, disposable polyethylene.	1 per student
Bleach, chlorine.	1 per class
Bunsen burner, std. nat. gas	1 per team
Gas lighter, flat file	1 per team
Gloves, disposable, medium	2 per student
Incubator, lab.	1 per class
Inoculating loop, nichrome	1 per team
Micrococcus luteus, culture	1 per team
Non-nutrient agar plates	6 per team
Paper towel, 100-sheet 2-ply roll. .	5 sheets per team
Pen, black wax marker for glass.	1 per team
Pipet bulbs	1 per team
Pipet, serological, disposable glass.	1 per team
Refrigerator, explosion proof	1 per class
Safety goggles, SG34 regular	1 per student
Soap, liquid antibacterial	1 per class
Swab applicator.	3 per team
Tape, masking 3/4 in. × 60 yd roll	1 per class
Tape, transparent w/dispenser	1 per team
Test tube with cap, 16 × 150 mm Pyrex	1 per team
Test-tube rack, 6-well LDPE.	1 per team
Tryptic soy agar base plates	1 per team
Washing bottle, fine stream 150 mL	1 per team

C32 Gram Staining of Bacteria

Required Materials	*Quantity Needed*
Apron, disposable polyethylene.	1 per student
Aquaspirillum serpens, culture	1 per class
Bacillus megaterium, culture.	1 per class
Beaker, low-form 250 mL Griffin	2 per team
Bleach, chlorine.	1 per class
Bunsen burner, std. nat. gas	1 per team
Coverslips, 22 mm plastic	2 per team
Forceps, Kirkbride slide	1 per team
Gas lighter, flat file	1 per team
Gloves, disposable, medium.	2 per student
Gram Stain Kit	1 per team

Inoculating loop, nichrome . . . . 1 per team
Lens tissue. . . . . 1 per team
Microscope slide, qual. precleaned . . . . . 2 per team
Microscope, the WARD'S Scope . . . . 1 per team
Paper towel, 100-sheet 2-ply roll. . 5 sheets per team
Pen, black wax marker for glass. . . . . 1 per team
Piccolyte II, 120 mL MSDS# 1490 . . 1 mL per team
Pipet, 3 in. glass dropping . . . . 1 per team
Safety goggles, SG34 regular . . . . 1 per student
Slide staining dish, slotted . . . . 1 per team
Stopwatch, digital . . . . 1 per team
Water, distilled . . . . 250 mL per team

C33 Gram Staining of Bacteria— Treatment Options

Required Materials	Quantity Needed
Apron, disposable polyethylene.	1 per student
Bacillus megaterium, culture.	1 per class
Beaker, low-form 100 mL Griffin	1 per team
Bleach, chlorine.	1 per class
Bunsen burner, std. nat. gas	1 per team
Coverslips, 22 mm plastic	1 per team
Forceps, Kirkbridge slide	1 per team
Gas lighter, flat file	1 per team
Gloves, disposable, medium.	2 per student
Gram Stain Kit	1 per team
Inoculating loop, disposable	1 per team
Microscope slide, qual. precleaned	1 per team
Microscope, the WARD'S Scope	1 per team
Paper towel, 100-sheet 2-ply	5 sheets per team
Pipet, 3 in. glass dropping	1 per team
Safety goggles, SG34 regular	1 per student
Slide staining dish, slotted	1 per team
Stopwatch, digital	1 per team
Water, distilled	100 mL per team

C34 Limiting Fungal Growth

Required Materials	Quantity Needed
Apron, disposable polyethylene.	1 per student
Bleach, chlorine.	1 per class
Bottle, spray mist dispenser 475 mL	1 per team
Corn starch, powder, LABgr	5 g per team
Gloves, disposable, medium.	2 per student
Jar caps, white metal.	8 per team
Jars, wide mouth glass	8 per team
Magnifier, dual 3× and 6×	1 per team
Orange peels.	2 per team
Paper towel, 100-sheet 2-ply roll. . 5 sheets per team
Pen, black wax marker for glass.	1 per team
Penicillium notatum, culture.	1 per class

Potato flakes, russet . . . . 5 g per team
Raisins. . . . . 20 per team
Safety goggles, SG34 regular . . . . 1 per student
Swab applicator. 8 per team
Washing bottle, fine stream 150 mL . . . . 1 per team

C35 Staining and Mounting Stem Cross Sections

Required Materials	Quantity Needed
Alcohol Pads, Sterile	1 rod per team
Apron, disposable polyethylene.	1 per student
Coverslips, 22 mm plastic	3 per team
Dicot stem CS, slide	1 per class
Dicot stem CS, vial	1 per team
Forceps, dissecting, medium	1 per team
Gloves, disposable, medium.	2 per student
Microscope slide, qual. precleaned	3 per team
Microscope, the WARD'S Scope	1 per team
Modified trichrome stain	1 mL per team
Monocot & Dicot Roots (CS), slide.	1 per team
Monocot stem CS, slide	1 per class
Monocot stem CS, vial	1 per team
Piccolyte II, 120mL MSDS# 1490 . .	1 mL per team
Safety goggles, SG34 regular	1 per student
Stopwatch, digital	1 per team
Wood splints, 5 1/2 in.	1 per team

Alternative Activity

Investigating Plant Cells Study. . . . 1 per 30 students

C36 Using Paper Chromatography to Separate Pigments

Required Materials	Quantity Needed
Apron, disposable polyethylene.	1 per student
Chromatography of Plant Pigments, kit . .	1 per class
Gloves, disposable, medium.	2 per student
Graduated cylinder, 10 mm PP.	1 per team
Pencil, Ticonderoga No. 2	1 per team
Ruler, 6 in. white vinylite.	1 per team
Safety goggles, SG34 regular	1 per student
Scissors, student econ. dissecting.	1 per team
Stopwatch, digital	1 per team

C37 Growing Plants in the Laboratory

Required Materials	Quantity Needed
Apron, disposable polyethylene.	1 per student
Brush, camel hair #1.	1 per team
Container, plastic	1 per team
Cupric sulfate, 20% aqueous sol.	1 per class
Forceps, dissecting, medium	1 per team
Gloves, disposable, medium.	2 per student

Labels, polypaper 1 1/2 × 3 in. . . . . 4 per team
Pen, black wax marker for glass. . . . . . . 1 per team
Pipet, 6 in. nonsterile . . . . 1 per team
Plant fertilizer . . . . 1 per class
Plant growth medium, Rapid Radish. . . . 1 per class
Plant stand with fluorescent light. . . . . . . 1 per class
Plant tray, nesting 13 × 15 × 3.5 in. . . . . 1 per class
Rapid Radish seeds . . . . 8 seeds per team
Ruler, 6 in. white vinylite . . . . 1 per team
Safety goggles, SG34 regular . . . . 1 per student
Stake, Rapid Radish . . . . 4 per team
Twist ties, Rapid Radish . . . . 4 per team
Tape, transparent w/dispenser . . . . 1 per class
Tray, Rapid Radish self-watering . . . . 1 per class
Water. . . . . 1 L per class
Alternative Activity
Plant Growth and Life Cycle Lab . . . 1 per 10 teams

C38 Growing Plants in the Laboratory—Fertilizer Problem

Required Materials	*Quantity Needed*
Apron, disposable polyethylene.	1 per student
Cupric sulfate, 20% aqueous sol.	1 per class
Fertilizer, nitrogen.	1 per class
Fertilizer, phosphorus	1 per class
Fertilizer, potash	1 per class
Labels, Polypaper 1 1/2 × 3 in.	8 per team
Pan, plate, or piece of plastic	1 per team
Pipet, 3 in. glass dropping.	2 per team
Plant fertilizer	1 per class
Plant growth medium, Rapid Radish. . . .	1 per class
Plant stand with fluorescent light.	1 per class
Plant tray, nesting 13 × 15 × 3.5 in.	1 per class
Rapid Radish Seeds	8 per team
Ruler, 6 in. white vinylite.	1 per team
Safety goggles, SG34 regular	1 per student
Stake, Rapid Radish	8 per team
Twist tie, Rapid Radish	8 per team
Tray, Rapid Radish self-watering	1 per class
Water.	1 L per class

Alternative Activity
Mineral Nutrition of Plants Lab . . . . 1 per 10 teams

C39 Response in the Fruit Fly

Required Materials	*Quantity Needed*
Aluminum foil, 12 in. wide roll	50 cm per team
Banana slices.	9 per team
Drosophila, wild type, culture.	40 flies per team
Ice	1 bucket per class
Lamp, clamp, with reflector	1 per team

Light bulb, clear 150 W, 120 V . . . . 1 per team
Pyrex Florence boiling flask-500 mL. . . . 4 per team
Tape, transparent w/dispenser . . . . 21 per team

C40 Conducting a Bird Survey

Required Materials	*Quantity Needed*
Audubon *Field Guide to North American Birds, Eastern Region*	1 per class
Audubon *Field Guide to North American Birds, Western Region*	1 per class
Binoculars, wide-angle Insta focus	2 per team
Bird identification form	5 per team
Index cards, 5 × 8 in., lined.	5 per team
Paper, white	10 sheets per team
Pencils, Ticonderoga No. 2.	2 per team

C41 Evaluating Muscle Exhaustion

Required Materials	*Quantity Needed*
Hand exerciser	1 per team
Stopwatch, digital	1 per team

C42 Blood Typing

Required Materials	*Quantity Needed*
Apron, disposable polyethylene.	1 per student
Blood Typing Lab	1 per class
Gloves, disposable, medium	2 per student
Microscope, the WARD'S Scope	1 per team
Pen, black wax marker for glass.	1 per team
Safety goggles, SG34 regular	1 per student

C43 Blood Typing—Whodunit?

Required Materials	*Quantity Needed*
Apron, disposable polyethylene.	1 per student
Blood Typing Whodunnit Lab	1 per class
Gloves, disposable, medium.	2 per student
Microscope, the WARD'S Scope	1 per team
Pen, black wax marker for glass.	1 per team
Safety goggles, SG34 regular	1 per student
Toothpicks, (additional)	170 per team
Water, distilled	50 mL per team

C44 Blood Typing—Pregnancy and Hemolytic Disease

Required Materials	*Quantity Needed*
Apron, disposable polyethylene.	1 per student
Blood Typing-Pregnancy & Hemolytic Disease, kit	1 per class
Gloves, disposable, medium.	2 per student
Safety goggles, SG34 regular	1 per student

C45 Screening Sunscreens

Required Materials	Quantity Needed
Agar plates, Sabouraud-dextrose	4 per team
Apron, disposable polyethylene	1 per student
Bleach, chlorine	1 per class
Dual-wave ultraviolet lamp	1 per class
Gloves, disposable, medium	2 per student
Incubator, lab	1 per class
Inoculating loop, disposable	8 per team
Paper towel, 100-sheet 2-ply roll	3 sheets per team
Paper, white	1 sheet per team
Pen, black wax marker for glass	1 per team
Saccharomyces, UV-sensitive, culture	1 per class
Safety goggles, SG34 regular	1 per student
Specimen container w/lid	1 per team
Sterile graduated pipet, 6 in.	2 per team
Stopwatch, digital	1 per team
Sunscreen	2 bottles per class
Swab applicator	2 per team
Tape, transparent w/dispenser	1 per class
Test tubes, with cap, 16 × 150 mm	1 per team
Water, distilled	40 mL per team

C46 Identifying Food Nutrients

Required Materials	Quantity Needed
Albumin, egg powder, LABgr	1 per class
Apron, disposable polyethylene	1 per student
Beaker, low-form 600 mL Griffin	1 per team
Benedict's solution, qualitative	2 mL per team
Biuret reagent, urea protein test	1 mL per team
Clamp, Stoddard test-tube	1 per team
Gloves, disposable, medium	2 per student
Glucose solution 15%	10 mL per team
Hot plate, 700 W single-burner	1 per team
Lugol's iodine solution	1 mL per team
Pen, black wax marker for glass	1 per team
Pipet, 6 in. nonsterile	8 per team
Safety goggles, SG34 regular	1 per student
Starch solution, 1%	10 mL per team
Sudan III (solvent red 23)	1 mL per team
Test-tube rack, 6-well LDPE	1 per team
Test tube with rim, 15 × 125mm	
Pyrex	10 per team
Vegetable oil	1 mL per team
Water	500 mL per team

C47 Identifying Food Nutrients— Food Labeling

Required Materials	Quantity Needed
Apron, disposable polyethylene	1 per student
Biuret reagent, urea protein test	3 mL per team
Burrito filling	2 mL per team
Gloves, disposable, medium	2 per student
Glucose test paper	1 strip per team
Graduated cylinder, 10 mL PP	1 per team
Lugol's iodine solution	3 mL per team
Micro spoon, 9 in. stainless steel	1 per team
Pen, black wax marker for glass	1 per team
Safety goggles, SG34 regular	1 per student
Scoop, laboratory	1 per team
Stirring rods, 6 in. glass, 150 × 5mm	4 per team
Sudan III (solvent red 23)	3 mL per team
Test-tube rack, 6-well LDPE	1 per team
Test tube with rim, 15 × 125 mm	
Pyrex	4 per team
Water, distilled	80 mL per team

C48 Urinalysis Testing

Required Materials	Quantity Needed
Apron, disposable polyethylene	1 per student
Beaker, low-form 600 mL Griffin	1per team
Clamp, Stoddard test-tube	1 per team
Gloves, disposable, medium	2 per student
Hot plate, 700 W single-burner	1 per team
Medicine cups, (additional)	20 per class
Microscope, the WARD'S Scope	1 per team
Pen, black wax marking for glass	1 per team
Pipet bulbs	1 per team
Pipet, 3 in. glass dropping	1 per team
Pipet, 6 in. nonsterile (additional)	30 per class
Safety goggles, SG34 regular	1 per student
Simulated Urinalysis Lab	1 per class
Stirring rod, 6 in. glass	1 per team
Stopwatch, digital	1 per team
Test-tube rack, 6-well LDPE	1 per team
Test tube with rim, 15 × 125 mm	
Pyrex	8 per team
Uric Acid (WM) slide (additional)	14 per class
Water	400 mL per class

Laboratory Assessment

As a teacher, only you know the best assessment methods to apply to student lab work in your classes.

After each lab, you may want students to prepare a lab report. A traditional lab report usually includes at least the following components:

- **title**
- **summary paragraph** describing the purpose and procedure
- **data tables** and **observations** that are organized and comprehensive
- **answers** to the Analysis and Conclusions questions

You may want students to use *Gowin's Vee*. Or, you may wish to use any of the other rubrics and checklists described below.

Teaching Resources CD-ROM Provides Customizable Rubrics, Checklists, and Gowin's Vees

The **Holt BioSources Teaching Resources CD-ROM** includes fully editable *scoring rubrics* and *classroom-management checklists* to expand your assessment options. These customizable assessment tools can provide a means to objective assessment or serve as a good starting point for designing your own specialized assessment tools. Laboratory-related rubrics and checklists include the following:

- Introduction to Scoring Rubrics
- Student-Designed Experiments Rubric
- Informal Assessment Direct Observations Checklist
- Evaluating Laboratory Work
- Scoring rubrics for *Quick Labs, Inquiry Skills Development* labs, *Laboratory Techniques* and *Experimental Design* labs, and *Biotechnology* labs

Using Gowin's Vee in the Lab contains complete instructions for using Gowin's Vee, for those who wish to experience the benefits of this powerful teaching tool.

- Vee Form Instructions—Teacher's Notes
- Teaching Students to Use the Vee
- Vee Form (a blank Vee form that students can use with any lab)
- Assessing Student-Constructed Vee Reports

The **Holt BioSources Teaching Resources CD-ROM** also includes the following worksheets, which can help you emphasize the importance of safe lab behavior:

- Student Safety Contract
- Safety Quiz

Using the Laboratory Techniques and Experimental Design Labs

This book contains two types of laboratory activities.

- **Laboratory Techniques** labs teach your students laboratory skills and procedures in the context of a real-world biological career.
- **Experimental Design** labs require your students to apply these skills to design their own experiments as they take on the role of an employee at a scientific consulting firm.

By using the Laboratory Techniques and Experimental Design labs, your students will do more than just get the answer. They will learn to:

- use techniques used by working biologists
- explore practical skills needed for successful achievement in the real world
- manage and allocate resources
- communicate effectively
- find information
- work as a team member
- select appropriate technology

Laboratory Techniques Labs

Each *Laboratory Techniques* lab includes a detailed, step-by-step procedure similar to many traditional labs. Unlike *Quick Labs* and *Inquiry Skills Development* labs, Laboratory Techniques labs are placed in the context of real-life occupational scenarios. Students learn practical lab skills, procedures, and fundamental concepts of biology while they discover the role of biology in various careers.

The parts of a Laboratory Techniques lab are described below.

- *Skills* identifies the techniques to be learned or practiced.
- *Objectives* describes the expected accomplishments.
- *Materials* is a list of the items required to perform the lab.
- *Purpose* describes the real-life setting.
- *Background* provides additional information needed for the lab.
- *Procedure* is a step-by-step explanation for doing the lab, including safety precautions and instructions for cleanup and disposal.
- *Analysis* questions ask students to examine the techniques and skills they used in the lab and to interpret their findings.
- *Conclusions* requires students to form opinions based on their observations and the data they collected.
- *Extensions* invites students to find out more about topics related to the subject of the lab.

Using Experimental Design Labs

Each *Experimental Design* lab invites students to join BioLogical Resources, Inc., a biological consulting firm. As employees of the firm, students are given a business letter from a "client" that outlines a problem, and a memo from their supervisor that offers suggestions and guidance. Based on the problem, they must create a plan for their procedure, select supplies and equipment, and conduct the work to earn a profit for the firm.

> To ensure best results and minimal safety hazards, do not perform any *Experimental Design* lab without first performing its prerequisite *Laboratory Techniques* lab, if there is one indicated at the beginning of the lab.

The parts of an Experimental Design lab are described below.

- *Prerequisites* identifies the Laboratory Techniques lab that introduces procedures used.
- *Review* lists concepts students will need to know for the lab.
- *Business letter* poses a problem for students to solve.
- *Memorandum* is a note from the supervisor directing the student to perform the work. The memorandum also contains the following:
    - *Proposal Checklist,* which spells out what students are to do before the lab and which must be approved by you
    - *Report Procedures,* which spells out what students should include in their final report
    - *Required Precautions,* which lists the safety procedures students must follow.
    - *Disposal Methods,* which provides instructions for the handling and removal of all materials used.
- *Materials and Costs* includes everything students should need to do the lab plus some nonessential items, forcing students to think carefully and to request only what they need.

Experimental Design labs require an approach different from most traditional labs.

If your students are new to working independently, they may initially feel uncomfortable when left on their own to propose a procedure for a self-directed lab. Be sure to take the following steps to help boost their confidence:

- **Perform the Laboratory Techniques** lab prerequisites, if there are any. Most Experimental Design labs are preceded by at least one prerequisite Laboratory Techniques lab.
- **Completely discuss the Laboratory Techniques lab prerequisites** with your students before they begin an Experimental Design lab.
- **Suggest that students use materials and equipment similar to those in the Laboratory Techniques** lab if they are unsure about how to begin.
- **Remind students to concentrate on what they need to know and on how they will measure it** so that they can remain focused.

- **Point out that it is more important to understand** what is going on in the lab than it is to perform the lab with excellent technique but no understanding.
- **Provide leading questions for students to consider** as they make their plans. Questions in the *Teacher's Notes and Sample Solutions* section that follows will help you underscore the applicability of techniques learned earlier.
- **Hold a question-and-answer session** before students begin the lab.
- **Hint (don't tell) that some of the items listed may be unnecessary.**
- **Remind students to check the Memorandum** in the Experimental Design lab for requirements for their proposal and final report.
- **Be thorough in helping students to develop their proposal,** as this will form the basis of their procedure. Each proposal should include:
    - the question(s) to be answered
    - the procedure to be used
    - at least one detailed data table
    - a list of proposed materials and costs
- **Emphasize that students cannot begin their procedure until you have approved their proposal and signed their *Proposal Checklist.***
- **Students can prepare their own invoice or use the *Materials and Costs* list of their lab as an invoice.** Give students the following tips for preparing an invoice:
    - An invoice should be generated from an itemized list of the costs incurred to perform the work requested by a client. Students should show the actual amount of time taken and the materials used, regardless of what they projected in their proposal.
    - An invoice should group itemized costs in meaningful categories, such as *Facilities and Equipment Use* and *Labor and Consumables.*
    - If you "fine" students for safety violations, have them subtract the fine when computing the *Subtotal* of their costs. Be sure that students notice the effect of the fine on the *Total Amount Due*—it will reduce the amount of profit made on the job.
    - To remain in business, the firm must show a profit. Students can compute the *Profit Margin,* which is the difference between the cost of the work and the amount the client is charged, by multiplying the *Subtotal* by a percentage, such as 0.25 (25 percent) or 0.30 (30 percent). Have students add the *Subtotal* and *Profit Margin* to compute the *Total Amount Due.*

Teacher's Notes and Sample Solutions ensure a safe, successful lab.

The pages that follow contain Student Objectives, Preparation Notes, Procedural Notes, and Sample Solutions for each of the Experimental Design labs.

EXPERIMENTAL DESIGN

C2 *Using a Microscope— Criminal Investigation*

Purpose
Students will apply the microscope skills learned from **Laboratory Techniques C1: Using a Microscope.**

Student Objective
Design an experiment to analyze trace evidence.

PREPARATION NOTES
Time Required: two to three 50-minute periods

Pre-Lab Discussion
Review the procedure used in **Laboratory Techniques C1: Using a Microscope,** and answer any questions students may have. Discuss with students cleanup procedures for this lab. Ask students the following questions to guide their thinking:
- How will you go about observing the evidence? *(The contents of each bag of evidence should be observed under the microscope and without the microscope.)*
- What materials will you need to complete this investigation? *(microscope, slides, coverslips, water, medicine dropper, lens paper)*
- How will you set up your data chart so that all observations can be recorded accurately? *(Students' data charts should include columns for the Crime Scene Evidence bag, Suspect A, Suspect B, and Suspect C. Rows could represent types of evidence.)*

Preparation Tips
- Prepare four large plastic bags labeled "Crime Scene Evidence," "Suspect A," "Suspect B," and "Suspect C." The bags should contain the items indicated in the sample data table shown in the Sample Data Table below.
- Provide separate containers for collecting broken glass and evidence that has not been destroyed.

Disposal
- Used evidence can be reused or thrown away.
- Evidence that has not been destroyed can be stored for reuse.

Results

PROCEDURAL NOTES
Safety Precautions
- Discuss all safety symbols and precautions with students.
- Remind students never to use direct sunlight as a light source when they are using a microscope with a mirror. Using direct sunlight will cause eye damage.

Procedural Tips
- Caution students never to use the coarse adjustment at high power.
- Tell students that they should always begin on low power when focusing the microscope.

SAMPLE SOLUTION

Question
Which suspects, if any, should be eliminated from the suspect list?

Proposed Procedure
1. Open the Crime Scene Evidence bag, and sort the materials. Record your observations in a data table.
2. Use the microscope to examine each type of evidence in the bag. Record your observations in the data table.
3. Repeat steps 1 and 2 for each of the remaining evidence bags.

Materials (per lab group)
- compound microscope
- 4 microscope slides with coverslips
- eyedropper
- 100 mL beaker
- crime-scene evidence bag
- evidence bags A, B, and C

Estimated Cost of the Job
$1,573.00, including a 30 percent profit (based on two students per group, two hours of labor, and two days of facility use). Refer to *Using Experimental Design Labs* to see how to prepare an invoice.

Conclusions
Students should conclude that Suspects A and C can be linked to the crime scene and should be investigated further. Suspect B had little matching evidence and may be dismissed as a suspect.

Sample Data Table

Material	Crime Scene	Suspect A	Suspect B	Suspect C
observations of material in bag	soil and sand, salt, red thread, green thread, flower petals, blond hair, black hair	soil, salt, green thread, flower petals, blond hair	grass and soil, sugar and cornmeal (yellow grains), leaves, cat hair	potting soil and sand, salt, red thread, flower petals, black hair

EXPERIMENTAL DESIGN

C3 *Using A Microscope—Slowing Protozoans*

Purpose

Students will observe the movements of single-celled organisms and test several dilutions of a protist-slowing agent to determine which is the most cost-effective for identifying unicellular organisms.

Student Objectives

- *Observe* two types of cellular movement and the effects of different concentrations of slowing agent on both types of movement.
- *Determine*, experimentally, the best concentration of slowing agent for identifying microorganisms.
- *Calculate* the cost-effectiveness of different concentrations of slowing agent.

PREPARATION NOTES

Time Required: one 50-minute period

Pre-Lab Discussion

Review the procedure used in **Laboratory Techniques C1: Using a Microscope,** and answer any questions students may have. Ask students the following questions to guide their thinking:

- What are the three main forms of cellular movement, and what are some common examples of unicellular organisms that move by these means? *(The euglena propels itself with its flagellum, a tail-like body extension that moves back and forth. Amoeba move using finger-like pseudopods. Also used to capture food, pseudopods are actually extensions of the cell membrane that are pushed out like fluid feet. Paramecia are covered with fine hairlike projections called cilia, which move against the fluid like thousands of little oars.)*
- How do you dilute a solution so that it has a 2:1 ratio of water to solution? *(Add two parts water to one part solution.)*

Preparation Tips

- Provide samples of *Paramecium* and *Euglena*.
- Provide separate containers for collecting broken glass and leftover solutions.

Disposal

- Slides and coverslips can be washed in the sink.
- Used culture water can be poured down the drain or stored for reuse.

PROCEDURAL NOTES

Safety Precautions

Discuss all safety symbols and precautions with students.

Procedural Tips

- Have students determine the direction of the organism's motion and *then* move the slide so that the organism is at the edge of the field of vision and moving across the field. It may take students several tries to get an accurate reading.
- Remind students to use low power at first and then adjust to a higher power if necessary.
- Students can dilute the slowing agent on the slide by increasing the ratio of culture water to slowing agent. For example, to produce a solution with a 2:1 culture-water-to-slowing-agent ratio, one would add two drops of culture water to one drop of slowing agent and mix.

SAMPLE SOLUTION

Question

What is the lowest concentration of slowing agent that will cause test subjects to move across the field of vision in no less than 10 seconds?

Proposed Procedure

1. Put on safety goggles and a lab apron.
2. Use a sterile pipet to prepare a wet mount of the *Euglena* culture water.
3. Time and record the number of seconds it takes for the organism to cross the field of vision.
4. Prepare a second wet mount using the 2 percent slowing agent. Do this by adding one drop of culture water and one drop of slowing agent to the slide and mixing the two substances with a toothpick. Use a separate sterile pipet for the *Euglena* culture water. Then repeat step 3.
5. Prepare a third and fourth wet mount in the same manner, increasing the amount of culture water by one drop each time. Repeat step 3 for each test.
6. Repeat steps 2–5 for the *Paramecium* sample using a new sterile pipet.
7. After completing all of the measurements, pool the data from all of the research groups. Determine the most dilute concentration of slowing agent that slows the fastest organism to a travel time at or above 10s.

Materials (per lab group)

- safety goggles
- lab aprons
- compound microscope
- 10 depression slides with coverslips
- 3 pipets

EXPERIMENTAL DESIGN C3 *continued*

- clock or watch with second hand
- 11 drops of *Paramecium* in culture water
- 11 drops of *Euglena* in culture water
- 8 drops of slowing agent
- 10 toothpicks

Estimated Cost of the Job
$898.30, including a 30 percent profit (based on two students per group and one hour of labor). Refer to *Using Experimental Design Labs* to see how to prepare an invoice.

Results
The data below represent a pool of all class data.

Conclusions
Answers may vary, but most students should conclude that Mr. Hayame could use a slowing-agent-to-culture-water ratio as low as 1:4 and still have enough time to identify the organisms.

Sample Data Table

Organism	Movement w/out slowing agent (in seconds)	Movement in 1:1 dilution (in seconds)	Movement in 1:2 dilution (in seconds)	Movement in 1:3 dilution (in seconds)	Movement in 1:4 dilution (in seconds)
Paramecium	4	>50	21	14	10
Euglena	8	>50	48	32	26

EXPERIMENTAL DESIGN

C6 Observing the Effect of Temperature on Enzyme Activity

Purpose
Students will measure the rate of an enzyme-catalyzed reaction and determine the effect of temperature on reaction rate.

Student Objectives
- *Design* an experiment that tests the effect of temperature on the rate of the reaction of hydrogen peroxide and catalase.
- *Relate* changes in temperature to changes in the rate of an enzyme-catalyzed reaction.

PREPARATION NOTES
Time Required: one 50-minute period

Pre-Lab Discussion
Review the procedure from **Laboratory Techniques C5: Observing the Effect of Concentration on Enzyme Activity,** and answer any questions students may have. Ask students the following questions to guide their thinking:
- What is a catalyst? *(a chemical that accelerates a reaction but does not change during the reaction)*
- What is the catalyst in this reaction? *(catalase)*
- What is a substrate? *(the reactant in any enzyme-catalyzed reaction)*
- What is the substrate in this reaction? *(hydrogen peroxide)*
- What are the products of this reaction? *(water and oxygen gas)*
- Why is this procedure useful for measuring the rate of this particular reaction? *(Oxygen forms bubbles; the more bubbles produced, the faster the disk rises.)*

Preparation Tips
- Prepare the catalase solution by adding 1 mL of catalase concentrate to 700 mL of distilled water at room temperature.
- Do not purchase or store hydrogen peroxide with a concentration greater than 3%. Prepare a dilute hydrogen peroxide solution by adding 33 mL of 3 percent hydrogen peroxide solution to 1 L of distilled or tap water and mix. Keep this solution away from light.
- Make filter paper disks by using a hole punch and #1 filter paper.
- Prepare a container of ice just before the beginning of the lab.
- Provide separate containers for collecting broken glass, unused hydrogen peroxide, and unused catalase.

Disposal
Unused catalase and dilute hydrogen peroxide can be stored for future use if kept refrigerated and away from light or they can each be flushed down the drain with copious amounts of water.

PROCEDURAL NOTES

Safety Precautions
- Discuss all safety symbols and precautions with students.
- Have students wear oven mitts when handling hot objects.

Procedural Tips
Remind students to rinse the graduated cylinder

EXPERIMENTAL DESIGN C6 *continued*

thoroughly between measurements of hydrogen peroxide and catalase solutions.

SAMPLE SOLUTION

Question
How does temperature change affect the reaction rate of hydrogen peroxide and catalase?

Proposed Procedure
1. Put on safety goggles, disposable gloves, and a lab apron.
2. Place a 50 mL beaker containing 50 mL of dilute hydrogen peroxide on a hot plate at medium heat. Heat the hydrogen peroxide to 80°C.
3. While the hydrogen peroxide is warming, use forceps to pick up a filter paper disk, and submerge it in a medicine cup filled with 10 mL of enzyme solution for 5 seconds.
4. Remove the disk, and blot it dry on a paper towel.
5. Using oven mitts, remove the hydrogen peroxide from the hot plate.
6. Drop the disk into the hydrogen peroxide. Begin timing as soon as the disk touches the surface of the hydrogen peroxide. Measure the depth in millimeters of the disk at its deepest position.
7. Stop timing when the disk rises to the surface. Use forceps to remove the disk and discard it.
8. Record the time, temperature, and distance traveled in a data table.
9. With the thermometer in the beaker of hydrogen peroxide, hold the beaker in a container of ice. Stir the hydrogen peroxide until the temperature drops 10°C from the last reading.
10. Repeat steps 3–9 for every 10°C drop in temperature. Continue until the hydrogen peroxide reaches 20°C.

Materials (per lab group)
- safety goggles
- disposable gloves
- lab aprons
- oven mitts
- 50 mL beaker
- thermometer
- hot plate
- forceps
- metric ruler
- 10 mL graduated cylinder
- container of ice
- watch or clock with second hand
- 50 mL of dilute hydrogen peroxide
- 10 mL of catalase solution
- medicine cup
- 4 paper towels
- 7 filter-paper disks

Estimated Cost of the Job
$914.29, including a 30 percent profit (based on two students per group and one hour of labor). Refer to *Using Experimental Design Labs* to see how to prepare an invoice.

Results
The reaction rate was slower when carried out at low and high temperatures. Students may recognize that between 20°C and about 40°C, increasing temperature increases reaction rate. After that, reaction rate declines.

Sample Data Table

Temperature (°C)	Time (in seconds)	Distance (mm)	Reaction Rate (mm/s)
80	18	46	2.6
70	14	46	3.3
60	10	46	4.6
50	7	43	6.1
40	6	43	7.2
30	11	43	4.2
20	13	46	3.5

Sample Graph

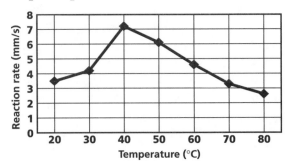

Conclusions
Students' data should support their conclusions. Students should find that the filter-paper disk assay can be used to test the effect of temperature on the reaction rate between catalase and hydrogen peroxide.

EXPERIMENTAL DESIGN

C8 *Measuring the Release of Energy—Best Food for Yeast*

Purpose
Students will determine the best sugar-based medium for yeast growth.

Student Objectives

- *Compare* three different sugar solutions for their effect on yeast growth.
- *Use* changes in temperature to monitor yeast growth.
- *Infer* the best food for promoting yeast growth.

PREPARATION NOTES

Time Required: one or two 50-minute periods

Pre-Lab Discussion

Review the procedure from **Laboratory Techniques C7: Measuring the Release of Energy from Sucrose,** and answer any questions students may have. Ask students the following questions to guide their thinking:

- What would a change in temperature indicate about the yeast? *(A temperature change may indicate a change in the rate of cellular respiration.)*
- What is the relationship between temperature and cellular respiration? *(Temperature increases as the rate of cellular respiration, or fermentation, increases.)*
- Why are yeast particularly abundant on flowers and ripening fruit? *(Yeast grow best in areas with high concentrations of sugar.)*
- What are the final products of yeast aerobic respiration? *(carbon dioxide and water)*
- What are the final products of yeast anaerobic fermentation? *(carbon dioxide and ethanol)*
- What is the purpose of using limewater in this experiment? *(Limewater reacts in the presence of CO_2, indicating that cellular respiration is indeed occurring.)*

Preparation Tips

- Prepare fructose solution by dissolving 100g of fructose in distilled water for a total volume of 500 mL. Repeat this procedure for lactose and sucrose.
- Assemble glass tubing, thermometers, and two-holed rubber stoppers according to directions in the teachers' notes for **Laboratory Techniques C5.**
- Provide a container for collecting broken glass.

Disposal

All solutions and mixtures may be poured down the drain with running water.

PROCEDURAL NOTES

Safety Precautions

Discuss all safety symbols and precautions with students.

Procedural Tips

- Have students design and follow a temperature-reading schedule. Each student should have a designated time to take temperature readings between classes.
- The best way to collect data for this experiment is to use an MBL or CBL temperature probe with a computer or graphing calculator.
- Because part of the thermometer will be covered by the stopper, it may be necessary for students to move the thermometer in order to read it. Caution them not to allow any air in the flask.
- You may wish to have students take an additional reading after 24 hours.

SAMPLE SOLUTION

Question

What is the best food for yeast growth?

Proposed Procedure

1. Put on safety goggles and a lab apron.
2. Pour half a package of dried yeast into each of four vacuum flasks. To the first flask, add 500 mL of distilled water. Label this flask "Control." Use a sterile stirring stick to stir until the yeast are completely suspended in the water. Stopper the flask with a prepared stopper assembly (two-holed rubber stopper with thermometer or temperature probe and glass tubing) so that the bulb of the thermometer is immersed in the water.
3. Connect one end of a 30 cm to 40 cm piece of rubber tubing to the glass tubing. Place the other end of the rubber tubing in a flask filled with limewater.
4. Repeat steps 2 and 3 for each of the sugar solutions, using 500 mL of each sugar solution and labeling each flask according to the solution.
5. Allow the flasks to sit for two days. Then take a temperature reading of each flask every hour for as many hours as possible. Record these readings and the time they were taken in a data table. Check the limewater for cloudiness, which indicates the presence of carbon dioxide. Record observations in a data table.

Materials (per lab group)

- safety goggles
- lab aprons
- 4 vacuum bottles
- 4 stopper assemblies w/ thermometers
- 4 flasks
- clock or watch
- 2 packages of dried yeast
- 80 mL of limewater
- 500 mL of distilled water

EXPERIMENTAL DESIGN C8 *continued*

- 500 mL of 20% fructose solution
- 500 mL of 20% lactose solution
- 500 mL of 20% sucrose solution
- 4 pieces of rubber tubing, 40 cm each
- wax pencil

Estimated Cost of the Job

$2,043.80, including a 30 percent profit (based on two students per group, two hours of labor, and two days of facility use). Refer to *Using Experimental Design Labs* to see how to prepare an invoice.

Results

Student data will vary. Reasons for lack of agreement between student data may include but not be limited to differences in starting temperature and different amounts of yeast used. Students should notice a reaction with the limewater for fructose and sucrose only.

Sample Data Table

Reading	Temperature			
	Control	Fructose	Lactose	Sucrose
1st hour	23°C	23°C	23°C	23°C
2nd hour	23°C	23°C	23°C	25°C
3rd hour	23°C	24°C	23°C	26°C
4th hour	23°C	24°C	23°C	27°C
5th hour	23°C	25°C	23°C	27°C

Graph

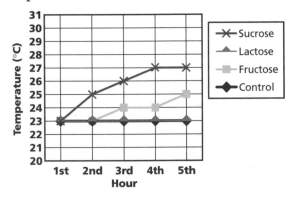

Conclusions

Students' data should support their conclusions. The fructose and sucrose yield the high rates of growth. Answers may vary, but students should find that both the fructose and sucrose make good mediums for growth.

EXPERIMENTAL DESIGN

C10 *Preparing a Root Tip Squash—Stopping Mitosis*

Purpose

Students will design an experiment to test the effect of caffeine on mitosis.

Student Objectives

- *Prepare* a stained slide of an onion root tip in the process of mitosis.
- *Identify* the stages of mitosis in onion root tip cells.
- *Evaluate* the effect of caffeine on mitosis.

PREPARATION NOTES

Time Required: two 50-minute class periods

Pre-Lab Discussion

Discuss with students the procedure from **Laboratory Techniques C9: Preparing a Root Tip Squash,** and answer any questions students may have. Review the stages of mitosis. Identify the name, events, and chromosome positions of prophase, metaphase, anaphase, and telophase. Show students a slide of a preserved onion root tip so that they will know what a well-prepared slide looks like. Model for students the techniques used in making a root-tip squash. Explain to students that only the very tip of the onion root will be stained. Ask students the following questions to guide their thinking:

- What is the purpose of staining the cells? (*to make chromosomes more visible*)
- What would be a suitable control for this experiment? (*Two controls are needed: an untreated slide to determine whether caffeine has any effect on mitosis, and a colchicine-treated slide to compare the effects of the caffeine with that of the colchicine.*)

Preparation Tips

- Purchase a prepared slide of an onion root tip treated with colchicine.
- Purchase onions at least five to six days before starting the lab. Green salad onions or white onions will suffice. Place the onions in water for two to three days or until the root tips are about 2 cm long. Each lab group will need six root tips.
- Soak half the root tips in a container of distilled water, and soak the other half in a container of 0.5% caffeine solution for 45–60 minutes. The root tips should still be attached to the onion.
- Prepare a fixative solution by mixing 25 mL of glacial acetic acid with 75 mL of 95 percent ethanol. Be sure to wear safety goggles, disposable gloves, and a lab apron, and work under a fume hood whenever handling glacial acetic acid. Also be sure

that there are no open flames in the room when handling ethanol.

- After the root tips have been soaked in caffeine, cut them from the onion stem and place each group of root tips in 100 mL of fixative solution. Let them sit inside a fume hood for 48 hours at room temperature. Label each group "distilled water" or "caffeine solution."
- Root tips can be stored in a 70% ethanol solution for two months or in distilled water for 48 hours after being fixed.
- Purchase 1 M HCl from WARD'S ahead of time.
- Provide separate containers for collecting broken glass, root tips, stain, and unused hydrochloric acid.

Disposal

- Root tips can be placed in a trash can.
- Dispose of all volumes of leftover HCl and fixative solutions less than 250 mL by neutralizing each with NaOH until the pH reaches a range of 5–9, and diluting it with 20 times as much water. Pour the resulting mixture down the sink with running water.
- To dispose of aceto-orcein stain, follow the directions on the supplier's MSDS.

PROCEDURAL NOTES

Safety Precautions

- Discuss all safety symbols and precautions with students.
- Remind students never to use direct sunlight as a light source if they are using a microscope with a mirror. Using direct sunlight will cause eye damage.
- Have students use plastic coverslips to avoid breakage.

Procedural Tips

- Not all cells will exhibit mitosis. Finding cells that do exhibit mitosis may require close and careful observation.
- Students should make at least three slides for each solution.
- Many students may have success with only one out of three slides. If they do not achieve a successful stain, encourage them to view a slide from another group.
- Warn students that pressing too hard when squashing the root tip could tear the cells apart or cause the coverslip to break.
- Have students wrap their slides in a paper towel before applying pressure and use a pencil eraser instead of their thumb. This will prevent fingerprints and broken coverslips.
- You may wish to have students use a mounting media to make permanent slides.

SAMPLE SOLUTION

Question

Does caffeine stop mitotic divisions in root tips?

Proposed Procedure

1. Put on safety goggles and a lab apron.
2. Using clean forceps, place three pieces of onion root tip that have been soaked in distilled water in a test tube with 1 M hydrochloric acid (HCl). Use enough to cover the root tip. Let the root tips soak in the HCl for eight minutes while the tubes are placed in a 60°C hot water bath.
3. Using a pipet, draw off as much HCl as possible, and dispose of the HCl in the container provided. Remove each root tip and pulverize it.
4. Cover the root tips in the test tube with aceto-orcein stain. Allow this to sit for about 10–15 minutes, or until the root tip turns bright pink.
5. Place one root tip in the center of a glass slide. Use a single-edged razor blade to cut the last 2 mm from the rest of the root tip.
6. Add a coverslip and squash the root tip.
7. Place the slide under the microscope and make and record observations at 100× and 400× magnifications.
8. Repeat steps 2–7 with other root tips from the same container until three successful slides are made and observed.
9. Repeat steps 2–8 for the root tips soaked in caffeine.
10. Compare the observations for the three types of root tips.

Materials (per lab group)

- safety goggles
- lab aprons
- compound microscope
- 6 microscope slides
- 6 plastic coverslips
- hot-water bath
- 500 mL beaker
- forceps
- pipet
- 2 test tubes
- test-tube rack
- single-edged razor blade
- 3 caffeine-treated root tips
- 3 untreated root tips
- prepared slide of onion root tip with colchicine
- 5 mL of 1 M HCl
- bottle of aceto-orcein stain
- wax pencil

EXPERIMENTAL DESIGN C10 *continued*

Estimated Cost of the Job

$1770.60, including a 30 percent profit (based on two students per group, two hours of labor, and two days of facility use). Refer to *Using Experimental Design Labs* to see how to prepare an invoice.

Results

Students should observe that the control cells were present in many stages of mitosis and that no cells treated with caffeine completed metaphase.

Sample Data Table

Slide	Observations	
	Control	With caffeine
1	cells in many phases	cells arrested in pro-metaphase
2	cells in many phases	cells arrested in pro-metaphase
3	cells in many phases	cells arrested in pro-metaphase

Conclusions

Students' results should support their conclusions. Answers will vary, but students should conclude that caffeine is an adequate substitute for colchicine in preventing cells from completing metaphase and thereby inducing polyploidy.

EXPERIMENTAL DESIGN

C12 *Analyzing Corn Genetics*

Purpose

Students will experimentally determine the germination rate and survival rate of corn from normal seeds and seeds containing the mutation for albinism.

Student Objectives

- *Determine* the germination rate of three lots of corn seeds.
- *Identify* any mutations in the corn that might reduce a farmer's yield.
- *Develop* a hypothesis to explain the poor yield of a corn crop.

PREPARATION NOTES

Time Required: This project will take three to four weeks to complete. About one 50-minute period will be needed for planting, and a few minutes of class time will be required periodically for watering plants and collecting data. This can be done during the first few minutes of class or between classes.

Pre-Lab Discussion

Discuss with students how the seeds will be divided and distributed to each lab team. Explain that the seeds should be allowed to grow until any obvious genetic problems are observed. Ask students the following questions to guide their thinking:

- How should the corn seeds be planted? (*Seeds should be sewn about 3 cm to 6 cm apart at a depth no greater than one seed's length.*)
- How often should the seeds be watered? (*Watering should be checked daily. The soil should remain damp but not drenched.*)
- What are the germination rate and survival rate, and how do they affect a farmer's crop? (*The germination rate is the percentage of planted seeds that sprout. The survival rate is the percentage of sprouted seeds that survive to maturity. Together, these two rates determine the total crop yield.*)
- How do you calculate the germination and survival rate? (*Students should use the following formulas: for germination rate,* number of germinated seeds/number of seeds planted $\times$ 100 = percent germination, *and for survival rate,* number of surviving plants/number of germinated seeds $\times$ 100 = percent survival.*)

After albino corn plants appear, ask students the following questions to guide their thinking:

- What is absent from the white (albino) corn plants? (*chlorophyll*)
- What caused the albino plants to occur? (*a mutation*)
- What will happen to the albino plants and why? (*They will die soon after germination, because without chlorophyll, these plants are unable to produce their own food.*)
- What are the genotypes of the parents of an albino individual? (*They would have to be heterozygous, because all albino, or homozygous, plants die before they can reproduce.*)
- What can the Punnett square reveal about crop yield? (*Nothing. In a population of corn plants, there are many different parents with a different genetic makeup. A Punnett square can only predict the offspring of two individual parents.*)

Preparation Tips

- Provide enough seeds for each student group to receive 10 seeds per lot. The plant tray or pot should be large enough to grow 30 seeds without overcrowding the seedlings.
- Designate one lot of seeds (in this case, lot B) as defective (albino). Purchase albino seeds for this lot. The seeds should come in a 3:1 green to albino ratio.
- Provide a container for collecting used potting soil for disposal or reuse.

EXPERIMENTAL DESIGN C12 *continued*

Disposal

Corn plants and potting soil can be thrown away or reused.

PROCEDURAL NOTES

Safety Precautions

- Discuss all safety symbols and precautions with students.
- Remind students to wash their hands with antibacterial soap before leaving the lab.

Procedural Tips

SAMPLE SOLUTION

Question

What is the germination rate of Mr. Clements's corn seed?

Proposed Procedure

1. Fill a plant tray about two-thirds full with potting soil. Dampen the soil.
2. Plant 10 seeds from each lot in separate rows (or pots). Make sure the seeds are labeled by lot. Place the seeds about 3 cm to 6 cm apart and at a depth of about 0.5 cm in the soil, or a depth no greater than the length of the seed.
3. Place the plant tray where it can receive light. Water the soil daily to keep the seedlings damp but not drenched.
4. Record in a data table the number of normal plants and the number of abnormal plants that germinate.
5. Allow the corn to grow for two to three weeks past germination.
6. Record the number of surviving corn plants in a data table.

7. Pool the class data, and determine the germination rate and survival rate for each lot.

Materials (per lab group)

- plant tray
- 3 kg of potting soil
- 10 corn seeds each from lots A, B, and C
- water

Estimated Cost of the Job

$14,216.80, including a 30 percent profit (based on two students per group, 22 days of facility use, and 2 1/2 total hours of labor). Refer to *Using Experimental Design Labs* to see how to prepare an invoice.

Results

See sample data tables below.

Conclusions

Students should determine that the germination rate for all lots is about 95 percent. The survival rate, however, varies from lot to lot. Students should conclude that the testing laboratory was accurate in reporting a 95 percent germination rate. But students should detect a survival rate of 74 percent in lot B. This number may vary, depending on the original number of albino seeds in lot B. Students may correctly conclude that the testing laboratory would be at fault only if it were asked to report the survival rate or predict crop yield. Students may recommend that Mr. Clements compare these results with the original data he received from the other company.

Sample Group Data Table

Lot number	Number of seeds	Number normal plants germinated	Number albino plants germinated	Number normal plants @ 4 wk	Number albino plants @ 4 wk
A	10	9	0	9	0
B	10	7	3	7	0
C	10	8	0	8	0

Sample Class Data Table

Lot number	Number of seeds	Number normal plants germinated	Number albino plants germinated	Number normal plants @ 4 wk	Number albino plants @ 4 wk	Total germination rate	Total survival rate
A	50	48	0	48	0	96%	100%
B	50	35	12	35	0	94%	74%
C	50	47	0	47	0	94%	100%

EXPERIMENTAL DESIGN

C15 *Karyotyping—Genetic Disorders*

Purpose
Students will complete a karyotyping form and interpret the results.

Student Objectives
- *Create* a human karyotype by arranging chromosomes in order by length, centromere position, and banding pattern.
- *Identify* a karyotype as normal or abnormal.
- *Describe* conditions that can be detected from abnormal karyotypes.

PREPARATION NOTES
Time Required: one 50-minute period for analysis, plus any time needed for additional research

Pre-Lab Discussion
Review the procedure from **Laboratory Techniques C14: Karyotyping,** and answer any questions students may have. Define the following terms for students: *amniocentesis,* a medical procedure in which cells are taken from a sample of fluid surrounding the fetus and analyzed for irregularities in the number or structure of the chromosomes; *chorionic villi sampling,* a procedure similar to amniocentesis in which a tiny piece of embryonic membrane is removed; and *smear,* a sample of cells that have been fixed to the surface of a microscope slide.

Explain that amniocentesis and chorionic villi sampling are done to detect chromosome disorders, such as Down syndrome, in the unborn fetus. Ask students the following questions to guide their thinking:

- Who usually receives amniocentesis? *(It is usually recommended to women over the age of 35 because the risk of chromosome disorders is greater in babies born to women in this age group; they are also recommended to younger mothers when problems are suspected.)*
- What is karyotyping, and how is it done? *(The karyotype of a cell is a classification of the arrangement, number, and structure of its chromosomes. Karyotyping is the process of analyzing a karyotype. This procedure involves arranging the chromosome pairs of a cell according to size and examining them for irregularities.)*
- What is the value of a double-blind study? *(If the conditions affecting the subjects are unknown to both the subject and the experimenter, the study is less likely to be biased by preconceived expectations on the parts of both the research team and the subjects.)*

Preparation Tips
- Prepare copies of karyotype forms in advance.
- You may wish to have students give oral presentations on their karyotype and the condition caused by the abnormal structure. If so, determine the requirements ahead of time.
- Karyotyping forms and karyotypes can be ordered from WARD'S.
- Provide separate containers for collecting paper scraps and chromosome cutouts.

Disposal
Paper scraps and chromosome cutouts can be recycled or reused.

PROCEDURAL NOTES

Safety Precautions
Discuss all safety symbols and precautions with students.

Procedural Tip
Have students review the normal karyotypes from **Laboratory Techniques C14** to help them determine the ordering of their chromosome pairs and to use as a comparison with abnormal karyotypes.

SAMPLE SOLUTION

Question
What genetic disorder, if any, is indicated by each karyotype?

Proposed Procedure
1. Cut apart the chromosomes on each "photomicrograph," leaving a slight margin around each chromosome.
2. Arrange the chromosomes in pairs. The members of each pair will be the same length and banding pattern and will have their centromeres located in the same area.
3. Arrange the pairs according to their length.
4. Tape each pair of chromosomes to a karyotyping form. Place the centromeres on the lines provided. Place the longest chromosome at position 1, and the shortest at position 22. (Place the two sex chromosomes at position 23.)
5. Analyze each karyotype, and complete the report.

Materials (per lab group)
- chromosome photomicrographs
- scissors
- 1 m of tape
- ruler
- karyotyping form

Estimated Cost of the Job
$769.60, including a 30 percent profit (based on two

EXPERIMENTAL DESIGN C15 *continued*

students per group and one hour of labor). Refer to *Using Experimental Design Labs* to see how to prepare an invoice.

Results

The data below represent a pool of all class data.

Sample Data Table

Smear letter	Result	Condition	Chromosome abnormality
A	abnormal female	Turner's syndrome	45 chromosomes XO
B	abnormal male	Klinefelter's syndrome	47 chromosomes XXY
C	abnormal female	Down syndrome	Trisomy 21, 47 chromosomes extra 21
D	abnormal male	Edwards' syndrome	Trisomy 18, 47 chromosomes extra 18
E	abnormal male	XYY syndrome	47 chromosomes XYY
F	abnormal female	Triple X	47 chromosomes XXX
G	abnormal female	Patau syndrome	Trisomy 13, 47 chromosomes, extra 13
H	normal male		
I	normal female		

Conclusions

Conclusions will vary, depending upon the karyotype of the individual student team. Each group should give a thorough explanation of how to perform a karyotype, what information is given in their karyotype, and what abnormality, if any, is indicated by their results. Students will need to pool their data in order to prepare a complete report.

EXPERIMENTAL DESIGN

C18 *Analyzing Blood Serum— Evolution of Primates*

Purpose

Students will use the Nutall agglutination technique to determine if a newly discovered primate is closely related to humans.

Student Objectives

- *Make* serial dilutions of blood serum.
- *Analyze* the amount of agglutination that occurs when human antiserum is added.
- *Relate* the amount of agglutination to the evolutionary relationships of these organisms.

- *Construct* an evolutionary tree to show the relationship of these organisms.

PREPARATION NOTES

Time Required: one 50-minute period

Pre-Lab Discussion

Review the procedure used in **Laboratory Techniques C17: Analyzing Blood Serum to Determine Evolutionary Relationships,** and answer any questions students may have. Discuss the procedures of the Nutall agglutination reaction, the making of evolutionary trees, and serial dilution. Ask students the following question to guide their thinking:

- How does the agglutination reaction relate to evolutionary relationships? (*If the antiserum of one subject is added to the blood of another subject, the greater the agglutination in the second subject's blood, the closer the subjects are on the evolutionary tree.*)

Preparation Tips

- Using a pipet and capped tubes, place approximately 25 mL of "human antiserum" into each vial for lab groups.
- Replace the label on the dropper bottle labeled "monkey serum" with one labeled "unknown primate serum."
- Provide separate containers for collecting broken glass, serum samples, and antiserum.

Disposal

Solutions can be washed down the sink with plenty of running water.

PROCEDURAL NOTES

Safety Precautions

- Discuss all safety symbols and precautions with students.
- Remind students that under no circumstances are they to test any blood other than the simulated blood provided.

Procedural Tips

SAMPLE SOLUTION

Question

What is the evolutionary relationship of the unknown primate to humans?

Proposed Procedure

1. Put on safety goggles, disposable gloves, and a lab apron.
2. Take one testing well tray and label it "human serum." Number the wells 1–8.
3. Use a clean dropper to place eight drops of serum

EXPERIMENTAL DESIGN C18 *continued*

in well 1. In well 2, place seven drops of serum and one drop of water. Use a separate dropper for the water. In well 3, place six drops of serum and two drops of water. With each consecutive well, continue decreasing the amount of serum and increasing the amount of water by one drop each so that well 8 has one drop of serum and seven drops of water.

4. Using a new plastic dropper, place eight drops of human antiserum in each well, and stir using a stirring stick.

5. Observe the wells for two minutes, occasionally stirring the well. After two minutes, record your results (ignore any reaction that may occur after these two minutes).

6. Record your results using the key below.
   +++ Heavy agglutination (fast and definite reaction)
   ++ Medium agglutination (definite but not immediate)
   + Slight agglutination (weak reaction; may take a longer amount of time)
   − No reaction

7. Repeat steps 2–6 for the chimpanzee and unknown primate using the other two trays.

8. Record all observations in a data table.

Materials (per lab group)
- safety goggles
- disposable gloves
- lab aprons
- 3 testing well trays
- 50 mL beaker
- watch or clock
- 1 vial of human antiserum
- 1 vial of human serum
- 1 vial of chimpanzee serum
- 1 vial of serum from unknown primate
- 20 mL of distilled water
- 5 plastic dropping pipettes
- 24 stirring sticks

Estimated Cost of the Job
$929.50, including a 30 percent profit (based on two students per group and one hour of labor). Refer to *Using Experimental Design Labs* to see how to prepare an invoice.

Results
Sample Data Table

Organism	1	2	3	4	5	6	7	8
Human	+++	+++	+++	+++	+++	+++	++	+
Chimp	+++	+++	+++	+++	+++	+++	+	−
Unknown primate	+++	+++	+++	+++	+++	++	−	−

Conclusions
Students should conclude that this unknown primate is not the closest relative to humans. The evolutionary tree should place the chimpanzee closer to the human than the unknown primate. The tree should look like this:

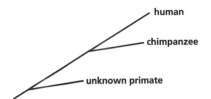

EXPERIMENTAL DESIGN

C21 Observing Animal Behavior—Grant Application

Purpose
Students will prepare a grant application based on results and suggestions from a preliminary study.

Student Objective
Write a grant application based upon a question formed from the preliminary research on an animal.

PREPARATION NOTES

Time Required: one or two 50-minute class periods for preparation; one to two weeks of out-of-class time for completion of the application

Pre-Lab Discussion
Review the procedure from **Laboratory Techniques C20: Observing Animal Behavior,** and answer any questions students may have. Review with students the requirements of the assignment, including research and grant requirements. Also, discuss where to find research on animal behavior. Ask students the following question to guide their thinking:
- What is the difference between an observation and an inference? *(An observation refers only to what is seen, heard, or otherwise perceived of an event. An inference is an interpretation of an observed event.)*

Materials (per lab group)
none

Preparation Tips
- This lab is designed as a continuation of **Laboratory Techniques C20: Observing Animal Behavior.** You may wish to have student groups trade papers or continue their own preliminary studies.
- You may wish to assign student groups to a particular animal ahead of time.
- Meet with students ahead of time to discuss and

EXPERIMENTAL DESIGN C21 *continued*

approve their procedures.
• You may wish to arrange some class time for library research.

Disposal
none

PROCEDURAL NOTES
Safety Precautions
none

Procedural Tips
• Advise students to seek (with their parents' permission) the advice of local animal experts such as pet store owners, zookeepers, and park rangers.
• Encourage students to work as a team and to divide the tasks of research, observation, and writing.

SAMPLE SOLUTION
Description of Procedure
Answers will vary, but students should describe in more detail the goals of the proposed research, procedures, and methods that will be used to test the hypothesis.

Evaluation
Answers will vary, but students should provide some measures for evaluating the accuracy of the study.

Budget
Answers will vary.

Project Staff Vitae
Answers will vary, but should provide complete information about both the "team members" and the "intern."

Materials (per lab group)
none

Estimated Cost of the Job
$320.00 (based on two students per group and four hours of labor). Refer to *Using Experimental Design Labs* to see how to prepare an invoice.

EXPERIMENTAL DESIGN

C23 *Examining Owl Pellets—NW vs. SE*

Purpose
Students will dissect the pellets of both Northwestern and Southwestern barn owls and identify the skulls of prey found within them.

Student Objectives
• *Dissect* owl pellets from two different geographic regions.
• *Sort* the contents of the owl pellets according to bone shape and size.
• *Identify* small mammal skulls found in each type of pellet using a dichotomous key.
• *Infer* regional differences in prey availability based on differences in pellet contents.
• *Predict* the effect that poisoning rats has on the owl population in each region.

PREPARATION NOTES
Time Required: two 50-minutes periods

Pre-Lab Discussion
Review with students the procedure from **Laboratory Techniques C22: Examining Owl Pellets,** and answer any questions students may have. Review the terms *incisors* (front teeth), *diastema* (gap between the front teeth and premolars), *infraorbital canal* (opening just below the eye socket), and *zygomatic arch* (ridge of bone on the side of the skull). Review with students how to calculate biomass and how to use a dichotomous key. Discuss how students will set up their data tables (see sample data table for reference). Ask students the following questions to guide their thinking:

• How are owl pellets of use to scientists? *(Because owls prey on a variety of small mammals, the contents of their pellets can be used to estimate the diversity of available prey in a given region. Also, as a direct indicator of what the owl has fed on, a pellet provides vital information for species management and protection.)*
• What is biomass? *(the total mass of all organisms in a group)*
• How would you find out how much of the owl's diet is composed of voles? *(For each type of prey, multiply the number of specimen found by the average prey mass. This is the prey biomass. Total the biomass for all prey found. Then divide the vole prey biomass by the total biomass and multiply by 100 to get the percentage of total biomass for voles.)*

Preparation Tips
• Each lab group will need a complete dichotomous key for identifying the contents of owl pellets and a set of diagrams illustrating skull characteristics of owl prey. Both the dichotomous key and the diagrams can be ordered from WARD'S.
• Provide students with a list of average body weights for various types of owl prey. These are listed in the Sample Class Data Table that follows.
• Provide separate containers for collecting mammal skeletons and aluminum foil wrappers.

- Purchase owl pellets in advance from WARD'S.

Disposal

- Aluminum foil wrappers can be recycled or thrown away.
- Mammal skeletons can be thrown away or collected for later use.

PROCEDURAL NOTES

Safety Precautions

Discuss all safety symbols and precautions with students.

Procedural Tips

- If the pellets do not come apart easily during dissection, soak them briefly in warm water until they soften.
- Students may find it useful to use egg cartons for storing bones while separating them into similar piles.
- Suggest that students begin sorting the bones first according to shape and then according to size. Remind students that some of the smaller skeletons may have come from immature animals.

SAMPLE SOLUTION

Question

What do the contents of an owl pellet reveal about its diet?

Proposed Procedure

1. Put on disposable gloves and a lab apron.
2. Place an owl pellet from the Northeastern barn owl on a dissecting tray. Remove the aluminum foil casing.
3. Use a dissecting needle and forceps to carefully break apart the owl pellet, taking care to avoid damaging the smaller bones.
4. As each bone is uncovered, use forceps to carefully place it on a piece of paper. Carefully remove all fur and feathers from each bone.
5. Place any undigested beetles or other arthropods in a separate area of the paper.
6. Assemble all the bone parts. Count and record the number of skulls found in the pellet as the number of prey ingested.
7. Sort all bone parts by shape and size.
8. Identify the skulls using the diagrams and a dichotomous key of owl prey skulls.
9. Record the identity, biomass, and number of each type of prey found in a data table.
10. Repeat steps 2–9 for the Southeastern barn owl pellet.
11. Try to reassemble the skeletons found in each owl pellet.
12. Pool all class data. For each type of pellet, multiply the number of each specimen found by prey weight to get the class total of biomass per prey. Then find the total biomass for all prey, and divide prey biomass by total biomass to get a percentage of total biomass. Compare the results for the Northwestern barn owl with those of the Southeastern barn owl.

Materials: (per lab group)

- disposable gloves
- lab apron
- dissecting tray
- dissecting needle
- forceps
- Northwestern barn owl pellet
- Southeastern barn owl pellet
- dichotomous key
- skull characteristics sheet
- 2 sheets of paper

Estimated Cost of the Job

$1,568.00, including a 30 percent profit (based on two students per group, two hours of labor, and two days of facility use). Refer to *Using Experimental Design Labs* to see how to prepare an invoice.

Results

Students will probably find 3–5 specimens in each pellet. Answers will vary, but class data should show that the pocket gopher makes up the largest percentage of the Northwestern barn owl's diet. For the Southeastern barn owl, data should show that the largest percentage of the owl's diet is composed of rats. See the sample class data table that follows.

Conclusions

Students' data should support their conclusions. Students should find that because the diet of the Southeastern barn owl is composed largely of rats, it would be more effected by the poisoning of rats than the Northwestern barn owl.

EXPERIMENTAL DESIGN C23 *continued*

Sample Class Data Table

Prey	Northwestern barn owl				Southeastern barn owl			
	Prey weight	Total number of prey found	Biomass of prey found	Percent of total mass	Prey weight	Total number of prey found	Biomass of prey found	Percent of total mass
Pocket gopher *Thomomys*	150g	7	1050g	44.6%	150g			
Rat								
Sigmodon	100g				100g	7	700g	31.9%
Oryzomys	80g				80g	5	400g	18.2%
Rattus	150g	2	300g	12.7%	150g	2	300g	13.7%
Vole *Microtus*	40g	9	360g	15.3%	40g	5	200g	9.1%
Mice								
Peromyscus	22g	4	88g	3.7%	22g	3	66g	3.0%
Mus	18g	7	126g	5.4%	18g	7	126g	5.7%
Reithrodontomys	12g	4	48g	2.0%	12g	4	48g	2.2%
Perognathus	25g	4	100g	4.2%	25g			
Mole								
Scapanus	55g	2	110g	4.7%	55g			
Scalopus	55g				55g	2	110g	5.0%
Shrew								
Blarina	20g				20g	5	100g	4.6%
Cryptotis	4g				4g	7	28g	1.3%
Sorex	4g	4	16g	0.7%	4g	1	4g	0.2%
Other Prey								
Bats	7g	4	28g	1.2%	7g	2	14g	0.6%
Birds	15g	5	75g	3.2%	15g	3	45g	2.1%
Insects	1g	4	4g	0.2%	1g	3	3g	0.1%
Crayfish	5g	2	10g	0.4%	5g	2	10g	0.5%
Small reptiles	20g	2	40g	1.7%	20g	2	40g	1.8%

EXPERIMENTAL DESIGN

C26 *Assessing and Mapping Factors in the Environment*

Purpose
Students will conduct a study of a local natural area.

Student Objectives
- *Identify* the dominant plant and animal species in a natural area.
- *Identify* and count the various populations of plants and animals in a given area.
- *Measure* wind speed, air temperature, soil temperature, and soil moisture.
- *Analyze* soil samples to determine the pH and levels of nitrate, phosphates, and potassium.
- *Relate* the abiotic factors in an environment to the biotic factors in an environment.
- *Draw* a map showing land characteristics and plant and animal species.
- *Create* a model sign illustrating the dominant plant and animal species.

PREPARATION NOTES
Time Required: three to five days

Pre-Lab Discussion
Discuss with students the procedures from **Laboratory Techniques C24: Mapping Biotic Factors in the Environment** and **Laboratory Techniques C25: Assessing Abiotic Factors in the Environment,** and answer any questions students may have. Review with students how to use field guides, anemometers, protractor apparati, thermometers, and soil test kits. Ask students the following question to guide their thinking:
- What limitations should be considered when mapping the area and suggesting trail locations? *(A trail should be safe and should disturb as little wildlife as possible. Ideally, a trail would show as many different plant and animal locations as possible without disturbing the wildlife.)*

Preparation Tips
- This is an outdoor activity. Be sure to communicate with administrators and obtain permission from parents if necessary.
- You may wish to present this lab activity as a field trip to a local natural area. Students can do field

work in one day and bring back samples for soil tests in the classroom. Alternatively, students could investigate a site on the school property.

- For this activity, select a vegetated area that is representative of the natural vegetation in your area.
- Find out whether there are poisonous plants or dangerous animals where you will be going. Learn to identify any hazardous species.
- Encourage students to be aware of the impact they are having on the environment they visit.
- This lab is designed for groups of four to eight students. Use any number that is convenient for your class size, but remember that too few quadrats may bias your data averages.
- The *Peterson Field Guide* series, *Audubon Field Guides,* and *Golden Guides* are excellent field guides.
- Prepare the protractor apparatus according to instructions in the teachers notes for **Laboratory Techniques C24**.
- Remind students to dress appropriately for field work. Students should wear comfortable clothing, long pants, and sturdy shoes with closed toes.
- Provide separate containers for collecting string, stakes, flags, and protractors.

Disposal

- Live organisms such as insects should be released into the study area after collection.
- Soil-test chemicals can be poured down the drain with running water.
- String, stakes, flags, and protractors can be stored for reuse.

PROCEDURAL NOTES

Safety Precautions

- Discuss all safety symbols and precautions with students.
- Have students wear safety goggles, disposable gloves, and lab aprons when using soil-test kits.
- Before going into the field, discuss with students some of the more dangerous organisms students might find and how they may avoid them. Show them pictures of poison ivy and other plants to avoid.
- Encourage students to bring sunglasses, sunscreen, and insect repellent if needed.
- Discuss field trip safety procedures with students. For example, have students remain within a distance that they can be seen or heard. You may wish to assign student partners for additional safety.

Procedural Tips

- Sketching, photographing, and writing field notes are generally more appropriate than collecting specimen for observation.
- Remind students not to leave garbage behind at the field site. Remind students to leave natural areas just as they found them.
- The plot of land students choose to study should be different from any previously studied.
- You may wish to decrease the quadrat size in areas where vegetation is particularly heavy.
- If there are very large numbers of small plants, such as grasses or forbs, students can estimate the total number for the quadrat by counting plants in a 1 m × 1 m area and multiplying by 25 or by the adjusted size of the quadrat if it has been decreased.
- The amount of shade in each quadrat will vary with the time of day. You may wish to have students record the time of day in their data table.

SAMPLE SOLUTION

Question

What are the abiotic and biotic factors of this environment and how do they interact?

Proposed Procedure

1. Using a meter stick and protractor, mark off a 5m × 5m plot, or quadrat, and divide it into 4 areas using stakes and string.
2. Use plant field guides to identify the plant types. Count each type of plant—(1) grasses, (2) forbs (nonwoody plants other than grasses), (3) vines, (4) shrubs, and (5) trees—in your group's quadrat. Record your data.
3. Select a few plants of each plant type and measure their height. Using a protractor and string, find the approximate height of any trees.
4. Add the total numbers for each plant type to determine the total number of plants in your quadrat.
5. Calculate the percent of all plants represented by the dominant, or most numerous, plant type in your quadrat. Record your results. Use the following formula to calculate percent:

$$\frac{\text{(number of dominant plants)}}{\text{(total number of plants)}} \times 100 = \% \text{ dominant plants}$$

For example:

$$\frac{\text{(300 grasses)}}{\text{(500 total plants)}} \times 100 = 60\% \text{ grasses}$$

6. Compare the vegetation of your quadrat with the class data. Record your observations in a data table.

7. Look for birds, flying insects, and insects that live on plants. Note any evidence of animal activity such as tracks and fecal material. Record your observations in a data table.

8. Using an anemometer, measure and record a wind speed reading for each of the four areas within the plot.

9. Observe your area for amount of shade, and record the approximate amount of shade in a data table.

10. In each of the four areas in your quadrat, measure temperature 4–5 cm above the ground, 1 m above the ground, and 1 cm into the ground. Record the temperature and the depth at which the measurement was taken in a data table.

11. Collect a soil sample from one of the four areas in the quadrat, place it in a sealable plastic bag, and take it to the lab. In the lab, measure 100 g of soil from your sample, and place it in a metal can with both ends removed and filter paper attached. Measure and record the mass of the container with the soil. Place your container in a drying oven set at 100°C for 24 hours.

12. Use oven mitts to carefully remove the container of soil from the oven. Let the contents cool. Then find the mass of the container with the soil, and record it.

13. Calculate and record the mass of the soil moisture in your sample using the following formula: *mass of container with soil before drying – mass of container with soil after drying = mass of soil moisture*

14. Calculate and record the percent of moisture in your soil using the following formula:

$$\frac{(mass\ of\ soil\ moisture)}{(mass\ of\ soil\ before\ drying)} \times 100 = \%\ of\ moisture\ in\ soil$$

15. Obtain a soil-test kit, and carefully read the safety rules included with the kit. Put on safety goggles, disposable gloves, and a lab apron. Following the directions for using the soil-test kit, test your dried soil for pH, nitrate content, phosphate content, and potassium content. Record your results.

16. Pool the class data.

17. Make recommendations about hiking trails, and complete a map and sign according to Mr. Grayson's instructions.

Materials (per lab group)

- safety goggles
- disposable gloves
- lab aprons
- oven mitts
- anemometer
- meter stick
- protractor
- thermometer
- garden trowel
- balance
- drying oven
- field guides
- metal can with both ends cut out
- one sheet of filter paper
- rubber band
- soil-test kit for pH
- soil-test kit for nitrogen
- soil-test kit for potassium
- soil-test kit for phosphorous
- 1 plastic zippered bag
- 8 stakes
- 4 colored flags
- 40 m of string
- nontoxic permanent marker

Estimated Cost of the Job

$2,050.36, including a 30 percent profit (based on four students per group, one day of facility use, four total days of study, and four total hours of labor). Refer to *Using Experimental Design Labs* to see how to prepare an invoice.

Results

Results will vary, depending on the area studied.

Sample Data Table

See **Laboratory Techniques C24** and **Laboratory Techniques C25** for sample data tables.

Conclusions

Answers will vary, depending on the area studied.

EXPERIMENTAL DESIGN

C28 *Studying an Algal Bloom— Phosphate Pollution*

Purpose

Students will test the effects of three levels of phosphate on the growth of algae in lake water.

Student Objectives

- *Design* an experiment to test the effects of various levels of phosphate that have been introduced into an ecosystem.
- *Infer* the effect of the addition of phosphate to a freshwater ecosystem.

PREPARATION NOTES

Time Required: one 50-minute period, plus 15–20 minutes per day for five to seven days

Pre-Lab Discussion

Review the procedure from **Laboratory Techniques C27: Studying an Algal Bloom,** and answer any questions students may have. Discuss with students the concepts of eutrophication and limiting factors. Review the technique for using a grid slide or for dividing the microscope field into quadrats for counting. Ask students the following question to guide their thinking:

- What is an algal bloom? *(An algal bloom is a rapid growth in the population of algae in a lake or pond.)*

Preparation Tips

- Prepare phosphate solution by mixing 0.55 g monosodium phosphate with 99.45 mL distilled water. Mix until thoroughly dissolved. This will make enough for several classes.
- Make "lake water" by mixing cultures of the following algae: *Spirogyra, Chlorella, Chamydomonus,* and *Closetrium.* Use tap water that has been left out to dechlorinate for at least 24 hours.
- Provide separate containers for collecting broken glass, phosphate, lake water, and containers of phosphate and lake water.

Disposal

- Unused lake water and phosphate may be collected and reused.
- Used lake water and phosphate mixtures may be flushed down the drain with copious amounts of water.

PROCEDURAL NOTES

Safety Precautions

- Discuss all safety symbols and precautions with students.

- Remind students never to use direct sunlight as a light source if they are using a microscope with a mirror. Using direct sunlight will cause eye damage.

Procedural Tip

Advise students to take drops of lake water from the surface of the lake water sample.

SAMPLE SOLUTION

Question

What is the effect of low levels of phosphate on algae in lake water?

Proposed Procedure

1. Put on safety goggles and a lab apron.
2. Using a grid slide, make a wet mount of the lake water by putting one drop of lake water on a slide with a coverslip. Count the number of algae in the field of vision by counting the algae in each grid space and then adding them.
3. Prepare three containers of lake water with three different levels of phosphate by adding a different amount of phosphate to lake water. Use 1 mL, 3 mL, and 5 mL of phosphate to 29 mL, 27 mL, and 25 mL of lake water, respectively.
4. Place these preparations in an area where there is plenty of light for the next five to seven days while observations are being made.
5. Use the procedures from step 1 to count the number of cells in slides made from each of the test samples and the original lake water. Make these observations every day for five to seven days.
6. Pool the class data to get a class average for each concentration.

Materials (per lab group)

- lab apron
- safety goggles
- compound microscope
- 24 grid slides with coverslips
- 25 mL graduated cylinder
- 3 small jars or 50 mL beakers
- medicine dropper
- grow lamp
- lake water
- 9 mL of phosphate solution
- 3 adhesive labels

Estimated Cost of the Job

$5,780.84, including a 30 percent profit (based on two students per lab group, seven days of facility use, and one hour of labor, plus 15 minutes a day for six additional days). Refer to *Using Experimental Design Labs* to see how to prepare an invoice.

EXPERIMENTAL DESIGN C28 *continued*

Results

Answers may vary. However, students should find that the greater the phosphate added, the larger the algal bloom.

Sample Data Table

Day	Lake water	Phosphate-to-lake-water ratios		
		1:29	3:27	5:25
1	32	28	30	33
2	34	31	34	37
3	36	39	41	49
4	40	48	53	62
5	43	53	64	77
6	50	63	75	88

Conclusions

Students' data should support their conclusions. Answers will vary, but students will probably infer that: (1) even a small amount of phosphate will increase the population of existing algae; (2) if the growth continues unchecked, the population may increase until the lake begins to show signs of eutrophication; and (3) the death of this algae will affect limiting factors such as dissolved oxygen and food for animals.

EXPERIMENTAL DESIGN

C29 *Classifying Mysterious Organisms*

Purpose

Students will explore the similarities and differences among four single-celled organisms and make recommendations about the identity and classification of each organism.

Student Objectives

- *Identify* four single-celled organisms.
- *Infer* the classification of an organism.
- *Sketch* various single-celled organisms.

PREPARATION NOTES

Time Required: one 50-minute period

Pre-Lab Discussion

Review procedure used in **Laboratory Techniques C1: Using a Microscope,** and answers any questions students may have. Ask students the following questions to guide their thinking:

- What are two major differences among various kinds of protists? *(presence or absence of chlorophyll; method of movement)*
- What are some of the problems biologists have

when they try to classify organisms? *(Some organisms are difficult to classify because they have traits that are common to more than one group. As a result, relationships among organisms are sometimes vague.)*

Preparation Tips

- Prepare the "pond water culture" ahead of time by mixing *Volvox, Paramecium, Amoeba,* and *Euglena* cultures.
- Provide separate containers for collecting broken glass, pond water, and slowing agent.

Disposal

You may wish to have students return live cultures to a separate container. Otherwise, the cultures can be washed down the drain with running water.

PROCEDURAL NOTES

Safety Precautions

- Discuss all safety symbols and precautions with students.
- Remind students never to use direct sunlight as a light source if they are using a microscope with a mirror. Using direct sunlight will cause eye damage.

Procedural Tips

- Remind students never to use the coarse adjustment at high power.
- Remind students that they should always begin on low power when focusing the microscope.

SAMPLE SOLUTION

Question

What is the identity and classification of the unknown organism?

Proposed Procedure

1. Put on safety goggles and a lab apron.
2. Obtain a culture dish with a pond water sample.
3. Place two or three drops of slowing agent on the well of a depression slide.
4. Add a few drops of the pond water sample to the slowing agent.
5. Observe it under the microscope, first under low power and then under high power. Record your observations.
6. Complete a scale drawing and include the magnification of each organism observed in the sample.
7. Identify the types of organisms observed.

Materials (per lab group)

- safety goggles
- lab aprons
- compound light microscope

EXPERIMENTAL DESIGN C29 *continued*

- depression slide with coverslip
- eyedropper
- pond water sample
- 3 drops of slowing agent

Results
Sample Data Table

Observed organism	Observations	Identity	Additional information
1	unicellular; nucleus present; no chloroplasts; moves by pseudopods	Amoeba	a heterotrophic protist (cannot make its own food); pseudopods formed by cytoplasmic streaming; engulfs food with pseudopods; reproduces by mitosis (asexual reproduction)
2	unicellular; nucleus is present; no chloroplasts; torpedo-shaped; moves by tiny hairs	Paramecium	moves with cilia; does not make its own food; ingests food though a gullet; reproduces by mitosis (asexual reproduction) and meiosis (sexual reproduction)
3	seems to be a group of cells; can been seen without microscope; many nuclei present; many chloroplasts	Volvox	autotrophic (can make its own food); moves with cilia (many); actually a colony of single-celled organisms; reproduces by mitosis (asexual reproduction) and meiosis (sexual reproduction)
4	nucleus is present; chloroplasts are present; moves with a flagellum; moves toward light	Euglena	semi-autotrophic; has an "eyespot" that detects the presence of light; reproduces by mitosis (asexual reproduction)

Estimated Cost of the Job

$902.10, including a 30 percent profit (based on two students per group and one hour of labor). Refer to *Using Experimental Design Labs* to see how to prepare an invoice.

Conclusions

Students' data should support their conclusions. The unknown organism is the *Euglena,* a protist.

EXPERIMENTAL DESIGN

C33 Gram Staining of Bacteria—Treatment Options

Purpose

Students will identify an unknown bacterium as Gram − or Gram + using the technique of Gram staining.

Student Objectives:
- *Make* a bacterial smear.
- *Identify* the bacterium as Gram − or Gram +.
- *Suggest* treatment for this bacterium.

PREPARATION NOTES

Time Required: one 50-minute period

Pre-Lab Discussion

Review the procedure from **Laboratory Techniques C32: Gram Staining of Bacteria,** and answer any questions students may have. Also review aseptic technique and safety procedures for using the Bunsen burner. You may wish to demonstrate correct aseptic technique, procedures for making a bacterial smear, and procedures for fixing a slide. Ask students the following questions to guide their thinking:

- How are eubacteria classified? *(Eubacteria are commonly classified by differences in their cell walls. A bacterium with a cell wall containing a large amount of a substance called peptidoglycan is classified as Gram positive, or Gram +. A bacterium with a cell wall containing a thin layer of peptidoglycan covered by an outer membrane is classified as Gram negative, or Gram −.)*
- How does a bacterium's response to Gram staining pertain to treatment? *(Gram-positive bacteria tend to be killed by antibiotics such as erythromycin and penicillin. Gram-negative bacteria tend to be resistant to penicillin but much more susceptible to streptomycin and tetracycline.)*
- Why should you use a decolorizer and a counterstain during the Gram-staining procedure? *(Peptidoglycan absorbs crystal violet stain and retains it through the decolorization process. As a result, Gram + bacteria show up purple. Gram − bacteria lose all purple color in the decolorization process and show up pink from the Safranin stain.)*

Preparation Tips

- Purchase live or freeze-dried cultures of *Bacillus megaterium*. Freeze-dried cultures should be rehydrated according to the directions on the package.
- Prepare the "unknown" bacterial culture by streaking *Bacillus megaterium* onto tryptic soy agar in a petri dish. Incubate at 37°C for 24 to 48 hours.
- Purchase premade crystal violet stain, or prepare it by dissolving 1.5 g crystal violet into 1 mL denatured alcohol. Then add 90 mL distilled water.

EXPERIMENTAL DESIGN C33 *continued*

- Purchase premade Gram's iodine stain, or prepare it by dissolving 1 g iodine and 2 g potassium iodide in 300 mL of distilled water.
- Purchase premade Safranin stain, or prepare it by dissolving 0.34 g Safranin in 90 mL of distilled water. Add distilled water until the volume reaches 100 mL.
- Purchase premade ethanol.
- Prepare a dilute household-bleach disinfectant solution by adding 50 mL household bleach to 950 mL water. Pour the disinfectant into squirt bottles labeled "disinfectant solution."
- Provide separate containers for collecting broken glass, inoculating loops, and contaminated slides.

Disposal

- Wearing disposable gloves, decontaminate spills and broken glass by covering them with paper towels and soaking the towels with a dilute household-bleach solution. Let this stand for 30 minutes. Then, using tongs, collect all waste materials, place them in a biohazard bag, and seal the bag. Wash your hands with antibacterial soap after this procedure.
- Used slides should be placed in a dilute solution of household bleach for 24 hours.
- All reusable materials that come in contact with bacteria should be autoclaved.
- All surfaces should be cleaned and decontaminated at the conclusion of each lab.
- Disinfectant solutions, staining solutions, and ethanol can be stored for future use.
- Solid trash that has been contaminated by biological waste must be collected in a separate and specially marked disposal bag. Package all sharp instruments in separate metal containers for disposal. No pipets should protrude from the disposal bag. It is recommended that disposal bags (following autoclaving) be placed in a sealed container, such as a plastic bucket with a lid. Liquids and gels must be absorbed by paper towel to minimize the risk of leakage.
- Contaminated materials that are to be decontaminated away from the laboratory must be placed in a durable, leakproof container that is closed prior to removal from the laboratory.

PROCEDURAL NOTES

Safety Precautions

- Discuss all safety symbols and precautions with students.
- Remind students to use a mechanical device—not their mouths—for pipetting.

Procedural Tips

- Remind students never to completely remove the top of the petri dish.
- Remind students to sterilize their work area before and after this lab.
- Students should take only a small amount from the stock culture and be sure to mix the bacteria thoroughly with the distilled water on the slide.

SAMPLE SOLUTION

Question

Is the unknown bacteria Gram + or Gram −?

Proposed Procedure:

1. Put on safety goggles, disposable gloves, and a lab apron. Use the aseptic technique throughout the lab. Spray the lab surface with disinfectant, and wipe the surface with a paper towel.
2. Place one drop of distilled water in the center of a microscope slide.
3. Using a disposable inoculating loop, pick up one bacterial colony from the stock culture dish. Take care not to gouge the agar. Do not remove the top of the petri dish; instead, hold it at a 45° angle.
4. Make a bacterial smear by stirring the bacteria into the drop of distilled water on the slide so that it is about 0.5 cm in diameter. The bacteria should be evenly spread throughout this smear. Allow the smear to air dry.
5. Once the smear has dried thoroughly, pick it up with the slide holder and quickly pass it through a burner flame three times with the bacterial smear facing up. Do not hold the slide in the flame. Place the slide onto a staining tray. Turn off the Bunsen burner.
6. Completely cover the smear with 10 drops of crystal violet stain, and let it sit for 60 seconds.
7. After 60 seconds, pour off the excess stain and gently rinse the smear in a clean beaker of distilled water. Tilt the slide to drain the remaining water drops, and blot it with a paper towel.
8. Place the slide back into the staining tray, cover it completely with 10 drops of Gram's iodine, and let it sit for 60 seconds.
9. Hold the slide at an angle, and gently rinse the slide with drops of ethanol until the ethanol flows clear.
10. Rinse the slide in a beaker of clean, distilled water, and drain the excess water.
11. Completely cover the smear with 10 drops of Safranin stain, and let it sit for 60 seconds.
12. Rinse the slide in a beaker of clean, distilled water, and drain.
13. Carefully blot dry using the clean paper towel.

14. Place the coverslip on the slide, and observe it on high power under the microscope.

Materials (per lab group)

- safety goggles
- disposable gloves
- lab aprons
- compound microscope
- microscope slide with coverslip
- Bunsen burner with striker
- slide holder
- staining tray
- disposable inoculating loop
- 100 mL beaker
- eyedropper
- clock or watch with second hand
- Fitz's culture
- one bottle of crystal violet stain
- one bottle of Gram's iodine
- one bottle of Safranin stain
- one bottle of 95% ethanol
- 100 mL of distilled water
- 5 paper towels
- disinfectant solution

Estimated Cost of the Job

$1,087.45, including a 30 percent profit (based on two students per group and one hour of labor). Refer to *Using Experimental Design Labs* to see how to prepare an invoice.

Results

The bacterium observed is purple in color, indicating that Fitz is suffering from a Gram + bacterium.

Conclusions

The purple color of the smear indicates that the cell walls of the bacteria have a thick layer of peptidoglycan and are therefore resistant to tetracycline. Students should suggest that Fitz has a Gram + infection and should therefore be treated with penicillin.

EXPERIMENTAL DESIGN

C34 Limiting Fungal Growth

Purpose

Students will investigate the characteristics and traits of mold, compare amounts of mold growth in a variety of growth media, and make recommendations about what conditions will deter mold growth.

Student Objectives

- *Apply* knowledge of molds to a situation not previously encountered.
- *Test* a variety of media for promotion of mold growth.
- *Infer* the media that will deter mold growth.

PREPARATION NOTES

Time Required: one 50-minute period and 1 to 2 weeks for incubation; during the incubation period, allow 5–10 minutes of class time each day for observation

Pre-Lab Discussion

Review the procedure from **Laboratory Techniques C31: Using Aseptic Technique,** and answer any questions students may have. Explain to students that fungi are heterotrophs and therefore need to acquire food. Explain that many of the human foods we see in the supermarket can also become food for mold and yeast. Also explain that fungi can be cultured in the same way as bacteria and are susceptible to contamination. Ask students the following questions to guide their thinking:

- What roles do fungi play in the environment? *(Fungi are valuable decomposers of organic material. However, many fungi are parasites on plants and animals and cause millions of dollars worth of damage each year, especially to grain and cereal crops. Some infections, such as ringworm and athlete's foot, are caused by fungi. Fungi are used to produce many useful products such as antibiotics, bread, and cheese.)*

- Why is it important to keep all of the test samples in a darkened area at 25°C? *(The fungi might be affected by light or fluctuations in temperature. Keeping these conditions constant for all subjects limits the number of variables and increases the validity of the experiment. Students may also realize that keeping the media darkened simulates the conditions inside a backpack.)*

Preparation Tips

- You or your students will need to wash the jars ahead of time to avoid contamination by other microorganisms. If possible, use a dishwasher to ensure that the jars are thoroughly cleaned.
- Prepare a dilute household-bleach disinfectant solution by adding 50 mL household-bleach to 900 mL water.
- Prepare a stock mold culture ahead of time.
- Provide separate containers for collecting broken glass, used cotton swabs, and mold samples.

Disposal

- Wearing gloves, decontaminate spills and broken glass by covering them with paper towels and soaking the towels with a dilute household-bleach solution. Let this stand for 30 minutes. Then, using tongs, collect all waste materials, place them in a biohazard bag, and seal the bag. Wash your hands thoroughly after this procedure.
- Used slides should be placed in a dilute solution of household bleach for 24 hours.
- All materials that have been in contact with fungi should be autoclaved.
- All surfaces should be cleaned and decontaminated at the conclusion of each lab.
- Disinfectant solutions, staining solutions, and ethanol can be stored for future use.
- Solid trash that has been contaminated by biological waste must be collected in a separate and specially marked disposal bag. Package all sharp instruments in separate metal containers for disposal. No pipets should protrude from the disposal bag. It is recommended that disposal bags (following autoclaving) be placed in a container with an outer seal, such as a plastic bucket with a lid. Liquids and gels must be absorbed by a paper towel to minimize the risk of leakage.
- Contaminated materials that are to be decontaminated away from the laboratory must be placed in a durable, leakproof container that is closed prior to removal from the laboratory.

PROCEDURAL NOTES

Safety Precautions

- Discuss all safety symbols and precautions with students.
- Caution students not to open the jars unless instructed.
- Students who have mold allergies should use caution throughout the lab to avoid exposure.
- If fungi spill onto the lab bench, sterilize the area with the disinfectant solution.

Procedural Tips

- Use a fresh cotton swab for each new sample.
- You may wish to have students keep a daily record of time and equipment use.
- Remind students to sterilize their work areas before and after this lab.
- To moisten food items, spray them 4 times with a spray bottle of distilled water.

SAMPLE SOLUTION

Question

Which environmental conditions will prevent mold growth, and which foods are less likely to grow mold?

Proposed Procedure

1. Put on safety goggles, disposable gloves, and a lab apron. Use aseptic technique throughout the lab. Spray the lab surface with disinfectant and wipe with a paper towel. Obtain a culture dish with a mold sample.
2. Label eight jars with the lab group, date, and name of the medium to be tested.
3. For food being tested for moisture, spray each item four times with distilled water.
4. Cover the bottom of each jar with its respective medium.
5. Rub a damp cotton swab over the surface of the mold culture, and then rub the swab over the surface of the contents of the jars. Use a fresh cotton swab each time to avoid contamination.
6. Cover each jar and let stand for 1 to 2 weeks in a cabinet or other darkened location at a temperature of approximately 25°C.
7. Clean up your work area and wash your hands before leaving the lab.
8. Use a hand lens or a microscope to observe the contents of each jar each day. On days 3, 7, and 10, note the number of mold colonies, if any, and record your observations about their appearance.

Materials (per lab group)

- safety goggles
- disposable gloves
- lab aprons
- hand lens
- *Penicillium notatum*
- 2 samples of potato flakes
- 2 samples of corn starch
- 2 samples of raisins
- 2 samples of orange peel
- 8 small jars with lids
- 8 cotton swabs (for inoculating)
- spray bottle of distilled water
- disinfectant solution in squirt bottle
- wax pencil
- paper towel

Estimated Cost of the Job

$6,709.17, including a 30 percent profit (based on two students per group, 10 days of facility use, and one total hour of labor). Refer to *Using Experimental Design Labs* to see how to prepare an invoice.

EXPERIMENTAL DESIGN C34 *continued*

Results

The molds tend to grow particularly well on the wet potato flakes.

Sample Data Table

Growth medium	Day 3		Day 7		Day 10	
	Observations	Number of colonies	Observations	Number of colonies	Observations	Number of colonies
dry potato flakes @ 25°C	no growth	0	no growth	0	no growth	0
moist potato flakes @ 25°C	some growth (white and fuzzy)	2–3	more growth (green)	10–15	most of medium covered (green)	difficult to count
dry cornstarch @ 25°C	no growth	0	no growth	0	no growth	0
moist cornstarch @ 25°C	some growth (white and fuzzy)	2–3	more growth (green)	8–14	approx. 1/2 covered (green)	difficult to count
dry raisins @ 25°C	no growth	0	no growth	0	little growth (white and fuzzy)	1–2
moist raisins @ 25°C	little growth (white and fuzzy)	1–2	some growth (green)	3–4	more growth (green)	5–7
dry orange peel @ 25° C	no growth 0	0	no growth	0	no growth	0
moist orange peel @ 25° C	little growth (white and fuzzy)	2–3	some growth (white and fuzzy)	5–10	more growth (green)	15–20

Conclusions

Students' data should support their conclusions. Students should find that the wet potato flakes and wet cornstarch make good media for growth. Students may correctly conclude that moisture seems to promote mold growth and that keeping food dry will probably help prevent mold growth.

EXPERIMENTAL DESIGN

C38 *Growing Plants in the Laboratory—Fertilizer Problem*

Purpose

Students will investigate the relationship between mineral nutrition and plant growth and make recommendations about an agricultural project based on their results.

Student Objectives

- *Apply* principles of plant growth and mineral nutrition to solving a problem not previously encountered.
- *Measure* relative growth rates of radish seeds in different fertilizers.
- *Infer* the best fertilizer to maximize plant growth.

PREPARATION NOTES

Time Required: 18–25 days; one 50-minute period for setup and 15 minutes every 3–5 days for data collection

Pre-Lab Discussion

Review the procedure from **Laboratory Techniques C37: Growing Plants in the Laboratory,** and answer any questions students may have. Ask students the following questions to guide their thinking:

- How can you quantitatively measure the effect of fertilizer on plant growth in an experiment? *(Measure the height of each plant, and compare fertilizers for their effect on plant height.)*
- What other observations might help you determine the best type of fertilizer to use for growing plants? *(color of plants, texture of leaves, size of leaves, health of leaves and stems as indicated by lack of spots or dead tissue, and sturdiness of stems)*

Preparation Tips

- Schedule the lab so that observation days fall on days that students are in school. This may require having students take one measurement early. For example, if the lab begins on a Monday, day 20 will fall on a Saturday.
- Provide reference books or brochures showing colored pictures of nutrient deficiency symptoms in plants.
- This lab can be done qualitatively (by looking at just the symptoms) rather than quantitatively.
- You may wish to have students take fresh and dry masses of plants harvested at the experiment's conclusion and examine the correlation of height and biomass.

- Prepare four fertilizer solutions using the packets of NPK fertilizer mix, NP fertilizer mix, NK fertilizer mix, and PK fertilizer mix. Mix the contents of each fertilizer packet with 1 L of water in a large container. Label the solutions according to their contents.
- Prepare an extra container of one of the four types of fertilizers, and label it "Fertilizer from Belize."
- Provide separate containers for collecting broken glass, extra soil, used solutions, unused fertilizer, and radish plants.

Disposal
- All solutions may be rinsed down the sink.
- Radish plants can be used in later experiments if desired.

PROCEDURAL NOTES

Safety Precautions
- Discuss all safety symbols and precautions with students.
- Have students wear protective equipment when planting seeds, transplanting seedlings, and preparing fertilizer solutions.

Procedural Tips
- Have each lab group prepare a growing container for one of the six treatments.
- Have students take the following measures to avoid contaminating the other growing containers: always fertilize away from the watering system, and do not return the containers to the watering system until the end of class; add the fertilizer slowly, putting on no more than the growth medium can accommodate; when returning the growing containers to the watering system, make sure no liquid is dripping from the sections; and carefully set the growing container back in its place.
- Caution students to avoid "burning" the plants by pouring the fertilizer solution onto the growth medium and not on the plants.
- A pot may become dry if it is separated from the mat by an air gap. If this occurs, have students record this observation, fill air gap with soil, and thoroughly water the pot.
- When students are measuring their plants, they should attempt to standardize their procedures.
- Remind students never to handle a plant by the stem.

SAMPLE SOLUTION

Question
What is the best kind of fertilizer for vegetable plants?

Proposed Procedure

1. Put on safety goggles, disposable gloves, and a lab apron.
2. Moisten the soils in a pan, plate, or piece of plastic so that they are about the same dampness.
3. Using an adhesive label and a waterproof pen or pencil, label a growing container with one of the following treatments: Treatment 1 (Belize fertilizer), Treatment 2 (NPK fertilizer), Treatment 3 (NP fertilizer), Treatment 4 (NK fertilizer), Treatment 5 (PK fertilizer), Treatment 6 (control).
4. Fill a growing container with growth medium. Tap the bottom of the growing container on the lab table to help the medium settle.
5. Place two radish seeds in each section, and carefully water the seeds. Place a pinch of growth medium on top of each seed.
6. Place the growing containers on the mat of a growing/watering system and turn on the lighting system.
7. On day 2 or 3, transplant seedlings from sections with two seedlings into empty sections, so that each section of the growing tray has one seedling.
8. Beginning on day 5, fertilize each container once a week by pouring 2.5 mL of the indicated fertilizer solution onto the soil around each plant.
9. Check the watering tray twice a week to be sure that it has plenty of water. Add about 2–2.5 mL (2 squirts with the plastic pipet) of the algae inhibitor to the watering tray when refilling to prevent algae growth.
10. Tie the plants to stakes when they begin to bend over.
11. Record the height of each plant every 5 days for 25 days. Note any changes in the appearance of the plants.
12. Record the average plant height for each day that measurements are taken.

Materials (per lab group)
- safety goggles
- lab apron
- lighting system
- watering system
- growing container
- metric ruler
- 1 kg of growth medium
- 8 Rapid Radish seeds
- 10 mL of fertilizer
- pan, plate, or piece of plastic
- 8 stakes
- 8 twist-ties

- adhesive plant label
- 2 plastic pipets
- 1 bottle of algae inhibitor

Estimated Cost of the Job

$17,252.17, including a 30 percent profit (based on two students per group, 25 total days of facility use, and two total hours of labor). Refer to *Using Experimental Design Labs* to see how to prepare an invoice.

Results

Plants that receive N, P, and K will be the tallest and most vigorous.

Sample Class Data Table

	Treatment					
Day	1 Belize	2 N, P, K	3 N, K	4 N, P	5 P, K	6 Control
0	0 cm	0 cm	0 cm	0 cm	0 cm	0 cm
5	?	2 cm	2 cm	2 cm	1 cm	1 cm
10	?	4 cm	4 cm	3 cm	3 cm	2 cm
15	?	7 cm	6 cm	4 cm	4 cm	3 cm
20	?	15 cm	13 cm	10 cm	7 cm	4 cm
25	?	23 cm	18 cm	17 cm	12 cm	5 cm

Conclusions

Students' data should support their conclusions. Students should find that the highest rate of growth (in cm) will occur with a fertilizer that contains N, P, and K. Students' recommendations regarding the fertilizer used by the vegetable growers in Belize will depend on the type of fertilizer provided.

EXPERIMENTAL DESIGN

C39 Response in the Fruit Fly

Purpose

Students will use skills in experimentation to set up a procedure and test for the response of fruit flies to two different types of stimuli.

Student Objective

Describe how fruit flies respond to two types of stimuli: light and scent.

PREPARATION NOTES

Time Required: two 50-minute periods. Part of the experiment will need to sit overnight.

Pre-Lab Discussion

Show students the stopper assembly that will be used in this lab activity. Place two flasks mouth-to-mouth, and connect them with transparent packing tape. Explain that this setup will allow the flies to move from one flask to the other without escaping. Ask students the following questions to stimulate their thinking:

- What do the terms *taxis*, *phototaxis*, and *chemotaxis* mean, and how do they relate to this experiment? *(The term* taxis *refers to a development or motion produced by a plant or animal in response to a stimulus. The response can be positive, such as moving toward the stimulus, or negative, such as moving away from it.* Phototaxis *is a response to light stimulus and* chemotaxis *is a response to a chemical substance.)*
- What is the purpose of putting the vials of *Drosophila* on ice? *(The lowered temperature will slow the organisms' movements so that it will be easier to place the flies in the flasks.)*
- How do you distinguish a positive response from a negative response? *(Flies will move toward the stimulus if there is a positive response and away from it if there is a negative response.)*

Preparation Tips

- Provide a container of ice for cooling the flies.
- Provide separate containers for collecting broken glass and aluminum foil.

Disposal

- Aluminum foil can be thrown away or recycled.
- Fruit flies can be cooled and returned to media vials for future use.

PROCEDURAL NOTES

Safety Precautions

- Discuss all safety symbols and precautions with students.

Procedural Tips

- *Drosophila* should stay on ice until all movement has come to a stop.
- The mouths of the flasks should be completely flush when they are taped so that no flies accidentally touch the sticky side of the tape.
- Before returning flies to the vial, first stop their movement by cooling the flasks on ice.

SAMPLE SOLUTION

Question

How do fruit flies respond to light and scent stimuli?

Proposed Procedure

1. Keep flies on ice until movement has stopped.
2. Label two flasks "phototaxis." For fly food, place three slices of banana into each of two flasks.
3. When flies have stopped moving, carefully drop 10 flies into one flask and 10 into the second flask.
4. Without inverting either flask, place the two

EXPERIMENTAL DESIGN C39 *continued*

flasks mouth-to-mouth, and connect them with wide transparent tape.

5. Using foil, completely cover one flask.
6. Carefully place the uncovered side under light, so that one flask is in the light and the other is darkened because of the foil.
7. Wait 24 hours and remove the foil. Immediately count the flies. Record the number of flies in each flask in your data table. Return the flies to the ice until movement has stopped.
8. Label two new flasks "chemotaxis." Place three banana slices in one flask.
9. Place 10 flies in each flask.
10. Connect the flasks with tape, and place the entire apparatus in a warm area as directed by your teacher.
11. Wait 24 hours and observe the flasks. Count the flies. Record the number of flies in each flask in your data table.

Materials: (per lab group)
- 2 vials of fruit flies (20 in each)
- 9 banana slices
- 50 cm of aluminum foil
- 4 500 mL Florence flasks
- 50 cm of wide transparent tape
- light source
- container of ice

Estimated Cost of the Job
$1,574.50, including a 30 percent profit (based on two students per group, two days of facility use, and one total hour labor). Refer to *Using Experimental Design Labs* to see how to prepare an invoice.

Results
Answers may vary, but sample data is listed below.

Sample Data Table: Phototaxis

Time of reading	Number of flies in flask with light	Number of flies in dark flask
Beginning	10	10
After 24 hours	17	3

Sample Data Table: Chemotaxis

Time of reading	Number of flies in flask with banana	Number of flies in flask without banana
Beginning	10	10
After 24 hours	19	1

Conclusions
Students' data should support their conclusions. Students should find that the flies prefer light to dark and that the scent of the banana attracts them. They should explain this to Mr. Nichols so that he can use this information to enhance his pesticide. They should use the proper terms of *phototaxis* and *chemotaxis*. Further, students may indicate that the behavior showed positive phototaxis and positive chemotaxis.

EXPERIMENTAL DESIGN

C40 *Conducting a Bird Survey*

Purpose:
Students will study birds found in the area in which they live. They will observe them, take notes on them, and give a pictorial representation of them.

INSTRUCTIONAL GOALS

Student Objectives
- *Observe* a bird or birds in a geographic area.
- *Identify* the bird or birds observed.

PREPARATION NOTES
Time Required: two 50-minute periods; time requirements will vary if the lab is assigned as a long-term, out-of-class project

Prelab Discussion
Prior to the start of this lab, meet with your students and clarify your expectations for this assignment. Discuss where students will conduct their observations, such as the school campus or a nearby park. Discuss how students should visually represent their subjects. You may wish to suggest any or all of the following methods: photographing, videotaping, sketching, or making a color copy of a photograph. Discuss how many birds will be observed by each group. This number should depend on the variety of local bird species. Five to ten species per group is recommended. If your observation area has a variety of habitats, you may wish to assign different groups to different habitats. If time permits, have each group present their findings to the class so that all students can see the variety of local bird species.

Preparation Tips
- You may wish to present this lab activity as a field trip or as a long-term, out-of-class project.
- If you choose to have a field trip, find out whether there are poisonous plants or dangerous animals where you will be going. Learn to identify any hazardous species.
- Encourage students to be aware of the impact they are having on the environment they visit.

EXPERIMENTAL DESIGN C40 *continued*

Disposal
none

PROCEDURAL NOTES

Safety Precautions
- Discuss all safety symbols and precautions with students.
- Before going into the field, discuss with students some of the more dangerous organisms students might find and how they may avoid them. Show them pictures of poison ivy and other plants to avoid.
- Encourage students to dress appropriately for field work. If possible, students should wear comfortable clothing, long pants, and sturdy shoes with closed toes.
- Encourage students to bring sunglasses, sunscreen, and insect repellent, if needed.
- Discuss field trip safety procedures with students. For example, have students remain within a distance that they can be seen or heard. You may wish to assign student partners for additional safety.
- Remind students that they are to be observers and that they should NEVER pick up a bird or an egg or touch any of the nests.

Procedural Tips
- Some information, such as nests, flight, and food gathering, may not be observable. If such a characteristic cannot be observed, have students indicate this on their index cards.
- Remind students not to leave garbage behind at the field site and to leave natural areas as they found them.

SAMPLE SOLUTION

Question
What are some of the bird species that live in the local area, and what characteristics do they have?

Proposed Procedure:
1. Choose a bird to study.
2. Use the criteria listed in the table to describe and identify the bird.
3. Refer to a field guide to describe and identify the bird.
4. Record observations about each of the characteristics in the observation table provided.
5. Photograph, videotape, or draw a sketch of the bird.
6. Complete an index card for each bird observed.
7. Repeat these procedures for each bird to be studied.

Materials: (per lab group)
- bird identification book
- 2 pairs of binoculars
- 5 bird identification forms
- five 5 × 8 index cards
- 2 pencils
- 10 sheets of paper

Estimated Cost of the Job
$307.13, including a 30 percent profit (based on two students per group, two hours of labor, two days of study, and 5 birds studied). Refer to *Using Experimental Design Labs* to see how to prepare an invoice.

Results
Answers will vary depending on the geographic location of your school. An example is given below.

1.	**Name of researcher(s):**	Juan Montoya and Alicia Firestein
2.	**Date of sighting:**	April 17, 1998
3.	**Location of sighting:**	Pena Adobe, Vacaville, California
	Description: park on the outskirts of Vacaville, CA; many oak and walnut trees; bird sighted sitting in oak tree; temperature about 25°C; park surrounded by open space with trees, native grasses, and hills.	
4.	**County of sighting:**	Solano County, California
5.	**Name of bird:**	Order Strigiformes; Western screech owl
6.	**Physical description:** Brown with some white, pointed small beak, rounded wing, short neck, short tail length, square tail shape, feet retracted in flight	
7.	**Social behavior:**	Solitary
8.	**Food acquisition technique:**	Flying
9.	**Nest type:**	None observed
10.	**Diet:**	Carnivorous, small mammals (mostly field mice and insects)

Conclusions
Conclusions will vary, but students should realize that there are a variety of birds in most geographic locations.

EXPERIMENTAL DESIGN

C41 *Evaluating Muscle Exhaustion*

Purpose
Students will design an experiment to test the effect of varying rest periods on muscle fatigue during repeated muscle use.

Student Objectives
- *Demonstrate* an understanding of experimental design by designing an experiment relating rest periods to muscle fatigue.
- *Infer* ways to avoid muscle fatigue in a factory.

PREPARATION NOTES
Time Required: one 50-minute period

EXPERIMENTAL DESIGN C41 *continued*

Pre-Lab Discussion

If necessary, show students how to set up the computer, or graphing calculator, and probe apparatus. Refer to the manual for instructions on how to set up the equipment. Ask students the following questions to guide their thinking:

• What is meant by the terms **muscle fatigue** and **muscle endurance?** (*Muscle fatigue is the exhaustion, or tiredness, of a muscle during exercise. Muscle endurance is the ability of a muscle to exercise for an extended period of time.*)

• What conditions in the candy plant might contribute to muscle fatigue? (*the number of chocolates that must be decorated before a break period is allowed, the length of the rest period between candies, constant use of the same hand*)

• What variables can contribute to muscle fatigue in this case? (*Examples include the pressure of the frosting bag, the number of rest periods, standing position, muscle development, and hand size.*)

• What might prevent muscle fatigue or increase muscle endurance? (*taking intermittent rest periods, alternating hands, using correct posture*)

Preparation Tip

A variety of hand exercisers that are similar to the probe exerciser are available at local sporting goods stores or department stores.

Disposal

PROCEDURAL NOTES

Safety Precautions

• Discuss all safety symbols and precautions with students.

• Students should not continue with this experiment if they experience any pain in their hand or arm.

Procedural Tips

• You may wish to have students use a computer or graphing calculator program designed to monitor muscle fatigue. In this case, the hand exerciser would be replaced by one that acts as both an exerciser and a probe.

• Students may choose to test another factor. Such tests may include alternating hands, varying time spent squeezing, and using one hand versus both hands.

• You may wish to bring in a frosting bag so that students can see the relationship between it and the hand exerciser.

• You may wish to have each lab group perform a different test sequence.

• You may wish to have another class complete steps 3–5 in reverse order and have the students compare the results from the two classes.

SAMPLE SOLUTION

Question

How does altering rest time affect muscle fatigue in repeated muscle exertion?

Proposed Procedure

1. Meet with group members and choose the factor that you will be testing (in this case, the length of the rest between squeezes).

2. One student should squeeze the exerciser while a partner counts the number of squeezes. The student should place the exerciser in the right hand. Begin timing and squeeze the exerciser as many times as possible until your hand exhausts. Make sure that the pressure used to squeeze the exerciser is consistent throughout the trial. Record the time it took to exhaust the muscle and the number of squeezes completed in a data table.

3. Rest for 2 minutes, then repeat steps 2–4, resting 1 second between squeezes. Record the time it took to exhaust the muscle and the number of squeezes completed in a data table.

4. Rest for 2 minutes, then repeat steps 2–4, resting 2 seconds between squeezes. Record the time it took to exhaust the muscle and the number of squeezes completed in a data table.

5. Rest for 2 minutes, then repeat steps 2–4, resting 3 seconds between squeezes. Record the time it took to exhaust the muscle and the number of squeezes completed in a data table.

6. Pool the class data, and graph or chart the results.

Materials: (per lab group)

• hand exerciser
• clock or watch with a second hand

Estimated Cost of the Job

$747.50, including a 30 percent profit (based on two students per group and one hour of labor). Refer to *Using Experimental Design Labs* to see how to prepare an invoice.

Results

Answers will vary depending on the experiment students choose to perform.

Sample Data Table

Trial	Total no. of squeezes	Total time spent	No. of squeezes per minute
no rest between squeezes			
1-second rest between squeezes	Answers will	vary.	
2-second rest between squeezes			
3-second rest between squeezes			

Conclusions

Students' data should support their conclusions.

C43 *Blood Typing—Whodunit?*

Purpose

Students will narrow a list of suspects by examining a sample of simulated blood to verify that the sample is "real" blood, determining the blood type of the sample, and comparing it with the blood types of five other samples.

Student Objectives

- *Examine* the properties of an alleged blood sample.
- *Identify* the substance as a blood sample.
- *Determine* the blood type of six blood samples.
- *Compare* the blood type of one sample with the blood types of five other samples to identify matching and nonmatching blood types.
- *Recognize* the limitations of using blood typing in criminal investigations.

PREPARATION NOTES

Time Required: one 50-minute period

Pre-Lab Discussion

Review the procedure from **Laboratory Techniques C42: Blood Typing,** and answer any questions students may have. Ask students the following questions to guide their thinking:

- What do A, B, and Rh stand for? *(These letters stand for different types of antigens found on red blood cells.)*
- What determines a blood type? *(Blood type is determined by the presence or absence of antigens and is encoded by genes. For example, type O blood has neither A nor B antigens, while type AB has both.)*
- How does the Rh antigen fit into blood typing? *(All blood types, A, B, AB, or O, either have or do not have the Rh antigen. In most cases, this is indicated by a plus or minus sign after the A, B, AB, or O.)*
- What is agglutination and how is it used in blood typing? *(Agglutination is a clumping of the blood that occurs when an antigen meets its antibody. When an anti-A serum is combined with type A blood, clumping occurs. The same thing would happen if type B blood were mixed with type A blood. Blood type is determined by introducing different types of antibodies into a sample of blood and observing any clumping.)*
- Can blood typing be used to positively identify someone as the criminal who committed a particular crime? *(No, it can only be used to identify any suspects who could not have left the blood stains. This is because there are many people who share the same blood type.)*

Preparation Tip

- Simulated blood samples can be purchased from WARD'S. Prepare these samples by relabeling them as needed.
- Prepare crime-scene samples by cutting 2 cm × 2 cm squares of cloth. Place them on a clean surface and dispense one drop of "crime scene" simulated blood onto each square, and allow them to dry. (To change the scenario from class to class, use a different suspect blood sample to stain the cloth squares.) Each lab group will need four cloth squares.
- Provide separate containers for collecting unused simulated blood, unused antiserums, used cloth squares, and broken glass.

Disposal

- Stained cloth squares can be thrown away or washed and used again.
- Simulated blood and anti-serums can be washed down the drain or stored for future use.

PROCEDURAL NOTES

Safety Precautions

- Discuss all safety symbols and precautions with students.
- Remind students never to use direct sunlight as a light source if they are using a microscope with a mirror. Using direct sunlight will cause eye damage.
- Remind students that under no circumstances are they to test any blood other than the simulated blood samples provided.

Procedural Tip

Students should use one of the cloth samples to identify the substance as blood by looking for simulated red blood cells in the cloth fibers. They should use the other three samples to determine the blood type. Adding water to these samples will rehydrate them to allow for blood typing.

SAMPLE SOLUTION

Question

Is the substance on the cloth blood, and if so, does its blood type match the blood type of the victim or any of the four suspects?

Proposed Procedure

1. Put on safety goggles, disposable gloves, and a lab apron.
2. Place one piece of stained cloth on a microscope slide, and place one drop of distilled water on the cloth. View the cloth under the microscope at low power and at high power to identify the stain. Record any observations, especially evidence that would prove that the substance in the stain is blood.

3. Using a wax pencil, label six blood typing trays as follows: Tray 1—Crime scene; Tray 2—Victim; Tray 3—Suspect 1; Tray 4—Suspect 2; Tray 5—Suspect 3; and Tray 6—Suspect 4.

4. Determine the type of blood in the cloth stains by placing a piece of cloth in each of the A, B, and Rh wells of blood typing tray 1. Add several drops of water to each well to dampen the blood stains. Add 3 to 4 drops each of the anti-A, anti-B, and anti-Rh serums into the A, B, and Rh wells, respectively.

5. Using separate, clean toothpicks, stir the contents of each well in Tray 1 so that the serums mix with the stains. Record any observed agglutination in a data table.

6. Place 3 to 4 drops of the victim's blood into each of the A, B, and Rh wells of Tray 2: Victim. Place 3 to 4 drops of anti-A serum into the A well of the victim tray, add 3 to 4 drops of anti-B serum into the B well of the tray, and add 3 to 4 drops of anti-Rh serum into the Rh well of the tray. Using separate, clean toothpicks, stir the contents of each well.

7. Repeat step 5 for each of the suspects' blood samples.

8. In each well of each tray, look for evidence of agglutination and record observations in the data table. Use this data to determine the blood types of the stains, the victim, and each suspect.

9. Compare the blood type of each suspect with that of the sample found at the crime scene. Eliminate the suspects whose blood does not match the crime scene blood.

Materials: (per lab group)

- safety goggles
- lab apron
- disposable gloves
- compound microscope
- 1 microscope slide
- 6 blood typing trays
- 4 stained cloth samples
- vial of simulated blood labeled "Victim"
- vial of simulated blood labeled "Suspect 1"
- vial of simulated blood labeled "Suspect 2"
- vial of simulated blood labeled "Suspect 3"
- vial of simulated blood labeled "Suspect 4"
- vial of anti-A typing serum
- vial of anti-B typing serum
- vial of anti-Rh typing serum
- 18 toothpicks
- 10 mL of distilled water
- wax pencil

Estimated Cost of the Job

$918.84, including a 30 percent profit (based on two students per group and one hour of labor). Refer to *Using Experimental Design Labs* to see how to prepare an invoice.

Results

Upon examining the crime-scene cloth sample under the microscope, students should observe simulated red blood cells that indicate the stain is indeed blood. In this case, the blood type of the crime scene stain is A, Rh+, a type that matches only that of Suspect 3.

Sample Data Table

Blood source	Anti-A serum	Anti-B serum	Anti-Rh serum	Blood type
crime scene	agglutination	no agglutination	agglutination	A, Rh+
victim	no agglutination	agglutination	no agglutination	B, Rh–
suspect 1	agglutination	agglutination	agglutination	AB, Rh+
suspect 2	no agglutination	agglutination	no agglutination	B, Rh–
suspect 3	agglutination	no agglutination	agglutination	A, Rh+
suspect 4	no agglutination	no agglutination	no agglutination	O, Rh–

Conclusions

Students' data should support their conclusions. Students should find that the stain on the cloth is indeed blood. They should suggest that, based on the results of the blood typing evidence, if the blood stain came from the burglar, only suspect 3 could have committed the burglary.

EXPERIMENTAL DESIGN
C44 **Blood Typing—Pregnancy and Hemolytic Disease**

Purpose

Students will design an experiment that uses the concepts of antigen/antibody reactions and blood factors to predict the potential for Rh hemolytic disease in the second Rh+ child born to an Rh− mother.

Student Objectives

- *Design* an experiment to test for the presence of Rh (+) antigens.
- *Interpret* the results and predict any risks for future pregnancies.

PREPARATION NOTES

Time Required: one 50-minute period

Pre-Lab Discussion

Review the procedure from **Laboratory Techniques C42: Blood Typing,** and answer any questions students may have. Review blood-typing procedures with students. Ask students the following questions to guide their thinking:

- What happens when antigens carried on the blood cell are transferred into the wrong system? (*This will cause antibodies to be formed that will cause agglutination. If agglutination occurs over a period of time, it may block small arteries and cause damage.*)
- What would be the value of knowing the Rh blood type of a married couple? (*If the couple planned to have children, knowing of a difference in Rh blood type could alert the couple to the potential for Rh hemolytic disease in their children.*)

Preparation Tips

- Simulated blood can be purchased as a kit from WARDS. Two vials of Rh− blood should be labeled "known Rh−" and "mother." Two vials of Rh+ blood should be labeled "known Rh+" and "fetal blood."
- You may wish to assign different blood samples to different lab groups, or have each lab group do all of the samples.
- Provide separate containers for collectiong broken glass, simulated blood, and antiserums.

Disposal

- Simulated blood and antisera can be washed down the drain or stored for future use.
- Trays containing simulated blood and antisera may be cleaned and the residue washed down the sink.

PROCEDURAL NOTES

Safety Precautions

- Discuss all safety symbols and precautions with students.
- Remind students that under no circumstances are they to test any blood other than the simulated blood samples provided.

Procedural Tips

SAMPLE SOLUTION

Question

Does the mother's blood type match her baby's blood type?

Proposed Procedure

1. Put on safety goggles, disposable gloves, and a lab apron.

2. Place 3–4 drops of the known Rh+ blood into one of the wells on a blood-typing tray.
3. Without touching the blood in the wells, drop in 3–4 drops of anti-Rh serum into this well. Stir using a clean toothpick.
4. Observe and record your observations in a data table.
5. Repeat steps 2–4 using the known Rh− blood sample, the mother's blood sample, and the baby's blood sample.

Materials: (per lab group)
- safety goggles
- disposable gloves
- lab aprons
- 5 eyedroppers
- blood-typing tray
- vial of anti-Rh serum
- blood sample: mother
- blood sample: first-born child
- blood sample: known Rh+ blood
- blood sample: known Rh− blood
- 4 toothpicks

Estimated Cost of the Job

$805.22, including a 30 percent profit (based on two students per group and one hour of labor). Refer to *Using Experimental Design Labs* to see how to prepare an invoice.

Results
Sample Data Table

Blood source	After anti-Rh serum	Positive or negative for Rh
known Rh+	agglutination	Rh+
known Rh−	no agglutination	Rh−
mother	no agglutination	Rh−
first-born child	agglutination	Rh+

Conclusions

Students should discuss in their conclusion that the mother is Rh−, that the first child is Rh+, and that a second pregnancy could pose dangers to an Rh+ child unless preventative action is taken.

EXPERIMENTAL DESIGN

C45 *Screening Sunscreens*

Purpose

Students will monitor the growth of UV-sensitive yeast to compare the effectiveness of two different sunscreens.

Student Objectives

- *Design* an experiment to monitor the effects and

EXPERIMENTAL DESIGN C45 *continued*

levels of solar UV on UV-sensitive yeast cells.
- *Investigate* the ability of sunscreen to prevent UV radiation exposure.
- *Determine* and compare the effectiveness of two sunscreens.

PREPARATION NOTES

Time Required: 3 days: 50 minutes the first day, 5–10 minutes the second day, and about 20 minutes the third day

Pre-Lab Discussion

Discuss with students the procedure for growing yeast cultures. Model the techniques for transferring yeast cells. Ask students the following questions to stimulate their thinking:
- When testing yeast response to sunlight, why is it important to make sure that the yeast are spread evenly in the petri dish? *(Since growth will be compared, it is important that an even layer of yeast cells is spread in each dish.)*
- When you transfer yeast cells to a petri dish, how can you make sure that the yeast are spread out evenly? *(Suspend the yeast in a liquid first, pour the suspension over the agar, and allow it to spread evenly and dry.)*

Preparation Tips

- Prepare agar according to the directions on the package, and prepare petri dishes with agar. Each lab group will need four petri dishes with agar.
- Prepare two sunscreen samples with sunscreens of different SPF ratings. Label the containers "original," and "test sample." The test sample should have a lower SPF rating than the original sunscreen. As a variation for this experiment, you may wish to provide a test sample with an SPF rating identical to or higher than the original.
- Prepare one yeast culture for each lab group. Open a yeast culture and remove a small amount of yeast cells with a sterile spreader. Replace the lid on the culture and refrigerate. Open a petri dish just enough so that you can reach into it without touching the side of the dish with the spreader. Gently make several streaks across the surface of the agar, remove the spreader, and close the lid. Incubate the culture overnight at 30°C or at room temperature for 1–2 days.
- Provide separate containers for collecting broken glass and contaminated materials.

Disposal

- Wearing gloves, decontaminate spills and broken glass by covering them with paper towels and soaking the towels with a dilute household bleach solution. Let this stand for 30 minutes. Then, using tongs, collect all waste materials, place them in a biohazard bag, and seal the bag. Wash your hands with antibacterial soap after this procedure.
- Used petri dishes should be placed in a dilute solution of household bleach for 30 minutes.
- All materials in contact with yeast should be autoclaved or soaked in bleach solution if an autoclave is not available.
- All surfaces should be cleaned and decontaminated at the conclusion of each lab.
- Disinfectant solutions can be stored for future use.
- Solid trash that has been contaminated by biological waste must be collected in a separate and specially marked disposal bag. Package all sharp instruments in separate metal containers for disposal. No pipets should protrude from the disposal bag. It is recommended that disposal bags (following autoclaving) be placed in an outer sealed container, such as a plastic bucket with a lid. Liquids and gels must be absorbed by paper toweling to minimize the risk of leakage.
- Contaminated materials that are to be decontaminated away from the laboratory must be placed in a durable, leakproof container that is closed prior to removal from the laboratory.

PROCEDURAL NOTES

Safety Precautions

- Discuss all safety symbols and precautions with students.
- Contaminates to the yeast used in this procedure may be pathogenic. Use aseptic techniques in handling and disposing of the materials in this experiment.
- If yeast spills onto the lab bench, sterilize the area with the disinfectant solution.
- Remind students to use a mechanical device—not their mouths—for pipetting.

Procedural Tips

- Have students incubate the cultures upside down to prevent condensation from dripping onto the culture.
- Have students use only a small amount of tape when securing the lids of the petri dishes, and advise them not to place any tape on the top surface of a lid. The tape can absorb some UV radiation.
- Have students position the petri dish perpendicular to the sun to provide the most direct sunlight.
- Students in different classes may get different results, depending on the angle of the sun during

the exposure of the yeast cells. Using a sun lamp will eliminate this discrepancy.

• If students decide to allow their cultures to sit at room temperature, be sure that the temperature does not fluctuate or get much colder than 25°C.

SAMPLE SOLUTION

Question

What effect does sunscreen have on the growth of UV-sensitive yeast cells?

Proposed Procedure

1. Put on safety goggles, disposable gloves, and a lab apron. Spray the surface of the work area with disinfectant, and wipe it with a paper towel. Label the bottoms of four petri dishes with a group name or number.

2. Using aseptic technique, scrape up some yeast cells from a yeast colony with a sterile disposable inoculating loop. Close the lid on the yeast culture and deposit the cells onto the wall of a jar filled with 25 to 30 mL of distilled water. Close the lid of the jar tightly and swirl the jar to suspend the cells in the distilled water. Repeat this procedure with a new loop each time until the water in the jar becomes cloudy.)

3. Using a small sterile pipet, measure 0.9 mL of distilled water into a small test tube. Place a cap on the test tube.

4. Using a new sterile pipet, add 0.1 mL of the yeast suspension from the jar into the test tube.

5. Place a cap on the test tube and swirl to mix.

6. Pour the contents of the test tube onto an empty agar dish.

7. Place the lid on the agar dish and tilt it to spread the liquid evenly over the agar.

8. Repeat steps 3–7 to prepare a total of four petri dishes.

9. Let the covered petri dishes sit for 10 minutes or until the liquid has disappeared.

10. With two small pieces of tape, secure the lid of each petri dish onto the bottom of the dish.

11. With a wax pencil, label the bottom of one prepared petri dish "no exposure." Cover this culture dish with thick paper so that little or no light reaches the yeast.

12. Label the bottom of a second prepared petri dish "original." Using a fresh cotton swab, evenly coat the inside top of the petri dish with sunscreen from the sunscreen sample labeled "original."

13. Label the bottom of a third prepared dish "test sample." Using a fresh cotton swab, evenly coat the inside top of the petri dish with sunscreen from the container labeled "test sample."

14. Label the bottom of a forth prepared dish "full exposure."

15. Expose the petri dishes to direct sunlight in a sunny outdoor location or to a sun lamp for 5–10 minutes.

16. Incubate these petri dishes for 2 days at 30°C or for 3 days at room temperature.

17. At the end of the incubation period, remove any obstructing materials and observe the level of yeast growth in each dish. Record these observations in the data table.

Materials (per lab group)

• safety goggles
• lab aprons
• disposable gloves
• clock or watch with a second hand
• 8 disposable inoculating loops
• sterile test tube with cap
• sterile 50 mL jar with lid
• incubator
• sun lamp
• 4 petri dishes with agar
• UV-sensitive yeast culture
• sample of original sunscreen
• test sample
• 2 sterile, calibrated 1 mL pipets
• 10 cm of transparent tape
• 2 sterile cotton swabs
• disinfectant solution
• 3 paper towels
• 1 sheet of thick paper
• 40 mL of distilled water
• wax pencil

Estimated Cost of the Job

$2,408.29, including a 30 percent profit (based on two students per group, 3 days of facility use, and 90 total minutes of labor). Refer to *Using Experimental Design Labs* to see how to prepare an invoice.

Results

Students should observe that the yeast culture with no exposure to sunlight has the most growth, while the culture with full exposure to sunlight had no growth. The original sunscreen should show more growth than the test sample.

Sample Data Table

Petri dish	Growth level
full exposure	no growth
no exposure	full growth
sunscreen sample 1	very little growth
sunscreen sample 2	some growth

Conclusions

Students should conclude that both of the samples absorb UV radiation, but the original product absorbs more UV radiation than the test product. Students will most likely recommend that the company continue to purchase the product they currently use. Students may recognize, however, that this experiment does not account for some other factors, such as how long the sunscreen will stay on in water.

EXPERIMENTAL DESIGN

C47 Identifying Food Nutrients—Food Labeling

Purpose

Students will test an unknown food for its nutrient content.

Student Objectives

• *Determine* the nutrient content of a frozen food.
• *Infer* whether the claims on a label are legitimate or fraudulent.

PREPARATION NOTES

Time Required: two or three 50-minute periods

Pre-Lab Discussion

Review the procedure from **Laboratory Techniques C46: Identifying Food Nutrients,** and answer any questions students may have.

Preparation Tips

• Prepare the burrito filling by mixing the following ingredients for each lab group: 50 g of mashed pinto beans or rehydrated bean flakes, 35 g of sour cream, 4 mL of salad oil, 4 g of soluble starch, and 5 mL of glucose. Thin the mixture with whole milk until it has a thick but wet consistency. You may wish to alter these ingredients from class to class.
• Lugol's iodine solution (or "iodine-iodide" solution), Biuret solution, and Sudan III are available ready-made from WARD'S. *Note: iodine reacts with metal, skin, and many other substances.*
• Provide a container for collecting broken glass.

Disposal

• Excess burrito filling can be placed in a trash can.
• Combine all wastes containing Benedict's solution and biuret solution. Slowly add 1 M sulfuric acid with stirring until the pH is between 5 and 6. Scour six 6d iron nails with steel wool until they are bright and shiny. Immerse the nails in the acidified solution. Let the nails remain immersed until all of the copper has precipitated (overnight should be sufficient). Remove the nails, and filter the solution. Heat the nails and any precipitate obtained from the filtrate sufficiently to convert the copper precipitate and copper on the nails into copper oxide. Let the nails and the precipitate cool, and then place them in the trash. Treat the filtrate with sufficient 1 M NaOH to bring the pH to from 8 to 10. Filter the liquid again. Let the precipitate dry, and place it in the trash. Pour the filtrate down the drain.

PROCEDURAL NOTES

Safety Precautions

• Discuss all safety symbols and precautions with students.
• Tell students that the chemicals used in the lab can cause serious damage to their eyes. If students get iodine in their eyes, they should immediately go to the laboratory eyewash station, flushing eyes with water for at least 15 minutes, and then see a doctor for further treatment.

Procedural Tips

• Students may need to use a chemical spatula to scoop the burrito mixture into the test tubes.
• Students may need to add more water to the burrito filling in order to see clear results.

SAMPLE SOLUTION

Question

What is the nutrient content of the sample burrito filling?

Proposed Procedure

1. Put on safety goggles, disposable gloves, and a lab apron.
2. Test for fat. Use a spatula to place a small amount of burrito filling in a test tube labeled "A." Add 20 mL of water, and stir with a sterile plastic stirring rod to dilute the filling. Add 3 mL of Sudan III solution to the substance. If a reddish orange color appears, lipids are present. Record observations in a data table.
3. Test for starch. Use a spatula to place a small amount of burrito filling in a test tube labeled "B." Add 20 mL of water, and stir with a sterile plastic stirring rod to dilute the filling. Add 3 mL of Lugol's iodine into the substance. If a dark blue color appears, starch is present. Record observations in a data table.
4. Test for glucose by using a spatula to place a small amount of burrito filling in a test tube labeled "C." Add 20 mL of water, and stir with a sterile plastic stirring rod to dilute the filling. Dip a glucose test strip into the substance for 3–5 seconds.

Then read the test tape to determine if glucose is present. Record observations in a data table.

5. Test for protein. Use a spatula to place a small amount of filling in a test tube labeled "D." Add 20 mL of water and stir with a sterile plastic stirring rod to dilute the filling. Add 3 mL of Biuret reagent to the test tube. Look for the faint violet color that indicates the presence of protein. Record observations in a data table.

Materials: (per lab group)

- safety goggles
- disposable gloves
- lab aprons
- 10 mL graduated cylinder
- 4 test tubes
- test-tube rack
- chemical spatula
- 4 plastic stirring rods
- 20 mL of burrito filling
- 3 mL of Lugol's iodine solution
- 3 mL of Biuret reagent
- 3 mL of Sudan III reagent
- 1 glucose test strip
- 80 mL of distilled water
- wax pencil

Estimated Cost of the Job

$1,595.88, including a 30 percent profit (based on two students per group, two hours of labor, and two days of facility use). Refer to *Using Experimental Design Labs* to see how to prepare an invoice.

Results

Students should determine that protein, starch, fat, and glucose are all present in the burrito filling.

Sample Data Table

Nutrient	Presence (+ or –)	Observation
lipids	+	reddish orange color forms
starch	+	dark blue color forms
glucose	+	reddish violet or black color forms
protein	+	violet color forms

Conclusions

Students' data should support their conclusions. Answers will vary, but students should conclude that the label is indeed fraudulent.

HOLT BIOSOURCES™
LAB PROGRAM

LABORATORY TECHNIQUES AND EXPERIMENTAL DESIGN

INCLUDES
LABS C1–C48

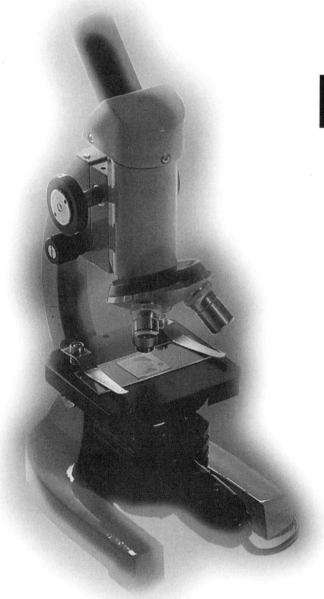

HOLT, RINEHART AND WINSTON
Harcourt Brace & Company

Austin • New York • Orlando • Atlanta • San Francisco • Boston • Dallas • Toronto • London

LABORATORY TECHNIQUES AND EXPERIMENTAL DESIGN

Staff Credits

Editorial Development

Carolyn Biegert
Janis Gadsden
Debbie Hix

Copyediting

Amy Daniewicz
Denise Haney
Steve Oelenberger

Prepress

Rose Degollado

Manufacturing

Mike Roche

Design Development and Page Production

Morgan-Cain & Associates

Acknowledgments

Contributors

David Jaeger
Will C. Wood High School
Vacaville, CA

George Nassis
Kenneth G. Rainis
WARD'S Natural Science Establishment
Rochester, NY

Suzanne Weisker
Science Teacher and Department Chair
Will C. Wood High School
Vacaville, CA

Editorial Development

WordWise, Inc.

Cover

Design—Morgan-Cain & Associates
Photography—Sam Dudgeon

Lab Reviewers

Lab Activities
Ted Parker
Forest Grove, OR

Mark Stallings, Ph.D.
Chair, Science Department
Gilmer High School
Ellijay, GA

George Nassis
Kenneth G. Rainis
Geoffrey Smith
WARD'S Natural Science Establishment
Rochester, NY

Lab Safety
Kenneth G. Rainis
WARD'S Natural Science Establishment
Rochester, NY

Jay Young, Ph.D
Chemical Safety Consultant
Silver Spring, MD

Printed in the United States of America

ISBN 0-03-051403-7

123456 022 00 99 98 97

LABORATORY TECHNIQUES AND EXPERIMENTAL DESIGN

Contents

Contents continued

Organizing Laboratory Data

Your data are all the records you have gathered from an investigation. The types of data collected depend on the activity. Data may be a series of weights or volumes, a set of color changes, or a list of scientific names. No matter which types of data are collected, all data must be treated carefully to ensure accurate results. Sometimes the data seem to be wrong, but even then, they are important and should be recorded accurately. Remember that nature cannot be wrong, regardless of what you discover in the laboratory. Data that seem to be "wrong" are probably the result of experimental error.

There are many ways to record and organize data, including data tables, charts, diagrams, and graphs. Your teacher will help you decide which format is best suited to the type of data you collect.

It is important to include the appropriate units when you record data. Remember that data are measurements or observations, not merely numbers. Data tables, graphs, and diagrams should have titles that are descriptive and complete enough to ensure that another person could understand them without having been present during the investigation.

Many important scientific discoveries have been made accidentally in the course of an often unrelated laboratory activity. Scientists who keep very careful and complete records sometimes notice unexpected trends in and relationships among data long after the work is completed. The laboratory notebooks of working scientists are studded with diagrams and notes; every step of every procedure is carefully recorded.

Data Tables and Charts

Data tables are probably the most common means of recording data. Although prepared data tables are often provided in laboratory manuals, it is important that you be able to construct your own. The best way to do this is to choose a title for your data table and then make a list of the types of data to be collected. This list will become the headings for your data columns. For example, if you collected data on plant growth that included both the length of time it took for the plant to grow and the amount of growth, you could record your data in a table like this:

Plant Growth Data

These data are the basis for all your later interpretations and analyses. You can always ask new questions about the data, but you cannot get new data without repeating the experiment.

Graphs

After data are collected, you must determine how to display them. One way of showing your result is to use a graph. Two types of graphs are commonly used: the line graph and the bar graph. In a line graph, the data are arranged so that two variables are represented as a single point. You could easily make a line graph of the data shown in the growth table. The first step is to draw and label the axes. Before you do this, however, you must decide which column of data should be represented on the *x*-axis (horizontal axis) and which should be represented on the *y*-axis (vertical axis).

Experiments have two types of variables, or factors that may change. Independent variables are variables that could be present even if other factors were not. For the example above, "Time" is an independent variable because time exists regardless of whether plants are present. An independent variable is, by convention, plotted on the *x*-axis of a graph. Dependent variables are variables that change because an independent variable changes. A dependent variable for this example would be "Height of plant." Dependent variables are plotted on the *y*-axis of a graph.

Next you must choose the scale for the axes of your graph. You want the graph to take up as much of the paper as possible because large graphs are much easier to read and make than small ones. For each axis, you must choose a scale that uses the largest amount of graph paper. Remember, once you choose the interval for the scale (the number of days each block represents on the *x*-axis, for example), you cannot change it. You cannot say that block one represents 1 day and block two represents 10 days. If you change the scale, your graph will not accurately represent your data.

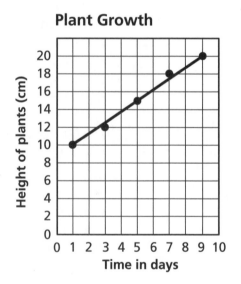

Plant Growth

The next step is to mark the points for each pair of numbers. When all points are marked, draw the best straight or curved line between them. Remember that you do not "connect the dots" when you draw a graph. Instead, you should draw a "best fit" curve—a line or smooth curve that intersects or comes as close as possible to your set of data points.

If you choose to represent your data by using a bar graph, the first steps are similar to those for the line graph. You must first choose your axes and label them. The independent variable is plotted on the *x*-axis, and the dependent variable is plotted on the *y*-axis. However, instead of plotting points on the graph, you represent the dependent variable as a bar extending from the *x*-axis to where you would have drawn the points. Using the sample data on plant growth, for example, on day 1, the height of the plant was 10 cm. On your graph, you would make a bar that extends to the height of the 10 cm mark on the *y*-axis.

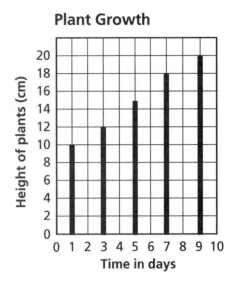

Plant Growth

Diagrams

In some cases, the data you must represent are not numerical. That means that they cannot be put into a data table or graphed. The best way to represent this type of information is to draw and label it. To do this, you simply draw what you see and label as many parts or structures as possible. This technique is especially useful in the biology laboratory, where many investigations involve the observation of living or preserved specimens. Remember, you do not have to be an artist to make a good laboratory drawing.

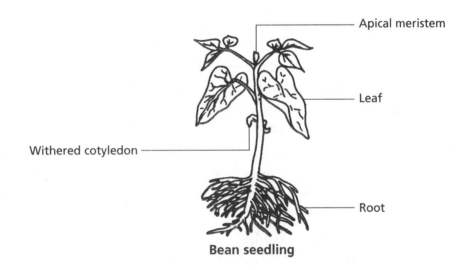

Bean seedling

There are several things you need to remember as you make your laboratory drawing. First, make the drawing large enough to be easily studied. Include all of the visible structures in your drawing. Second, drawings should also show the spacing between the parts of the specimen in proportion to its actual appearance. Size relationships are important in understanding and interpreting observations. Third, in order for your drawing to be the most useful to you, you need to label it. All labels should be clearly and neatly printed. Lines drawn from labels to the corresponding parts should be straight, so be sure to use a ruler. Label lines should never cross each other. Finally, be sure to title the drawing. Someone who looks at your drawing should be able to identify the specimens. Remember, neatness and accuracy are the most important parts of any laboratory drawing.

Laboratory Safety

Your biology laboratory is a unique place where you can learn by doing things that you couldn't do elsewhere. It also involves some dangers that can be controlled if you follow these safety notes and all instructions from your teacher.

It is your responsibility to protect yourself and other students by conducting yourself in a safe manner while in the laboratory. Familiarize yourself with the printed safety symbols—they indicate additional measures that you must take.

While in the Laboratory, at All Times . . .

- **Familiarize yourself with a lab activity—especially safety issues—before entering the lab.** Know the potential hazards of the materials, equipment, and the procedures required for the activity. Ask the teacher to explain any parts you do not understand before you start.

- **Never perform any experiment not specifically assigned by your teacher.** Never work with any unauthorized material.

- **Never work alone in the laboratory.**

- **Know the location of all safety and emergency equipment used in the laboratory.** Examples include eyewash stations, safety blankets, safety shower, fire extinguisher, first-aid kit, and chemical-spill kit.

- **Know the location of the closest telephone,** and be sure there is a posted list of emergency phone numbers, including poison control center, fire department, police, and ambulance.

- **Before beginning work: tie back long hair, roll up loose sleeves, and put on any personal protective equipment as required by your teacher.** Avoid or confine loose clothing that could knock things over, ignite from a flame, or soak up chemical solutions.

- **Report any accident, incident, or hazard—no matter how trivial—to your teacher immediately.** Any incident involving bleeding, burns, fainting, chemical exposure, or ingestion should also be reported to the school nurse or physician.

- **In case of fire, alert the teacher and leave the laboratory.**

- **Never eat, drink, or apply cosmetics.** Never store food in the laboratory. Keep your hands away from your face. Wash your hands at the conclusion of each laboratory activity and before leaving the laboratory. Remember that some hair products are highly flammable, even after application.

- **Keep your work area neat and uncluttered.** Bring only books and other materials that are needed to conduct the experiment.

- **Clean your work area at the conclusion of the lab as your teacher directs.**

• **When called for, use the specific safety procedures below.**

Eye Safety

• **Wear approved chemical safety goggles as directed.** Goggles should always be worn whenever you are working with a chemical or chemical solution, heating substances, using any mechanical device, or observing a physical process.

• **In case of eye contact**
  (1) Go to an eyewash station and flush eyes (including under the eyelids) with running water for at least 15 minutes.
  (2) Notify your teacher or other adult in charge.

• **Wearing contact lenses for cosmetic reasons is prohibited in the laboratory.** Liquids or gases can be drawn up under the contact lens and into direct contact with the eyeball. If you must wear contact lenses prescribed by a physician, inform your teacher. You must wear approved eye-cup safety goggles—similar to goggles individuals wear when swimming underwater.

• **Never look directly at the sun through any optical device or lens system, or gather direct sunlight to illuminate a microscope.** Such actions will concentrate light rays that will severely burn your retina, possibly causing blindness!

Electrical Supply

• **Never use equipment with frayed cords.**

• **Ensure that electrical cords are taped to work surfaces** so that no one will trip and fall and so that equipment can't be pulled off the table.

• **Never use electrical equipment around water or with wet hands or clothing.**

Clothing Protection

• **Wear an apron or lab coat when working in the laboratory to prevent chemicals or chemical solutions from coming in contact with skin or contaminating street clothes.** Confine all loose clothing and long jewelry.

Animal Care

• **Do not touch or approach any animal in the wild.** Be aware of poisonous or dangerous animals in any area where you will be doing outside fieldwork.

• **Always obtain your teacher's permission before bringing any animal (or pet) into the school building.**

• **Handle any animal only as your teacher directs.** Mishandling or abuse of any animal will not be tolerated!

Sharp Object Safety

• **Use extreme care with all sharp instruments, such as scalpels, sharp probes, and knives.**

• **Never use double-edged razor blades in the laboratory.**

• **Never cut objects while holding them in your hand.** Place objects on a suitable work surface.

Chemical Safety

- **Always wear appropriate personal protective equipment.** Safety goggles, gloves, and an apron or lab coat should always be worn when working with any chemical or chemical solution.

- **Never taste, touch, or smell any substance or bring it close to your eyes, unless specifically told to do so by your teacher.** If you are directed by your teacher to note the odor of a substance, do so by waving the fumes toward you with your hand. Never pipet any substance by mouth; use a suction bulb as directed by your teacher.

- **Always handle any chemical or chemical solution with care.** Check the label on the bottle and observe safe-use procedures. Never return unused chemicals or solutions to their containers. Return unused reagent bottles or containers to your teacher. Store chemicals according to your teacher's directions.

- **Never mix chemicals** unless specifically told to do so by your teacher.

- **Never pour water into a strong acid or base.** The mixture can produce heat and can splatter. Remember this rhyme:

  > "Do as you oughta—
  > Add acid (or base) to water."

- **Report any spill immediately to your teacher.** Handle spills only as your teacher directs.

- **Check for the presence of any source of flames, sparks, or heat (open flame, electric heating coils, etc.) before working with flammable liquids or gases.**

Plant Safety

- **Do not ingest any plant part used in the laboratory (especially seeds sold commercially).** Do not rub any sap or plant juice on your eyes, skin, or mucous membranes.

- **Wear protective gloves (disposable polyethylene gloves) when handling any wild plant.**

- **Wash hands thoroughly after handling any plant or plant part (particularly seeds).** Avoid touching your hands to your face and eyes.

- **Do not inhale or expose yourself to the smoke of any burning plant.**

- **Do not pick wildflowers or other plants unless directed to do so by your teacher.**

Proper Waste Disposal

- **Clean and decontaminate all work surfaces and personal protective equipment as directed by your teacher.**

- **Dispose of all sharps (broken glass and other contaminated sharp objects) and other contaminated materials (biological and chemical) in special containers as directed by your teacher.**

Hygienic Care

- Keep your hands away from your face and mouth.

- Wash your hands thoroughly before leaving the laboratory.

- Remove contaminated clothing immediately; launder contaminated clothing separately.

- When handling bacteria or similar microorganisms, use the proper technique demonstrated by your teacher. Examine microorganism cultures (such as petri dishes) without opening them.

- Return all stock and experimental cultures to your teacher for proper disposal.

Heating Safety

- When heating chemicals or reagents in a test tube, never point the test tube toward anyone.

- Use hot plates, not open flames. Be sure hot plates have an "On-Off" switch and indicator light. Never leave hot plates unattended, even for a minute. Never use alcohol lamps.

- Know the location of laboratory fire extinguishers and fire blankets. Have ice readily available in case of burns or scalds.

- Use tongs or appropriate insulated holders when heating objects. Heated objects often do not look hot. Never pick up an object with your hands unless you are certain it is cold.

- Keep combustibles away from heat and other ignition sources.

Hand Safety

- Never cut objects while holding them in your hand.

- Wear protective gloves when working with stains, chemicals, chemical solutions, or wild (unknown) plants.

Glassware Safety

- Inspect glassware before use; never use chipped or cracked glassware. Use borosilicate glass for heating.

- Do not attempt to insert glass tubing into a rubber stopper without specific instruction from your teacher.

- Always clean up broken glass by using tongs and a brush and dustpan. Discard the pieces in an appropriately labeled "sharps" container.

Safety With Gases

- Never directly inhale any gas or vapor. Do not put your nose close to any substance having an odor.

- Handle materials prone to emit vapors or gases in a well-ventilated area. This work should be done in an approved chemical fume hood.

Using Laboratory Techniques and Experimental Design Labs

You will find two types of laboratory exercises in this book.

1. *Laboratory Techniques* labs help you gain skill in biological laboratory techniques.
2. *Experimental Design* labs require you to use the techniques learned in *Laboratory Techniques* labs to solve problems.

Working in the World of a Biologist

Laboratory Techniques and Experimental Design labs are designed to show you how biology fits into the world outside of the classroom. For both types of labs, you will play the role of a working biologist. You will gain experience with techniques used in biological laboratories and practice real-world work skills, such as creating a plan with available resources, working as part of a team, developing and following a budget, and writing business letters.

Tips for success in the lab

Preparation helps you work safely and efficiently. Whether you are doing a Laboratory Techniques lab or an Experimental Design lab, you can do the following to help ensure success.

- **Read a lab twice** before coming to class so you will understand what to do.
- **Read and follow the safety information** in the lab and on pages viii–xi.
- **Prepare data tables** before you come to class.
- **Record all data and observations immediately** in your data tables.
- **Use appropriate units** whenever you record data.
- **Keep your lab table organized** and free of clutter.

Laboratory Techniques Labs

Each Laboratory Techniques lab enables you to practice techniques that are used in biological research by providing a step-by-step procedure for you to follow. You will use many of these techniques later in an Experimental Design lab. The parts of a Laboratory Techniques lab are described below.

1. *Skills* identifies the techniques and skills you will learn.
2. *Objectives* tells you what you are expected to accomplish.
3. *Materials* lists the items you will need to do the lab.
4. *Purpose* is the setting for the lab.
5. *Background* is the information you will need for the lab.
6. *Procedure* provides step-by-step instructions for completing the lab and reminders of the safety procedures you should follow.
7. *Analysis* items help you analyze the lab's techniques and your data.
8. *Conclusions* items require you to form opinions based on your observations and the data you collected in the lab.

9. *Extensions* items provide opportunities to find out more about topics related to the subject of the lab.

Experimental Design Labs

Each of these labs requires you to develop your own procedure to solve a problem that has been presented to your company by a client. The procedures you develop will be based on the procedures and techniques you learned in previous Laboratory Techniques labs. You must also decide what equipment to use for a project and determine the amount you should charge the client. The parts of an Experimental Design lab are described below.

1. *Prerequisites* tells you which Laboratory Techniques labs contain procedures that apply to the Experimental Design lab.
2. *Review* tells you which concepts you need to understand to complete the lab.
3. The *Letter* contains a request from a client to solve a problem or to do a project.
4. The *Memorandum* is a note from a supervisor directing you to perform the work requested by the client, and providing clues or directions that will help you design a successful experiment or project. The *Memorandum* also contains the following: a *Proposal Checklist,* which must be completed before you start the lab; *Report Procedures,* which tells you what should be included in your lab report; *Required Precautions,* which indicate the safety procedures you should follow during the lab; and *Disposal Methods,* which tells you how to dispose of the materials used.
5. *Materials and Costs* is a list of what you might need to complete the work and the unit cost of each service and item.

What you should do before an Experimental Design lab

Before you will be allowed to begin an Experimental Design lab, you must turn in a proposal that includes the question to be answered, the procedure you will use, a detailed data table, and a list of all the proposed materials and their costs. Before you begin writing your proposal, follow these steps.

- **Read the lab thoroughly,** and jot down any clues you find that will help you successfully complete the lab.
- **Consider what you must measure or observe** to solve the problem.
- **Think about** *Laboratory Techniques* labs you have done that required similar measurements and observations.
- **Imagine working through a procedure,** keeping track of each step and of the equipment you will need.

What you should do after an Experimental Design lab

After you finish, prepare a report as described in the *Memorandum.* The report can be in the form of a one- or two-page letter to the client, plus an invoice showing the cost of each phase of the work and the total amount you charged the client. Carefully consider how to convey the information the client needs to know. In some cases, graphs and diagrams may communicate information better than words can.

Name _____

Date _____ Class _____

C1 *Using a Microscope*

Skills

- using a compound light microscope
- making a wet mount
- using a stereomicroscope

Objectives

PREPARATION NOTES

Time Required: three 50-minute periods

- *Demonstrate* the proper use and care of a compound light microscope.
- *Focus* the compound light microscope at low power and at high power.
- *Make* a wet-mount slide to examine under the microscope.
- *Demonstrate* the proper use and care of a stereomicroscope.
- *Compare* the movement of the images seen through a compound light microscope and a stereomicroscope.

Materials

Materials

• Materials for this lab activity can be purchased from WARD'S. See the *Master Materials List* for ordering instructions.

- compound light microscope
- prepared slide
- lens paper
- glass microscope slide (2)
- coverslip (2)
- scissors
- newspaper
- medicine dropper
- water
- forceps
- dissecting needle or pencil
- thread
- stereomicroscope
- paper towel
- small plants, such as moss
- leaves

Purpose

You are a laboratory assistant who has been selected to be part of a newly formed research team. Throughout the year, you will be observing many organisms and performing many tests. To perform these tasks, you will use a variety of tools. Today you will be trained on the use of a compound light microscope and a stereomicroscope. You will also learn how to make a wet-mount slide.

Background

• Most standard student compound light microscopes are 100× on low power and 400× on high power. If your microscopes have a scanning lens and/or an oil immersion objective, tell students not to use them in this lab.

In almost every type of biological research, the microscope plays a fundamental role. Biologists use it to study the fine structures of cells and tissues—things too small to be seen with the unaided eye. The microscope used most often is the **light microscope,** which uses light to form an enlarged image of a specimen. Two types of light microscopes are the **compound light microscope** and the **stereomicroscope.** Compound light microscopes are used to view tiny living organisms as well as preserved cells mounted on glass—a **microscope slide**—and covered with a **coverslip.** A slide that is prepared with water is called a **wet mount.** Stereomicroscopes are used to study larger specimens and provide a three-dimensional view of the specimen's surface.

Under the compound light microscope, most objects and microorganisms are observed in a drop of water. If you think of that drop of water as a pond and the objects and microorganisms as fish in the pond, you will begin to see why it is important to be able to focus at different depths. **Depth-of-field** focusing is always done under high power with the fine adjustment.

Procedure

Preparation Tip

Provide a variety of prepared slides of cells or tissues that are easy for students to focus on, and allow them to look at cells of different shapes and sizes.

Disposal

• Instruct students to place used newspaper and plants in the laboratory trash can and to wash and dry slides and coverslips and return them to a designated place.

• Provide a separate container for the disposal of broken glass.

PROCEDURAL NOTES

Safety Precautions

• Discuss all safety symbols and caution statements with students.

• Instruct students to follow the proper procedures for carrying and using the compound microscope.

Procedural Tips

• Remind students to always begin on low power when focusing the microscope.

• This lab activity is meant to give students new skills and confidence in using the microscope. Ideally there should be plenty of specimens and microscopes available. The more time spent on this lab, the more success students will have with future labs using the microscope. Other lab activities rely heavily on the premise that students have mastered the basic techniques and have acquired the right attitude toward use and care of the microscope, including safety.

Part 1—Learning the Compound Light Microscope

1. Complete the following data table as you do Part 1.

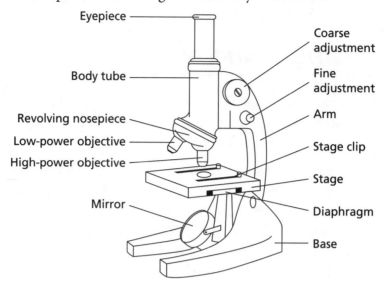

Function of the Parts of a Compound Light Microscope

Microscope part	Function
Eyepiece (magnification: _____)	where the eye looks through the lens
Body tube	directs light from the objective to the eye
Arm	holds the base of the microscope with the objectives and eyepiece
Stage	supports the slide
Coarse adjustment	used to bring an object into focus under low power
Fine adjustment	Used to fine-tune the focus under low power; used to focus under high power
Lamp or mirror	light source
Revolving nosepiece	holds the objectives
Low-power objective (magnification: _____)	lens with the least magnification
High-power objective (magnification: _____)	lens with the greatest magnification
Diaphragm	controls the amount of light passing through the object to be viewed
Base	holds the light or mirror; connects to the arm to hold the other parts of the scope together

- Explain what the microscope is—how it works and what it is used for. It will take the students a while to get a sense of the size of the objects they will be observing. Provide examples to aid in their understanding of magnitude.

- Caution students to watch carefully as they change to high power because occasionally slides (such as those of whole mounted specimens) are thicker than normal. If this is the case, the high-power objective may hit the slide when it is moved into place.

- Acquaint students with both iris and disc diaphragms, and show them how to change the setting on each type.

- Warn students not to touch the lamps; they get very hot. If the microscopes have a plastic or vinyl cover, caution students to make sure the lamp is cool before replacing the cover. A hot lamp may melt the cover.

- Have available prepared slides of a variety of microorganisms and cellular structures for students to observe and practice their microscope skills.

2. Carry a microscope to your lab table as shown by your teacher. *Note: A microscope is expensive and fragile. It is important to use it correctly to avoid damaging it and avoid breaking slides or destroying specimens. When you use a microscope, be sure it rests securely on your lab table away from the edge.*

3. Locate each microscope part listed in the data table and shown in the diagram on the previous page. Observe the magnification power (a number followed by an $\times$) of the eyepiece and the low- and high-power objectives. Record these numbers in your data table.

4. If your microscope has a built-in lamp, plug it in and turn it on. If your microscope has a mirror, adjust the mirror to reflect light through the hole in the center of the stage. **CAUTION: If your microscope has a mirror, never use direct sunlight as a light source. Direct sunlight will damage your eyes.**

5. Raise the objectives (or lower the stage) as far as possible by turning the coarse-adjustment knob toward you. Secure a prepared slide to the stage using the stage clips. Turn the low-power objective into position over the stage. While observing the stage from eye level, use the coarse-adjustment knob to position the objective as close to the slide as it will go without touching the slide.

6. Look through the eyepiece. Always keep both eyes open as you look into the eyepiece. Keeping both eyes open avoids eye strain. If the lens is dirty, ask your teacher to demonstrate the correct way to clean it. *Note: Never use anything other than lens paper to clean the lenses of the microscope.* Focus with the coarse-adjustment knob by turning it away from you. *Note: Never focus objectives downward. You may run the objective into the slide and break the slide or damage the objective.*

7. Complete focusing by slowly turning the fine-adjustment knob back and forth. When the object you are viewing is in focus and exactly in the middle of your field of vision, switch to high power. *Note: Never use the coarse-adjustment knob at high power.*

Part 2—Making a Wet Mount

8. Use scissors to cut out a capital letter *R* from a piece of newspaper. *Note: Do not use one from a headline.* **CAUTION: Handle scissors carefully. Notify your teacher of any cuts.**

9. With a medicine dropper, place one drop of water in the middle of a clean glass microscope slide. **CAUTION: Glassware is fragile. Notify your teacher promptly of any broken glass or cuts. Do not clean up broken glass or spills unless your teacher tells you to do so.** With forceps, place the letter *R* in the drop of water as seen in the diagram on the next page.

10. Hold a coverslip at a 45° angle to the slide at the edge of the drop of water as seen in the diagram on the next page. Lower the coverslip slowly to avoid forming air bubbles. Under the microscope, air bubbles look round and have dark edges.

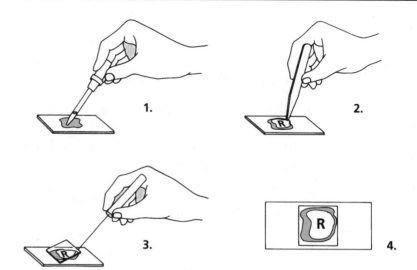

11. Place your wet mount on the microscope stage with the letter *R* facing you. Using the low-power objective, center and focus the microscope on the letter *R*. Then switch to high power.

♦ What happens to the image of the letter *R* as you go from low to high power?

Answers will vary. You see less of the *R*; you see more detail; the image becomes fuzzy.

12. As you look through the eyepiece, slowly adjust the diaphragm to obtain the appropriate light for viewing.

♦ What happens as you adjust the diaphragm?

The field of view becomes darker or lighter.

13. As you look into the microscope, use your fingers to move the slide to the right and then to the left.

♦ What happens to the image as you move the slide to the right?

The image in focus moves to the left.

♦ What happens to the image as you move the slide to the left?

The image in focus moves to the right.

♦ Move the slide away from you, and record what happens to the image.

The image in focus moves toward you.

Part 3—Depth-of-Field Focusing

14. Make a wet mount slide of two threads by crossing the threads in the center of a clean glass microscope slide. Use a medicine dropper to add a drop of water. Add a coverslip to the slide.

15. Place your wet mount on the stage of the microscope. Under low power, adjust the slide on the microscope stage so that the point where the threads cross is in the center of your field of vision. Bring the threads into focus.

♦ Can you see both threads in focus at the same time?

Yes.

16. Switch to high power.

♦ Using the fine adjustment, can you see both threads in focus at the same time? Why?

No, the threads are at different depths. The high-power objective has less depth of field.

17. Slowly turn the fine-adjustment knob back and forth, and practice focusing on different parts of the two threads.

Part 4—Comparing the Stereomicroscope With the Compound Light Microscope

18. Look at the stereomicroscope, and identify the parts labeled in the diagram below. Determine how its uses differ from those of the compound light microscope. Compare the working distance (the space between the objective and the stage) of the stereomicroscope to that of the compound light microscope.

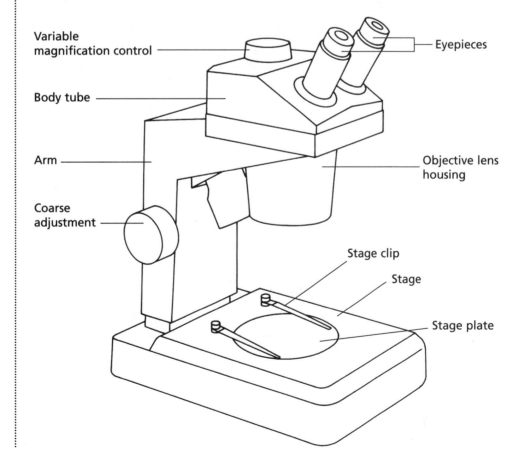

◆ How does the working distance affect the size of objects that can be viewed under the stereomicroscope?

The working distance of the stereomicroscope is much larger than that of the compound light

microscope so that larger specimens can be viewed as three-dimensional objects.

19. Lay a clump of moss or a leaf on a paper towel, and place it on the stage of the stereomicroscope. Focus on part of the moss or leaf. Move the paper towel to the left.

◆ Compare the movement of the plant seen through the eyepieces to the direction you moved the plant.

It appeared to move to the left, the same direction you moved the plant.

◆ How does the image move when seen with a stereomicroscope compared with the way an image moves when seen with a compound light microscope?

The image seen with a stereomicroscope moves in the same direction as the object; the image seen

with a compound light microscope moves in the opposite direction from the object.

20. Compound light microscopes work by having light pass through the objects to be viewed. However, not all objects are translucent. Some are opaque; light cannot pass through them. Make wet mounts of a moss plant and a small piece of a leaf. Observe the wet mounts under low power.

◆ How does your view of the moss plant compare with your view of the leaf?

The parts of the moss plant can be seen, while the parts of the leaf cannot be seen because it is too

thick to allow much light to shine through it.

◆ If an opaque object is viewed through a compound light microscope, what will you see?

Only the silhouette of the object will be visible.

21. Place a leaf under the stereomicroscope. Adjust the light source so that light shines on the leaf not through it.

◆ How does this way of lighting an object allow you to observe objects that cannot be viewed under a compound light microscope?

It uses reflected light to produce an image of the surface of opaque objects.

22. Dispose of your materials according to the directions from your teacher.

23. Clean up your work area and wash your hands before leaving the lab.

Analysis

24. What does the magnification number on the eyepiece mean?

The eyepiece magnification number is how many times larger the eyepiece makes the image

coming from the objective appear to be.

25. Calculate the total magnification of your compound light microscope at low power and at high power. Remember, you look at a specimen through both the objective lens and the eyepiece (Hint: Multiply the eyepiece magnification by the objective magnification.) Show your work below.

Total magnification at low power: Eyepiece magnification (10×) × low-power objective

magnification (10×) = 100×. Total magnification of high power: Eyepiece magnification (10×)

× high-power objective magnification (40×) = 400×.

26. Under which power, high or low, is the largest field of view seen?

low power

27. What will happen to the field of view when you switch from low power to high power?

The field of view will become darker and smaller as you switch from low power to high power.

28. When making a wet mount, why must you always use a coverslip?

Using a coverslip on a wet mount protects the objectives from coming into contact with water and

specimens and keeps them clean.

29. Why is it necessary to be able to focus at different depths?

to analyze all parts of what you are looking at or to be able to see all parts of a slide

30. Why do objects look three-dimensional under the stereomicroscope?

There is a lens for each eye, and depth perception requires visual input from both eyes.

Conclusions

31. If a microorganism were swimming from right to left across your field of view under a compound light microscope, which way would you move the microscope slide to keep it in view? Why?

You would move the slide to the right because the apparent direction of motion of an object under

a compound light microscope is opposite from the object's actual direction.

32. If the same microorganism as above were swimming from right to left across your field of view under a stereomicroscope, which way would you move the microscope slide to keep it in view? Why?

You would move the slide to the left because the apparent direction of motion of an object under a

stereomicroscope is the same as the object's actual direction.

33. When is the microscope a useful tool for biologists?

The microscope is useful to biologists whenever they need to examine objects that are smaller than

the unaided eye can see. Microscopes also enable biologists to magnify larger objects to see them in

more detail.

34. A student brings in a piece of moldy bread. Which microscope would you use to view the mold? Why?

The stereomicroscope would give a good overall picture of the mold on the bread, while the

compound light microscope would reveal the fine structure of the mold on a slide prepared from it.

Extensions

35. *Research technicians* use microscopes to study and identify many things. Find out about the training and skills required to become a research technician.

36. In addition to light microscopes, biologists today use many different kinds of microscopes in their everyday activities. Find out the differences and the amount of magnification for the following microscopes: electron microscope, scanning electron microscope, transmission electron microscope, and scanning tunneling electron microscope.

HOLT

BIOSOURCES

LAB PROGRAM

EXPERIMENTAL DESIGN

C2 | *Using a Microscope—Criminal Investigation*

Prerequisites | • **Laboratory Techniques C1: Using a Microscope** on pages 1–8

Review | • procedures for using a microscope
| • procedures for making a wet mount

CITY OF OAKWOOD
POLICE DEPARTMENT
Oakwood, Missouri 65432-1221

September 20, 1997

Caitlin Noonan
Research and Development
BioLogical Resources, Inc.
101 Jonas Salk Dr.
Oakwood, MO 65432-1101

Dear Ms. Noonan,

As you may have heard, we have recently had a string of burglaries here in Oakwood. We had not been able to gather much evidence until yesterday evening, when we found several types of trace evidence at the scene of the latest burglary. Three people were identified as being near the business at the time of the burglary. One of the three suspects is a relative of a high-ranking member of our city staff, so it is extremely important that our facts are correct. We need a party not associated with the city to test the trace evidence that has been gathered.

The City of Oakwood Police Department would like to begin a working relationship with your research company. Enclosed is evidence gathered from the site of the burglary. I have also included the trace evidence collected from each of the three suspects. Please study the evidence to see if any of the suspects can be linked to the scene of the crime. The city is willing to pay your fee. Please let me know if you can take the job.

Sincerely,

Roberto Morales

Roberto Morales
Chief of Police
City of Oakwood Police Department

BioLogical Resources, Inc. Oakwood, MO 65432-1101

M E M O R A N D U M

To: Team Leader, Forensics Dept.

From: Caitlin Noonan, Director of Research and Development

As we are a relatively new company, this account from the city is *very* important. I want your team to make thorough observations with and without a microscope. Please be extremely thorough and organized in your procedures, and make your written observations as detailed as possible. If we are called to present our evidence in court, we will need to be able to discuss our procedures, results, and conclusions in a detailed, organized, and logical manner.

Proposal Checklist

Before you start your work, you must submit a proposal for my approval. **Your proposal must include the following:**

_____ • the **question** you seek to answer

_____ • the **procedure** you will use

_____ • a detailed **data table** for recording observations

_____ • a complete, itemized list of proposed **materials** and **costs** (including use of facilities, labor, and amounts needed)

Proposal Approval: _____
                                                           (Supervisor's signature)

Report Procedures

When you finish your tests, prepare a report in the form of a business letter to Chief Morales. **Your report must include the following:**

_____ • a paragraph describing the **procedure** you followed to examine the evidence

_____ • a complete **data table** showing your observations of the evidence

_____ • your **conclusions** about which, if any, of these suspects can be linked to the crime scene

_____ • a detailed **invoice** showing all materials, labor, and the total amount due

Safety Precautions

- Glassware is fragile. Notify your teacher promptly of any broken glass or cuts. Do not clean up broken glass or spills unless your teacher tells you to do so.

- Never use electrical equipment around water, nor with wet hands or clothing. Never use equipment with frayed cords.

- Wash your hands before leaving the laboratory.

- If you are using a microscope with a mirror, do not use direct sunlight as a light source.

Disposal Methods

- Dispose of waste materials according to instructions from your teacher.

- Place used paper towels in a trash can.

- Place broken glass and evidence that has not been destroyed in the separate containers provided.

- Wash reusable materials such as glassware and lab utensils, and return them to the supply area.

FILE: City of Oakwood Police Department

MATERIALS AND COSTS (Select only what you will need. No refunds.)

I. Facilities and Equipment Use

Item	Rate	Number	Total
facilities	$480.00/day		
personal protective equipment	$10.00/day		
compound microscope	$30.00/day		
microscope slide with coverslip	$2.00/day		
depression slide with coverslip	$2.00/day		
beaker	$5.00/day		
eyedropper	$2.00/day		
clock or watch with second hand	$10.00/day		
scissors	$1.00/day		
transparent ruler	$1.00/day		
test tube rack	$5.00/day		
test tubes	$3.00/day		

II. Labor and Consumables

Item	Rate	Number	Total
labor	$40.00/hour		
crime scene evidence bag	provided		
evidence bags A, B, and C	provided		
wax pencil	$2.00 each		
pH paper	$2.00/strip		
lens paper	$1.00/sheet		

Fines

OSHA safety violation	$2,000.00/incident		
		Subtotal	
		Profit Margin	
		Total Amount Due	

Name _____

Date _____ Class _____

HOLT
BIOSOURCES
LAB PROGRAM
EXPERIMENTAL DESIGN

C3 Using a Microscope— Slowing Protozoans

Prerequisites • **Laboratory Techniques C1: Using a Microscope** on pages 1–8

Review • procedures for making a wet mount
• procedures for using the microscope

AQUA ANALYSIS
Mobile, Alabama

September 27, 1997

Karl Smith
Microbiology Division
BioLogical Resources, Inc.
101 Jonas Salk Dr.
Oakwood, MO 65432-1101

Dear Mr. Smith,

I am the owner of Aqua Analysis, a small business that identifies microorganisms in water samples. For a variety of reasons, my clients need to know what microorganisms can be found in the lakes, creeks, and streams in their area.

Many of my customers have limited budgets. I try to keep my costs low, but over the last two years the slowing agent that I use has doubled in price. The slowing agent is essential to my work; without it, most of the protozoans would move too quickly to be observed. I cannot continue to use this product without raising my fee, but some of my clients simply cannot afford to pay a higher rate. I must come up with another solution.

I noticed recently that the agent that I use slows the organisms down more than necessary. I am curious to know whether diluting the slowing agent would allow me to cut costs without reducing my accuracy. In order to identify an organism, I need to observe it clearly for about 10 seconds. I would like to know how much I can dilute the slowing agent and still be able to identify the organisms. I would do a study myself, but I have a backlog of orders to handle and do not have any extra time. I have included a sample of the slowing agent. Please let me know what you find.

Sincerely,

Mikako Hayame

Mikako Hayame
Owner, Aqua Analysis

BioLogical Resources, Inc. Oakwood, MO 65432-1101

M E M O R A N D U M

To: Team Leader, Limnology Dept.

From: Karl Smith, Director of Microbiology

Your department is best equipped to handle Ms. Hayame's request. Please get started as soon as possible. In a recent phone conversation, Ms Hayame gave me some additional information that should help you in planning your experiments.

The slowing agent Ms. Hayame uses comes from the factory as a 2% solution. When preparing her slides, she adds one drop of slowing agent to one drop of culture water on a slide. She has suggested that you test several ratios of slowing agent to culture water. These ratios are 1:1, 1:2, 1:3, and 1:4. Have your team members conduct an experiment to test *Paramecium* and *Euglena* for these ratios.

Because Ms. Hayame is limited financially, we will be conducting this study at the lowest possible cost. As always, please avoid any unnecessary expenditures.

Proposal Checklist

Before you start your work, you must submit a proposal for my approval. **Your proposal must include the following:**

_____ • the **question** you seek to answer

_____ • the **procedure** you will use

_____ • a detailed **data table** for recording measurements

_____ • a complete, itemized list of proposed **materials** and **costs** (including use of facilities, labor, and amounts needed)

Proposal Approval: _____
(Supervisor's signature)

Report Procedures

When you finish your tests, prepare a report in the form of a business letter to Ms. Hayame. **Your report must include the following:**

_____ • a paragraph describing the **procedure** you followed to examine the effectiveness of each dilution of slowing agent

_____ • a complete **data table** showing time taken to cross the microscope field for each organism in each solution

_____ • your **conclusions** about which dilution best fits Ms. Hayame's needs and how much money, if any, will be saved

_____ • a detailed **invoice** showing all materials, labor, and the total amount due

Safety Precautions

• ◇◇ Wear safety goggles and a lab apron.

• ◇ Glassware is fragile. Notify your teacher promptly of any broken glass or cuts. Do not clean up broken glass or spills unless your teacher tells you to do so.

• ◇ Never use electrical equipment around water, nor with wet hands or clothing. Never use equipment with frayed cords.

• ◇ Wash your hands before leaving the laboratory.

• If you are using a microscope with a mirror, do not use direct sunlight as a light source.

Disposal Methods

• ◇ Dispose of waste materials according to instructions from your teacher.

• Toothpicks and paper towels used to clean lab tables can be thrown into a trash can.

• Place broken glass and leftover solutions in the separate containers provided.

• Wash reusable materials such as glassware and lab utensils, and return them to the supply area.

FILE: Aqua Analysis

MATERIALS AND COSTS (Select only what you will need. No refunds.)

I. Facilities and Equipment Use

Item	Rate	Number	Total
facilities	$480.00/day	_____	_____
personal protective equipment	$10.00/day	_____	_____
compound microscope	$30.00/day	_____	_____
depression slide with coverslip	$2.00/day	_____	_____
pipet	$2.00/day	_____	_____
clock with second hand	$10.00/day	_____	_____
beaker	$5.00/day	_____	_____
scissors	$1.00/day	_____	_____
ruler	$1.00/day	_____	_____

II. Labor and Consumables

Item	Rate	Number	Total
labor	$40.00/hour	_____	_____
Euglena in culture water	$2.00/drop	_____	_____
Paramecium in culture water	$2.00/drop	_____	_____
slowing agent	provided	_____	_____
tape	$0.25/m	_____	_____
paper towels	$0.10/sheet	_____	_____
toothpicks	$0.10 each	_____	_____

Fines

OSHA safety violation	$2,000.00/incident	_____	_____

Subtotal	_____
Profit Margin	_____
Total Amount Due	_____

HOLT
BIOSOURCES
LAB PROGRAM
LABORATORY TECHNIQUES

C4   *Testing for Vitamin C*

Skills
- comparing vitamin C content in foods with a standard
- organizing data on a bar graph

Objectives
- *Determine* the vitamin C content of selected fruit and vegetable juices.
- *Relate* food processing methods to vitamin C content.

Materials

PREPARATION NOTES

Time Required: one 50-minute period

Materials
Materials for this lab activity can be purchased from WARD'S. See the *Master Materials List* for ordering instructions.

- safety goggles
- lab apron
- test tubes, 13 mm × 100 mm (11)
- test-tube rack
- medicine droppers (4)
- indophenol solution, 0.1%
- ascorbic acid solution, 0.1%
- ascorbic acid solution, 0.05%
- orange juice, freshly squeezed

- orange juice, frozen
- orange juice, canned
- orange juice, boiled
- orange juice, squeezed and exposed to air at room temperature for 1 hour
- lemon juice
- pineapple juice
- tomato juice
- carrot juice

Purpose

In your job as a food technologist, you analyze the content of various foods and beverages. You also study the effect of preparation and storage methods on the nutrient content of foods. You are currently conducting a study of the vitamin C content of fruit and vegetable juices. Today you will use an indicator to test for the vitamin C content of various types of juices and determine how preparation and storage methods affect the vitamin C content of orange juice.

Background

Preparation Tips
In buying juice for this investigation, avoid those labeled "Vitamin C added." The vitamin C content found naturally in the juice is the data the student will be asked to find.

Vitamins are organic compounds that are needed by the body in small amounts. Many vitamins, including vitamin C, act as coenzymes in reactions taking place during metabolism. **Coenzymes** are compounds that must be present in order for enzymes to react. Vitamin C, also known as **ascorbic acid,** plays an important role in protein metabolism and is vital to the maintenance of the blood vessels, bones, teeth, and gums. Vitamin C also acts as an antioxidant in the body. **Antioxidants** defend tissue against *free radicals,* byproducts of metabolism that attack healthy tissues.

Disposal
- Ascorbic acid solutions and the juices can be poured down the sink after use.
- Dispose of indophenol as directed by the supplier's MSDS.

A healthy diet should include foods rich in vitamin C. Foods containing vitamin C come from plants or plant products. Fresh fruits, vegetables, and fruit juices contain significant amounts of vitamin C. The vitamin content of foods is affected by the way the food is processed and stored. Chemicals called **indicators** detect the presence of specific substances by changing color. Vitamin C can be detected by the indicator **indophenol,** which turns from blue to colorless in the presence of vitamin C.

Procedure

Part 1—Establishing a Standard

1. Put on safety goggles and a lab apron.

PROCEDURAL
NOTES

Safety Precautions
• Discuss all safety sym-
bols and caution state-
ments with students.
• Warn students not
to ingest any of the
juices, especially those
combined with the
indicator.

Procedural Tips
• Review with students
how to set up the cal-
culation in step 6. Note
that the calculation is
an inverse relationship.
• If possible, put the
indicator solution in
bottles that have a
dropper in the cover.
This will help prevent
contamination of the
indicator solution.

2. A standard is a positive test for a known substance that is used to compare with the result of a test for that substance. To establish a standard, place 15 drops of indophenol in a test tube. Use a clean medicine dropper to add 0.1% ascorbic acid solution one drop at a time. After each drop, swirl the test tube. Continue to add drops of ascorbic acid until the indophenol solution becomes colorless. Note the number of drops required.

♦ How many drops of ascorbic acid solution are needed to turn the indophenol colorless?

Answers will vary; students should note that this step establishes their standard.

3. Repeat step 2 using a 0.05% ascorbic acid solution and a clean dropper.

♦ How many drops of ascorbic acid solution are now needed to turn the indophenol colorless? Why do the results differ from those of step 2?

The number should double from step 2. The ascorbic acid solution is half as strong as the previous one.

Part 2—Performing the Test

4. Place 15 drops of indophenol in one test tube. Use a clean medicine dropper to add fresh orange juice one drop at a time. After each drop, swirl the test tube. Continue to add drops of juice until the indophenol solution becomes colorless. Record the number of drops of juice required in the table below.

Ascorbic Acid (Vitamin C) Content of Various Juices

Juice	Number of drops added	Percentage of vitamin C
orange juice, freshly squeezed		
orange juice, frozen		
orange juice, canned		
orange juice, boiled		
orange juice, squeezed and exposed to air	Answers will vary.	
lemon juice		
pineapple juice		
tomato juice		
carrot juice		

5. Repeat step 4 for the other juices listed in the table on the previous page.
Note: Be sure to use a separate test tube for each juice test, and rinse the dropper well between each juice test.

6. With the results obtained using a 0.1% ascorbic acid solution as the standard, or basis for comparison, calculate the percentage of ascorbic acid in each type of juice tested. Record the percentages in the table on the previous page. Set up your calculations based on the example that follows. In the example, *x* equals the percentage of ascorbic acid in the test sample.

If it takes 10 drops of 0.1% ascorbic acid to neutralize the indophenol solution and 7 drops of a solution with an unknown percentage of ascorbic acid, then

$$\frac{x}{0.1\%} = \frac{10 \text{ drops}}{7 \text{ drops}}$$

$$x = \frac{(10)(0.1\%)}{7}$$

$$x = \frac{1\%}{7}$$

$$x = 0.14\%$$

7. Use the grid below to draw a bar graph of the data from the Percentage of vitamin C column of the table.

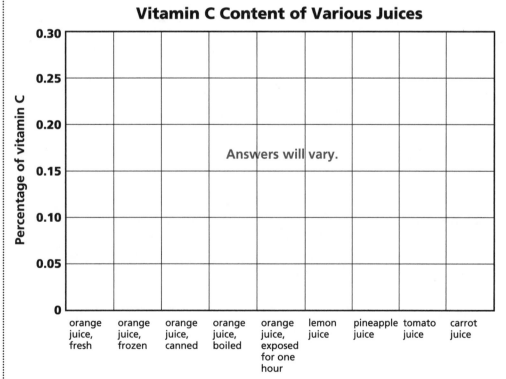

Vitamin C Content of Various Juices

8. Dispose of your materials according to the directions from your teacher.

9. Clean up your work area and wash your hands before leaving the lab.

Analysis **10.** What is the control in this investigation?

the ascorbic acid solution of known concentration

11. What procedure or procedures seem to reduce the vitamin C content of juices?

Boiling and exposure to air reduce vitamin C content the most, followed by freezing and canning.

Conclusions **12.** To ensure sufficient intake of vitamin C, what juices can be included in your diet? List them from highest vitamin C content to lowest.

Orange juice (highest), lemon, pineapple, tomato, and carrot (most likely lowest); individual test

results may vary.

13. From the results of this investigation, suggest a storage procedure that will maintain the vitamin C content of orange juice after it has been removed from a freezer.

It should be stored in a refrigerator in a sealed container.

Extensions **14.** Test fresh vegetables and fruits for their vitamin C content. Compare the vitamin C content of a teaspoon of the crushed fruit or vegetable with the vitamin C content of the same fruits that have been sliced up and left exposed to air for an hour.

15. Test juices or juice drinks that are labeled "vitamin C added." Compare these products with the juices tested in this investigation.

HOLT

BioSources

LAB PROGRAM
LABORATORY TECHNIQUES

C5 *Observing the Effect of Concentration on Enzyme Activity*

Skills

PREPARATION NOTES

- making a dilution series of a stock enzyme solution
- determining the rate of a reaction
- collecting and graphing data

Objectives

Time Required: one 50-minute period

- *Prepare* a series of dilutions of a 100% enzyme solution to make enzyme solutions of various concentrations.
- *Determine* how enzyme concentration affects reaction rate.

Materials

Materials
Materials for this lab activity can be purchased from WARD'S. See the *Master Materials List* for ordering instructions.

- safety goggles
- lab apron
- medicine cups for dilutions (6)
- marking pen or pencil
- 100% catalase (stock) solution (30 mL)
- 10 mL graduated cylinder
- water
- stirring rod

- clear vial
- dilute hydrogen peroxide solution (20 mL)
- forceps
- filter-paper disks (18)
- paper towels
- watch or clock with second hand
- metric ruler

Purpose

You are a research assistant who is testing enzyme activity in various body tissues. You need a simple test to measure the effect of enzyme concentration on reaction rate, the speed of a chemical reaction. You decide to conduct an experiment in which you soak disks of filter paper in different concentrations of the enzyme catalase and measure the speed at which the catalase breaks down hydrogen peroxide.

Background

Preparation Tips
• Hydrogen peroxide can be obtained from a pharmacy or grocery store. Do not purchase or have on hand hydrogen-peroxide solutions that have a concentration greater than 3%. If purchased as 3%, dilute 33 mL of the hydrogen-peroxide solution to 1 L with distilled water. Keep this solution away from light. The lab requires about 20 mL per group.

Enzymes are proteins that speed up the rate of chemical reactions that would otherwise happen more slowly. The enzyme may be altered, but it is not consumed by the completed reaction. You have hundreds of different enzymes in each of your cells. Each of these enzymes is responsible for one particular chemical reaction that occurs in the cell.

Hydrogen peroxide is a chemical used to treat wounds. It is an effective antiseptic because it is deadly to cells. Of course, the cells that you want to destroy are bacterial cells that may enter an open wound, not your own cells.

You may be surprised to find out that hydrogen peroxide is produced as a waste product in every cell of your body. The enzyme **catalase,** which is found in the cells of many living things, speeds up the reaction that breaks down hydrogen peroxide into two substances—water and oxygen. The reaction is as follows:

$$2\ H_2O_2 \xrightarrow{\text{catalase}} 2\ H_2O + O_2$$

In this reaction, hydrogen peroxide (H_2O_2) is the **substrate,** the reactant in the

- To prepare the 100% enzyme solution, add 1 mL catalase concentrate to 700 mL of room temperature distilled water. If distilled water is not available, tap water can be used. Make the solution the same day you will use it.
- Disks can be made by using a hole punch and #1 filter paper.

Procedure

Disposal
- Enzyme solutions may be flushed down the drain with copious amounts of water.
- Dilute hydrogen-peroxide solutions can be poured down the drain.
- Filter-paper disks can be placed in the trash.

PROCEDURAL NOTES

Safety Precautions
- Have students wear safety goggles and a lab apron.
- Discuss all safety symbols and caution statements with students.
- Caution students to be careful when working with chemicals. Remind them to notify you immediately of any chemical spills. Also caution students never to taste, touch, or smell any substance or bring it close to their eyes, unless specifically directed to do so.

Procedural Tips
- Make sure students rinse the graduated cylinder thoroughly with water after they measure the catalase and before they measure the hydrogen peroxide solution. If the cylinder is not clean, the reaction will take place in the cylinder instead of in the vials.

chemical reaction. In the presence of catalase, two molecules of hydrogen peroxide yield two molecules of water and one molecule of oxygen. In the equation on the previous page, catalase is placed on top of the arrow to show that catalase speeds up the reaction but is not consumed by the reaction.

The rate of this reaction can be determined by measuring the amount of oxygen produced. Filter-paper disks soaked in different concentrations of catalase solutions react at different speeds. In all cases, however, the catalase reacts with the hydrogen peroxide and causes it to break down into water and oxygen. The oxygen builds up on the disk, causing it to float. If one concentration is catalyzing more rapidly than another, the oxygen will build up faster and the disk will float sooner.

1. Put on a safety goggles and a lab apron.

2. Determine how you will make a series of dilutions of the enzyme catalase. One way to make a dilution is to start out with a 100% enzyme solution. To make 10 mL of a 50% enzyme solution, for example, you would mix 5 mL of water and 5 mL of the 100% enzyme solution. Complete the following table to determine how you will mix each of the enzyme dilutions needed in this lab. *Note: The first two lines are filled in for you.*

Enzyme Dilutions

Final quantity needed	Concentration of final solution	mL of catalase	mL of water
10 mL	100%	10	0
10 mL	80%	8	2
10 mL	60%	6	4
10 mL	40%	4	6
10 mL	20%	2	8
10 mL	0%	0	10

3. Use a marking pencil to label six medicine cups for the enzyme solutions as follows: 100%, 80%, 60%, 40%, 20%, and 0%.

4. Obtain a cup containing 30 mL of the 100% enzyme solution.

5. Use a graduated cylinder to measure 10 mL of the 100% enzyme solution into the medicine cup labeled 100%. **CAUTION: Glassware is fragile. Notify your teacher promptly of any broken glass or cuts. Do not clean up broken glass or spills unless your teacher tells you to do so.**

6. Prepare and label 10 mL of each of the dilutions according to the measurements in the enzyme dilution table on the previous page. Mix each dilution thoroughly with a stirring rod. *Note: Be sure to rinse the stirring rod with water after making each dilution.*

7. Fill the clear vial with 20 mL of hydrogen peroxide solution. *Note: If you are measuring the hydrogen peroxide solution in a graduated cylinder, clean the cylinder very carefully when you are finished. You do not want to mix catalase and hydrogen peroxide in the cylinder.*

8. Use forceps to pick up one of the filter-paper disks. Hold the forceps as close to the edge of the disk as possible. Submerge the disk in the 100% catalase solution for five seconds. Continue to hold the disk with the forceps.

9. Remove the disk from the solution, and blot it dry on the paper towel for five seconds.

10. Drop the disk into the hydrogen peroxide. Measure the time it takes for the disk to rise to the surface of the hydrogen peroxide. Begin timing as soon as the disk touches the surface of the hydrogen peroxide. Use the metric ruler to measure the distance the disk sinks into the hydrogen peroxide. Multiply this measurement by two to determine the distance traveled. Enter the time and the distance traveled in the column for Trial 1 in the data table below.

Reaction Rate of Enzyme Dilutions

% Catalase	Time in seconds (s)				Distance in millimeters (mm)				Reaction rate (mm/s)
	Trial 1	Trial 2	Trial 3	Avg.	Trial 1	Trial 2	Trial 3	Avg.	
100				9				48	5.33
80				11				48	4.36
60				13	Answers will vary. Sample averages and rates are given.			48	3.69
40				16				48	3.00
20				24				48	2.00
0				no reaction				0	0

11. Repeat steps 8 through 10 two more times, using a clean filter-paper disk each time. Enter the times and distances traveled in the columns for Trial 2 and Trial 3. Find the average time and average distance traveled. Then calculate the reaction rate by dividing the average distance by the average time.

$$\text{Reaction rate} = \frac{\text{average distance}}{\text{average time}}$$

12. Repeat steps 8 through 11 for the 80%, 60%, 40%, and 20% solutions. Complete the data table on the previous page for each trial for each solution. *Note: Be sure to use a clean filter-paper disk and a clean paper towel for each trial to avoid contamination.*

13. Repeat steps 8 through 11 for the 0% solution. *Note: If the disk has not risen to the surface within three minutes, write "no reaction" in the data table.*

14. Dispose of your materials according to the directions from your teacher.

15. Clean up your work area and wash your hands before leaving the lab.

Analysis

16. Which concentration of catalase had the fastest reaction time?

100%

17. Which concentration of catalase had the slowest reaction time?

0%

18. On the grid below, plot a line graph of the reaction rate versus the enzyme concentration. Place enzyme concentration on the *x*-axis and the reaction rate on the *y*-axis.

Reaction Rate v. Enzyme Concentration

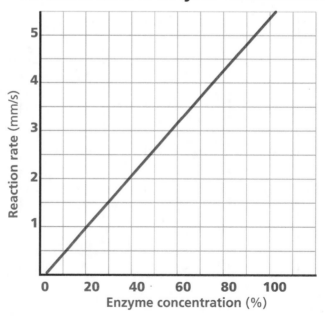

19. Why did you measure the distance traveled by the disks to determine reaction rate?

Reaction rate measures the speed of the reaction. Speed is normally calculated by dividing distance

by time.

Conclusions

20. Based on the graph and the overall slope of the line, what can you conclude about the effect of enzyme concentration on reaction rate?

Since it takes longer for the disk to rise when the enzyme concentration is lower, it can be

concluded that as enzyme concentration increases, the rate of the reaction increases.

21. Is this procedure a good way to test enzyme activity? Explain your answer.

This procedure is a good way to test enzyme activity because the results are quantitative instead of

qualitative. The distance traveled and the time can be measured, and their rates (speed) can then

be calculated and plotted against enzyme concentration.

22. In the lab, the term *100% enzyme* is only relative—it is merely the concentration of enzyme your teacher mixed. In other words, the enzyme concentration could have been made much higher. Do you think that the trend noted in the graph above would continue if the enzyme samples were even more concentrated than those in this lab? Explain your answer.

Reaction rate would have increased as long as there was enough substrate to occupy all enzyme

active sites. If the concentration of enzyme exceeds the amount of available substrate, the reaction

rate will remain constant and will not continue to increase.

Extensions

23. A student wishes to know if the amount of substrate (in this case hydrogen peroxide) can change the reaction rate. The student's teacher gives the student a beaker of 100% catalase and hydrogen peroxide. Using the above lab model, discuss how the student would set up the lab to test the effect of varying substrate concentration.

24. For many reactions, the rate cannot be determined easily, only the time. Use your data from the above investigation to construct a graph that shows the time of reaction versus enzyme concentration. What information is given by this graph?

C6 *Observing the Effect of Temperature on Enzyme Activity*

Prerequisites
- **Laboratory Techniques C5: Observing the Effect of Concentration on Enzyme Activity** on pages 21–26

Review
- the terms *substrate, catalyst, enzyme,* and *catalase*

SRE SIMPSON RESEARCH ENTERPRISES ——— Boston, MA

January 15, 1998

Sam Ashike
Health Division
BioLogical Resources, Inc.
101 Jonas Salk Dr.
Oakwood, MO 65432-1101

Dear Mr. Ashike,

Our company will soon begin a study on the roles of an enzyme called catalase and hydrogen peroxide in the aging of the body. Hydrogen peroxide helps the body fight infection. However, it can also become toxic to the body, damaging mitochondria and impairing cell function. This can lead to tissue and organ damage, as well as a decrease in the body's ability to cope with environmental changes. Fortunately, the enzyme catalase acts as a catalyst to break down hydrogen peroxide into its harmless components.

As you know, the rate of a chemical reaction typically increases with increasing temperature. While temperature inside the body remains relatively constant, the temperature of the skin is more susceptible to changes in outside temperature. We suspect that the rate of reaction of catalase and hydrogen peroxide changes more in the skin than elsewhere in the body. Our study will investigate whether there is a relationship between climate and the aging of the epidermis.

For this study, we will need a simple assay (test) of the effect of temperature on catalase activity. I would like your research company to design an experiment that investigates the effect of temperature on the reaction rate of hydrogen peroxide and catalase. We are willing to pay your normal fee. Please keep me posted.

Sincerely,

Dr. Everett Simpson

Dr. Everett Simpson
Owner and President
Simpson Research Enterprises

BioLogical Resources, Inc. Oakwood, MO 65432-1101

M E M O R A N D U M

To: Team Leader, Physiology Dept.

From: Sam Ashike, Director of Health

I am confident that your teams will be able to develop an experimental design for the study described in Dr. Simpson's letter. Doing preliminary research for another company requires exemplary record keeping. Please be very thorough in recording your data and preparing your report.

I want your teams to design and complete an experiment that will test catalase activity in hydrogen peroxide at varying temperatures. I suggest that you begin at 80°C and then cool the liquid down at increments of 10 degrees. Given the supplies you will be using, it will be easier to control temperature change this way.

You may find it useful to do some research on the effect of temperature changes on enzyme activity. This may give you a better idea of how to interpret your results.

Proposal Checklist

Before you start your work, you must submit a proposal for my approval. **Your proposal must include the following:**

_____ • the **question** you seek to answer

_____ • the **procedure** you will use

_____ • a detailed **data table** for recording measurements

_____ • a complete, itemized list of proposed **materials** and **costs** (including use of facilities, labor, and amounts needed)

Proposal Approval: _____

(Supervisor's signature)

Report Procedures

When you finish your analysis, prepare a report in the form of a business letter to Dr. Simpson. **Your report must include the following:**

_____ • a paragraph describing the **procedure** you followed to test the effect of temperature on the rate of reaction of hydrogen peroxide and catalase

_____ • a complete **data table** showing time, distance traveled, and reaction rate

_____ • a **graph** showing change in reaction rate against change in temperature

_____ • your **conclusions** about the effect of temperature on the rate of reaction of hydrogen peroxide and catalase

_____ • a detailed **invoice** showing all materials, labor, and the total amount due

Safety Precautions

• Wear safety goggles, disposable gloves, and a lab apron.

• Wear oven mitts when handling hot objects.

• Glassware is fragile. Notify your teacher promptly of any broken glass or cuts. Do not clean up broken glass or spills unless your teacher tells you to do so.

• Never use electrical equipment around water, nor with wet hands or clothing. **Never** use equipment with frayed cords.

• Wash your hands before leaving the laboratory.

Disposal Methods

• Dispose of waste materials according to instructions from your teacher.

• Place used filter-paper disks, cups, and paper towels in a trash can.

• Place broken glass, unused hydrogen peroxide, and unused catalase in the separate containers provided.

• Wash reusable materials such as glassware and lab utensils, and return them to the supply area.

FILE: Simpson Research Enterprises

MATERIALS AND COSTS (Select only what you will need. No refunds.)

I. Facilities and Equipment Use

Item	Rate	Number	Total
facilities	$480.00/day		
personal protective equipment	$10.00/day		
compound microscope	$30.00/day		
microscope slide with coverslip	$2.00/day		
forceps	$5.00/day		
metric ruler	$1.00/day		
thermometer	$5.00/day		
hot plate	$15.00/day		
laboratory tray	$5.00/day		
50 mL beaker	$5.00/day		
light source	$15.00/day		
10 mL graduated cylinder	$5.00/day		
watch or clock with second hand	$10.00/day		
container of ice	$5.00/day		

II. Labor and Consumables

Item	Rate	Number	Total
labor	$40.00/hour		
dilute hydrogen peroxide	$1.00/mL		
catalase solution	$2.00/mL		
medicine cups	$0.50 each		
paper towels	$0.10 each		
filter-paper disks	$0.20 each		

Fines

Item	Rate	Number	Total
OSHA safety violation	$2,000.00/incident		

Subtotal		
Profit Margin		
Total Amount Due		

Name _____

Date _____ Class _____

C7 | *Measuring the Release of Energy from Sucrose*

Skills
- measuring change in temperature
- conducting a controlled experiment

Objectives

PREPARATION NOTES

Time Required: 2 days

- *Predict* how the temperature of a solution will change as a chemical process occurs in the solution.
- *Relate* a change in temperature to the release of energy from sucrose.

Materials

Materials

Materials for this lab activity may be purchased from WARD's. See the *Master Materials List* for ordering instructions.

- safety goggles
- lab apron
- 500 mL insulated bottle (2)
- wax pencil
- 1000 mL beaker
- 150 g of sucrose
- 800 mL of lukewarm water

- glass stirring rod
- dried yeast package
- 2-hole stopper with a thermometer inserted in one hole and a 10 cm piece of glass tubing inserted in the other hole (2)
- 250 mL flask with 100 mL of limewater (2)
- 40 cm piece of rubber tubing (2)

Purpose

You are a student teacher who has been asked to give a lecture on the processes of fermentation and cellular respiration. To show that these processes release energy, you set up the following demonstration using yeast and sucrose.

Background

The carbohydrate **sucrose,** or table sugar, is made by plants to store energy. All living things must have energy to power their cells. That energy is obtained from organic molecules such as sucrose through the processes of **fermentation** and **cellular respiration.** Fermentation, which occurs in the absence of oxygen, releases a relatively small amount of energy. Cellular respiration, which occurs in the presence of oxygen, releases a great deal of energy. Since energy conversions are not 100 percent efficient, some of the energy released is in the form of thermal energy. The addition of thermal energy to a system results in an increase in temperature.

Another product of both fermentation and cellular respiration is carbon dioxide. A solution called **limewater** can be used to indicate the presence of carbon dioxide. Limewater, which is normally clear, turns cloudy in the presence of carbon dioxide.

Procedure

1. Put on safety goggles and a lab apron.

2. Using a wax pencil, label two insulated bottles with the initials of your group's members. Label one of the bottles "Control" and the other "Experimental."

Preparation Tips

• Assemble glass tubing, thermometers, and two-hole stoppers. To insert glass tubing into the stopper, first be

3. In a 1000 mL beaker, dissolve 150 grams of sucrose in 800 mL of lukewarm water. Pour half of the sucrose solution into the insulated bottle marked "Control." **Caution: Glassware is fragile. Notify your teacher promptly of any broken glass or cuts. Do not clean up broken glass or spills unless your teacher tells you to do so.**

sure that the ends of the tubing are fire-polished and that the outside diameter of the tubing is only slightly larger than the diameter of the hole in the stopper. Lubricate the holes in the stopper with glycerin or water. Then, wearing thick leather gloves or with both hands protected by multiple layers of cloth, gently insert the tubing into a hole without twisting or turning the tube or stopper. Do not force the glass tubing; it can shatter. Repeat for the thermometer.

• Provide a container for disposal of any broken glass.

Disposal

• Solutions containing yeast and sucrose may be poured down the drain.

PROCEDURAL NOTES

Safety Precautions

• Discuss all safety symbols and caution statements with students.

Procedural Tips

• Provide opportunities for students to come to the lab to take additional data readings.

• If enough oxygen is dissolved in the solution and the temperature is high enough, cellular respiration may occur in addition to fermentation. The occurrence of cellular respiration will be marked by a steeper rise in temperature in the "Experimental" bottle.

• Microcomputer Based Laboratory (MBL) probes and Calculator Based Laboratory (CBL) probes can be used to measure changes in tempera-

4. Add one package of fresh yeast to the sucrose solution remaining in the beaker, and stir. Pour the sucrose-yeast mixture into the bottle marked "Experimental."

5. Set up the two insulated bottles, two 250 mL flasks with limewater, and two 40 cm pieces of rubber tubing as shown in the diagram at right. Adjust the thermometers until they extend down into the solutions in the bottles and the liquid column can be seen above the stopper. *Note: Be sure that the glass tubing does not touch the solutions in the bottles.*

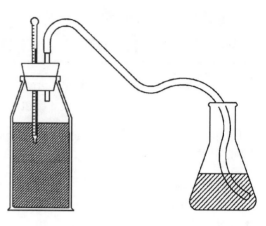

6. Record the time and the initial temperature of the solution in each bottle in the data table below.

Temperature of a Sucrose Solution in Insulated Bottles

Experimental bottle		Control bottle	
Time	Temperature	Time	Temperature

◆ What do you predict will happen in each bottle over the next 48 hours?

Answers will vary. Students should predict that the temperature in the "Control" bottle will not

change while the temperature in the "Experimental" bottle will rise as the yeast breaks down

sucrose and releases energy stored in the sucrose.

7. Clean up your work area and wash your hands before leaving the lab.

ture. Have students insert a thermister (temperature probe) into one of the holes of the two-hole stopper. Provide modeling clay to seal the hole so that air cannot enter and so that heat will remain inside.

8. Continue to record the temperature in each bottle for 48 hours. Take turns with your lab partner during the school day to take temperature data between class periods. Record as many readings as possible during the next 48 hours.

◆ What happened in each flask of limewater?

The limewater in the flask paired with the "Experimental" bottle turned cloudy. The limewater in

the flask paired with the "Control" bottle stayed clear.

9. 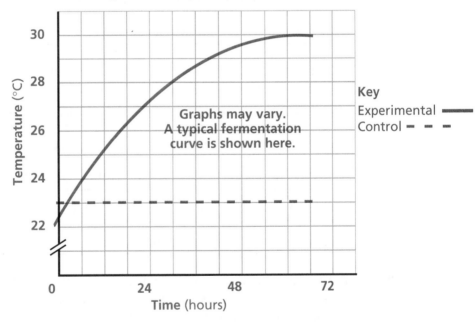 Dispose of your materials according to the directions from your teacher.

10. Make a line graph of the data in your data table using the grid below. Plot time on the *x*-axis (horizontal axis) and temperature on the *y*-axis (vertical axis). Plot a point on the graph that corresponds to the temperature at each time. Complete the graph by drawing a curve through the plotted points. The curve will show the yeast's energy production. Use two different colors to graph the results from each bottle.

Change in Temperature in Bottles

Graphs may vary.
A typical fermentation
curve is shown here.

Key
Experimental ▬▬▬
Control ▬ ▬ ▬

Analysis

11. What are some indications that a chemical reaction was taking place inside one of the vacuum bottles?

Bubbles appeared in the limewater, which turned cloudy, and there was a rise in temperature.

12. What is the purpose of the glass tube and rubber tubing in this experiment?

The glass tube enables any gases produced to be released from the bottles and prevents the rubber

stopper from popping off due to pressure buildup. The rubber tubing attached to the glass tube

enables the gas to bubble through limewater, showing that the gas is carbon dioxide.

Conclusions

13. How would you explain the change in temperature observed during the experiment to a class of biology students?

Temperature rose in the "Experimental" bottle because energy was released by yeast as they broke

down sucrose. Temperature varied very little in the "Control" bottle because there were no yeast to

break down sucrose.

14. How could you alter this experiment to show that the energy is released from sucrose?

Answers will vary. Students should suggest adding a third setup, in which there is yeast but no

sucrose, to the experiment.

15. How would you explain what happens to the limewater in the two setups to a class of biology students?

The limewater turns cloudy in the "Experimental" setup because carbon dioxide gas is released by

the yeast during fermentation. The limewater remains clear in the "Control" setup because there

are no yeast in the bottle and no carbon dioxide is produced.

16. Which process, fermentation or cellular respiration, caused the changes observed in this experiment? How could you test your hypothesis?

Answers will vary. Students should realize that oxygen cannot enter the systems in the bottles and

that any oxygen present at the outset would be depleted rapidly. Therefore, fermentation is the

most likely cause of the changes observed during the experiment. This hypothesis could be tested

by testing the solution in the "Experimental" bottle for the presence of alcohol, which is produced

by fermentation but not by cellular respiration.

Extensions

17. Soon after the yeast is placed in the sucrose solution, take one drop of the solution and place it on a slide. Observe the yeast with a microscope under high power, and estimate the number of cells seen. Repeat the above process on the second and third days. Relate your findings to the energy-releasing process.

18. Look up the meanings of the terms *exothermic, endothermic, exergonic,* and *endergonic.* Write a paragraph that describes the processes of fermentation and cellular respiration using the appropriate terms from the list above.

Name _____

Date _____ Class _____

HOLT
BIOSOURCES
LAB PROGRAM
EXPERIMENTAL DESIGN

C8   *Measuring the Release of Energy— Best Food for Yeast*

Prerequisites	• Laboratory Techniques C7: Measuring the Release of Energy from Sucrose on pages 31–34
Review	• cellular respiration

art's
everage Company   Springfield, Illinois

March 17, 1998

Sam Ashike
Health Division
BioLogical Resources, Inc.
101 Jonas Salk Dr.
Oakwood, MO 65432-1101

Dear Mr. Ashike,

Here at Bart's Beverage Company we are always striving to improve our a products. We would like your firm to assist us in our latest study on yeast. Yeast is of great importance to us. We use it in developing our nonalcoholic beverages. Although many companies that produce nonalcoholic adult beverages leave yeast out of the brewing process, we have found that the yeast adds a unique flavor preferred by our customers. We aerate the solution to prevent anaerobic fermentation, and what little alcohol is produced is taken out by way of dialysis or a low-temperature vacuum distillation process.

We are interested in discovering which type of sugar produces the highest rate of cellular respiration. We would like your firm to investigate a variety of products potentially suitable for our yeast cultures. Please conduct a study as soon as possible to determine the best sugar for yeast respiration, and bill us when the work is complete.

Sincerely,

Wayne Zaslovsky

Wayne Zaslovsky
Director of Research
Bart's Beverage Company

BioLogical Resources, Inc. Oakwood, MO 65432-1101

M E M O R A N D U M

To: Team Leader, Food Testing Dept.
From: Sam Ashike, Director of Health

Attached is a copy of a letter I received yesterday from a beverage company in Illinois requesting that we test different sugars for their effects on yeast growth. Please design an experiment to accomplish this task.

This morning I received a phone call from Mr. Zaslovsky. He wants the following solutions tested: a 20% fructose solution, a 20% lactose solution, and a 20% sucrose solution. He has sent samples of the yeast used by Bart's Beverage Company. I suggest that you use temperature as an indicator of the rate of cellular respiration. Since heat is a product of fermentation, an increase in temperature should correspond with an increase in cellular respiration. You may also want to verify that cellular respiration is occuring by testing for carbon dioxide.

As always, it is important that we provide quick and accurate results. Make sure that your work is thorough. Doing quality work is the best way to keep a client.

Proposal Checklist

Before you start your work, you must submit a proposal for my approval. **Your proposal must include the following:**

_____ • the **question** you seek to answer

_____ • the **procedure** you will use

_____ • a detailed **data table** for recording temperature readings

_____ • a complete, itemized list of proposed **materials** and **costs** (including use of facilities, labor, and amounts needed)

Proposal Approval: _____
(Supervisor's signature)

Report Procedures

When you finish your analysis, prepare a report in the form of a business letter to Mr. Zaslovsky. **Your report must include the following:**

_____ • a paragraph describing the **procedure** you followed to determine the best sugar-based medium for yeast growth

_____ • a complete **data table** showing changes in temperature for each solution

_____ • a **graph** showing changes in temperature over time for each solution

_____ • your **conclusions** about which medium Bart's Beverage Company should use to grow the yeast

_____ • a detailed **invoice** showing all materials, labor, and the total amount due

Safety Precautions

- Wear safety goggles and a lab apron.

- Glassware is fragile. Notify your teacher promptly of any broken glass or cuts. Do not clean up broken glass or spills unless your teacher tells you to do so.

- Wash your hands before leaving the laboratory.

- Do not taste or ingest any food item in the laboratory.

Disposal Methods

- Dispose of waste materials according to instructions from your teacher.

- Place broken glass in the container provided.

- Wash reusable materials such as glassware and lab utensils, and return them to the supply area.

FILE: Bart's Beverage Company

MATERIALS AND COSTS (Select only what you will need. No refunds.)

I. Facilities and Equipment Use

Item	Rate	Number	Total
facilities	$480.00/day		
personal protective equipment	$10.00/day		
scissors	$1.00/day		
ruler	$1.00/day		
hot plate	$20.00/day		
500 mL beaker	$10.00/day		
vacuum bottle	$10.00/day		
stopper assembly w/ thermometer	$10.00/day		
flask	$10.00/day		
clock or watch	$10.00/day		

II. Labor and Consumables

Item	Rate	Number	Total
labor	$40.00/hour		
2 packages of dried yeast	provided		
limewater	$0.05/mL		
distilled water	$0.10/mL		
20% fructose solution	$0.20/mL		
20% lactose solution	$0.20/mL		
20% sucrose solution	$0.20/mL		
rubber tubing	$0.20/cm		
wax pencil	$2.00 each		
tape	$0.25/m		

Fines

OSHA safety violation	$2,000.00/incident		

Subtotal	
Profit Margin	
Total Amount Due	

Name _____

Date _____ Class _____

C9 *Preparing a Root Tip Squash*

Skills

PREPARATION NOTES

Time Required: one 50-minute period

- preparing and staining a wet mount of onion *(Allium sativum)* root tips
- making a root tip squash
- identifying the stages of mitosis
- calculating the percentage of cells in mitosis

Objectives

- *Demonstrate* the proper technique in preparing and staining a slide of onion root tips to observe cells in the process of mitosis.
- *Identify* the stages of mitosis.
- *Determine* the percentage of cells in mitosis.

Materials

Materials

Materials for this lab activity can be purchased from WARD'S. See the *Master Materials List* for ordering instructions.

- safety goggles
- lab apron
- vial of specially treated onion root tips
- forceps
- treating dish
- 1 M HCl in dropping bottle (6 mL)
- watch or clock with second hand
- distilled water
- microscope slide (3)
- paper towels
- aceto-orcein stain in dropping bottle (30 mL)
- wooden macerating stick
- coverslip (3)
- compound light microscope

Purpose

You are a botanist who is studying the effects of a new fertilizer. To determine how effective the new fertilizer is, you decide to study the mitotic activity in the roots of onion plants. You also want to determine the percentage of cells that are undergoing mitosis. To observe mitosis in the onion root cells, you make a squash of onion root tips, which contain actively dividing cells.

Background

Preparation Tips

• If you do not want students to handle hydrochloric acid in step 3, you can do this step for them. Dispense the onion root tips into the treating dish. Pour off the holding fluid in the treating dish. Flood the root tips with 1 M HCl, and allow them to stand for 10 minutes. After 10 minutes, remove the hydrochloric acid and refill the dish with distilled water. The tips are now ready for student use.

All cells undergo a process of growth and division called the **cell cycle,** which consists of five phases. The G1, S, and G2 phases are collectively called **interphase.** Cell growth and DNA replication occur during this period. Most of a cell's life is spent in interphase.

Mitosis, or the M phase, is divided into four stages: **prophase, metaphase, anaphase,** and **telophase.** The following events occur in each stage.

Prophase: In this stage, the chromatin coils into thick threads. These are the chromosomes. Each chromosome now consists of paired chromatids, which are replicated chromosomes that are held together by a centromere. In late prophase, the nuclear membrane breaks down very rapidly and nucleoli disappear. The breakdown of the nuclear membrane frees the coiled chromosomes and allows the attachment of spindle fibers, which move the chromosomes into the next stage of mitosis. Chromosomes are distinct in the central region of the cell.

• If you choose to prepare the onion root tips yourself instead of buying them, do the following. Purchase green salad onions (scallions) five or six days before the start of this lab. Place onions in water for two to three days or until root tips are approximately 2 cm long. Cut off root tips and place in 100 mL fixative solution for 48 hours. Remove root tips from fixative solution, and store in a labeled container with 100 mL of 70% ethanol solution until ready for use. Root tips can be stored in this way for about 2 months. When ready to use, place root tips in a large test tube, and add enough 1 M HCl to cover the root tips. Place the tube into a 60°C water bath for 15 minutes. Use a bulb pipet to draw off as much HCl from the tube as possible. Add enough aceto-orcein stain to the test tube to cover the root tips completely. After about 15 minutes, the root tips will turn bright pink and are ready for students to use to prepare a squash.

Procedure

• To prepare 1 M hydrochloric acid, dilute 8 mL of concentrated HCl in 92 mL of distilled water. Place in labeled dropper bottles for student use.

• To prepare fixative, mix 25 mL of glacial acetic acid with 75 mL of 95% ethanol. Fix roots in fume hood for 48 hours. Be sure there are no sources of flame or ignition in the room.

Metaphase: The chromosomes appear at the center of the spindle apparatus with each centromere attached to two spindle fibers, one from each side of the cell.

Anaphase: The centromere of each chromosome divides, separating the chromatids from each other. The chromatids of each pair are then free to move along the spindle fibers toward opposite poles. They continue to migrate toward the poles until they form distinctive clumps at each end.

Telophase: This is the final stage of mitosis. A new nuclear membrane forms around each clump of chromosomes, which have started to uncoil and return to the chromatin network seen prior to the beginning of mitosis.

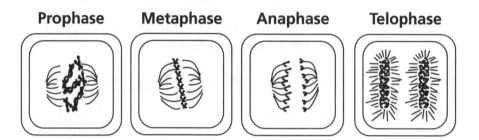

Mitosis results in two identical nuclei. Following nuclear division, the cell undergoes **cytokinesis,** or the C phase, which is the division of the cytoplasm. The formation of two complete cells occurs in the C phase of the cell cycle.

In plants, mitosis occurs in special growth regions called **meristems.** Meristems can be found at the tips of the roots and stems in vascular plants.

To observe mitosis in stem and root meristems, biologists prepare special slides called a **squash.** This preparation is just what it sounds like. Actively dividing cells from a root or stem meristem are removed and treated with hydrochloric acid to fix the cells, or stop them from dividing. The cells are then stained, made into a wet mount, and squashed and spread into a single layer by applying pressure to the coverslip.

Part 1—Preparing a Cell Squash

1. Put on safety goggles and a lab apron.

2. Use forceps to carefully remove three onion root tips from their vial and place them in a treating dish.

3. Flood the root tips with 1 M hydrochloric acid (HCl). Allow the root tips to stand in the HCl for 10 minutes. **CAUTION: If you get an acid or a base on your skin or clothing, wash it off at the sink while calling to your teacher. If you get an acid or a base in your eyes, promptly flush it out at the eyewash station while calling to your teacher. Notify your teacher in the event of an acid or base spill.**

- To prepare 70% ethanol, mix 70 mL of 95% ethanol with 30 mL of distilled water. Be sure there are no flames in the room.

- Prepare separate containers for the disposal of root tips, acid, stain, and broken glass.

Disposal

- To dispose of volumes of hydrochloric acid less than 250 mL, do the following. Always wear personal protective equipment, and work near an eye-wash station. Neutralize the acid by adding small amounts of 1 M NaOH as required. Dilute the volume of waste materials with 20 times as much water. Place a beaker containing this diluted mixture in the sink, and run water to overflowing for 10 minutes, flushing to a known sanitary sewer.

- To dispose of aceto-orcein stain, follow the directions on the supplier's MSDS.

PROCEDURAL NOTES

Safety Precautions

- Discuss all safety symbols and caution statements with students.

- Review rules for carrying and using the compound microscope.

Procedural Tips

- Prior to the start of the lab, review the process of mitosis. Be sure to identify the name, events, and chromosome positions of prophase, metaphase, anaphase, and telophase. Make sure students can identify each phase of mitosis in the prepared slide.

4. Remove the hydrochloric acid from the treating dish, and dispose of it as instructed by your teacher. Refill the dish with distilled water.

5. Place a microscope slide on a piece of paper towel. **CAUTION: Glassware is fragile. Notify your teacher promptly of any broken glass or cuts. Do not clean up broken glass or spills unless your teacher tells you to do so.** Add three drops of aceto-orcein stain to the center of the slide. **CAUTION: Aceto-orcein stain is flammable. It will also stain your skin and clothing. Promptly wash off spills to minimize staining.**

6. Use forceps to transfer a prepared root tip from the treating dish to the drop of stain on the microscope slide.

7. Pulverize the tissue by gently but firmly tapping the root tip with the end of a wooden macerating stick. *Note: Move the stick in a straight up-and-down motion.*

8. Allow the root tip to stain for 10 to 15 minutes. *Note: Do not let the stain dry. Add more stain if necessary.*

9. Place the slide on a smooth, flat surface. Add a coverslip to the slide to make a wet mount. Place the wet mount between two pieces of paper towel.

10. Use the eraser end of a pencil to press down on the coverslip. Apply only enough pressure to squash the root tip into a single cell layer. *Note: Be very careful not to move the coverslip while you are pressing down with the pencil. Also be very careful not to press too hard. If you press too hard, you might break the glass slide and tear apart the cells in the onion root tip.*

♦ Why do you squash and spread out the root tip?

to make the tissue one cell layer thick

11. Repeat steps 5 through 10 for the remaining root tips.

Part 2—Observing the Onion Root Tip Squash

12. View one slide at a time with a compound light microscope under both low and high power. *Note: Remember that your mount is fairly thick, so be careful not to change to the high-power objective too quickly. Doing so could shatter the coverslip and destroy your preparation. You will need to focus carefully with the fine-adjustment knob to better see the structures under study.* View all three slides you have made. Select the slide that shows the most cells undergoing mitosis. Use this slide to complete steps 13 through 15.

♦ What is the shape of the onion root tip cells?

Most cells will appear rectangular, with some cells almost square in shape.

◆ What color did the aceto-orcein stain the chromosomes?

dark red

13. Observe the slide under high power. Without moving the slide, count or estimate the number of cells in the viewing area.

◆ What is the total number of cells you see in the viewing area?

Student answers will vary.

14. Now, mentally divide the viewing area into three viewing sections. *Note: Do not move the slide.* In each area, count the number of cells in prophase, metaphase, anaphase, and telophase. Record the numbers in the data table below.

Stages of Mitosis

Cell stage	Viewing area			Total number of cells per stage
	1	**2**	**3**	
Prophase	6	4	8	18
Metaphase	2	4	3	9
Anaphase	3	2	2	7
Telophase	1	2	1	4

◆ What are the most common stages of mitosis you observed?

Most students will observe prophase and anaphase, but all stages of mitosis will likely be present.

15. For the next observation, move the slide to a new viewing area near the one you just looked at. Repeat steps 13 and 14 in this new area. Repeat steps 13 and 14 for a third viewing area.

16. Dispose of your materials according to the directions from your teacher.

17. Clean up your work area and wash your hands before leaving the lab.

Analysis **18.** Why do specimens have to be thin to be viewed through the microscope?

Internal structures cannot be identified if the specimen is more than one cell layer thick. Also, light

must be able to pass through the specimen.

19. How many cells were in mitosis?

Student answers will vary.

Teacher's Notes (side column):

• Students need to be very patient and diligent when viewing their prepared slides. Students are asked to prepare more than one slide to increase their chances of successfully observing all four stages of mitosis. Because success rates for making good quality slides are low, share good slides with the class so that all students have the opportunity to view and count the cells. Students tend to feel greater success in viewing the slide if they look first for chromosomes in anaphase and prophase, followed by metaphase and telophase.

• Remind students to always begin on low power when focusing the microscope and to use only the fine adjustment when using high power.

• As an optional extension, you may want to have students make their slides permanent. To make a permanent slide, remove the coverslip from the temporary mount, add a drop of mounting medium (Piccolyte II), and replace the coverslip. Place the slide on a flat surface, and allow it to dry for several days.

20. In the space provided, sketch one cell to represent each stage of mitosis you observed on your slide. Label each stage.

Prophase	Metaphase	Anaphase	Telophase
			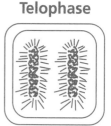

21. What percentage of cells on your slide were in the process of mitosis? Use the following formula to figure out your answer. Show your calculations in the space provided.

$$\% \text{ cells in mitosis} = \frac{\text{total number of cells in all phases of mitosis}}{\text{total number of cells}} \times 100$$

Student answers will vary but should resemble the following: number of cells in mitosis = 38, total

number of cells = 856, % of cells in mitosis = 38/856 × 100 = 4.4%.

22. What percentage of cells were in each phase of mitosis? Use the following formula to figure out your answer.

$$\% \text{ cells in phase} = \frac{\text{total number of cells in phase}}{\text{total number of cells}} \times 100$$

Student answers will vary but should approximate the following: prophase = 18/856 = 2%,

metaphase = 9/856 = 1.1%, anaphase = 7/856 = 0.8%, telophase = 4/856 = 0.5%.

23. Why did you make more than one slide in this investigation?

Some slides have a "better" group of cells to observe than others. As you compare the different

slides, you can choose the "best" slide for observation purposes.

Conclusions **24.** The aceto-orcein stain contains acetic acid, which causes the cells to swell. What is the purpose of allowing the stain to remain on the root tips for 10 to 15 minutes before preparing the squash?

The acetic acid in the stain causes the cell cytoplasm to gradually swell and helps spread apart the

chromosomes so they can be studied.

25. How do you explain the percentage of cells undergoing mitosis?

The number of cells undergoing mitosis seems low because most of the cell's life is spent at rest or

in interphase.

26. What are some reasons why you may have observed a low number of cells undergoing mitosis?

Not all cells are undergoing mitosis even though there is fast growth in the root tip. The cell rests

and enlarges in size following mitosis. Some stages of mitosis (especially prophase and telophase)

were more difficult to find than others (metaphase and anaphase).

27. Why did you (the botanist) choose to work with the root tip?

The root tip is one of the areas that is the most rapidly growing. The cells will be undergoing

mitosis at a higher rate in this area than in other regions of the plant. Since the root tip contains no

chlorophyll, cells undergoing mitosis are easy to see.

Extensions **28.** Design an experiment to test the effect of different kinds of fertilizer on the percentage of cells undergoing mitosis in an onion root tip. *Note: Be sure to control all variables other than the fertilizer.*

29. *Botanists* are biologists who study plants. They study plant structure, function, growth, and reproduction. Because there are so many things to know about so many different kinds of plants, many botanists specialize in studying just one function of just one kind of plant. Find out about the training and skills required to become a botanist.

Name _____

Date _____ Class _____

HOLT
BIOSOURCES
LAB PROGRAM
EXPERIMENTAL DESIGN

C10 *Preparing a Root Tip Squash*

Prerequisites Review

- **Laboratory Techniques C9: Preparing a Root Tip Squash** on pages 39–44
- the stages of mitosis
- position of the chromosomes during each phase of mitosis

Paula's Perfect Plants
Riverside, California

March 11, 1998

Karl Smith
Microbiology Division
BioLogical Resources, Inc.
101 Jonas Salk Dr.
Oakwood, MO 65432-1101

Dear Mr. Smith,

I am a plant breeder who develops new varieties of ornamental plants for nurseries around the country. For the past 10 years, I have used a chemical called colchicine to produce polyploid plants, or plants that have more than the diploid number of chromosomes. Polyploid plants tend to be larger than diploid plants. Colchicine prevents a cell undergoing mitosis from completing metaphase, causing the number of chromosomes in a plant cell to double. The resulting condition is known as polyploidy.

Recently I read an article about colchicine that disturbed me. Apparently colchicine is more toxic than I had thought. I have always used care when handling chemicals, and fortunately, I have not had any problems. If necessary, I will now take the necessary precautions and continue to use the colchicine. However, if it is possible, I would like to find an alternative procedure. I would like your company to develop a new, safer procedure for inducing polyploidy. I am willing to pay any reasonable fee. Please let me know what you decide.

Sincerely,

Paula Bates-Pierce

Paula Bates-Pierce
Plant Breeder
Paula's Perfect Plants

BioLogical Resources, Inc. Oakwood, MO 65432-1101

M E M O R A N D U M

To: Team Leader, Cytology Dept.

From: Karl Smith, Director of Microbiology

Your research teams should find the problem described in the attached letter interesting and challenging. I have taken the liberty of doing a little research on the effects of caffeine on mitosis. I was able to find an obscure reference that suggested that caffeine in high doses acts as an inhibitor to mitotic division. Have your research teams design and conduct an experiment to test caffeine for its effect on mitosis.

Mitosis is easily observed in onion root tip cells. I have had my assistant prepare two sets of root tips by soaking them in either caffeine or water and fixing them. These will be available to you in two separate beakers. Follow the standard procedures for preparing a root tip squash. Be sure to prepare more than one root tip to ensure a successful slide. Use the root tips soaked in water as a control. Compare the caffeine-soaked root tips with the control root tips and with a prepared slide of a root tip that has been soaked in colchicine.

Proposal Checklist

Before you start your work, you must submit a proposal for my approval. **Your proposal must include the following:**

_____ • the **question** you seek to answer

_____ • the **procedure** you will use

_____ • a detailed **data table** for recording observations

_____ • a complete, itemized list of proposed **materials** and **costs** (including use of facilities, labor, and amounts needed)

Proposal Approval: _____
(Supervisor's signature)

Report Procedures

When you finish your analysis, prepare a report in the form of a business letter to Ms. Bates-Pierce. **Your report must include the following:**

_____ • a paragraph describing the **procedure** you followed to prepare the root tip squash and to test the effect of caffeine on mitosis

_____ • a complete **data table** showing your observations for each control and experimental slide

_____ • your **conclusions** about the effectiveness of caffeine as a substitute for colchicine

_____ • a detailed **invoice** showing all materials, labor, and the total amount due

Safety Precautions

• Wear safety goggles and a lab apron.

• Glassware is fragile. Notify your teacher promptly of any broken glass or cuts. Do not clean up broken glass or spills unless your teacher tells you to do so.

• Never use electrical equipment around water, or with wet hands or clothing. Never use equipment with frayed cords.

• Wash your hands before leaving the laboratory.

• If you are using a microscope with a mirror, do not use direct sunlight as a light source.

Disposal Methods

• Dispose of waste materials according to instructions from your teacher.

• Place used paper towels in a trash can.

• Place broken glass, root tips, and unused hydrochloric acid in the separate containers provided.

• Wash reusable materials such as glassware and lab utensils, and return them to the supply area.

FILE: Paula's Perfect Plants

MATERIALS AND COSTS (Select only what you will need. No refunds.)

I. Facilities and Equipment Use

Item	Rate	Number	Total
facilities	$480.00/day	_____	_____
personal protective equipment	$10.00/day	_____	_____
compound microscope	$30.00/day	_____	_____
microscope slide	$1.00/day	_____	_____
plastic coverslip	$1.00/day	_____	_____
forceps	$5.00/day	_____	_____
test tube	$2.00/day	_____	_____
test-tube rack	$5.00/day	_____	_____
stirring rod	$2.00/day	_____	_____
500 mL beaker	$5.00/day	_____	_____
hot-water bath	$15.00/day	_____	_____
pipet	$2.00/day	_____	_____
thermometer	$5.00/day	_____	_____
prepared slide of onion root tip treated with colchicine	$5.00/day	_____	_____

II. Labor and Consumables

Item	Rate	Number	Total
labor	$40.00/hour	_____	_____
single-edged razor blade	$3.00 each	_____	_____
ethanol	$0.10/mL	_____	_____
cotton ball	$0.10 each	_____	_____
caffeine-treated root tips	provided	_____	_____
untreated root tips	provided	_____	_____
1 M HCl	$0.20/mL	_____	_____
aceto-orcein stain	$30.00/bottle	_____	_____
mounting medium	$0.10/cm	_____	_____
wax pencil	$2.00 each	_____	_____

Fines

OSHA safety violation	$2,000.00/incident	_____	_____

Subtotal	_____
Profit Margin	_____
Total Amount Due	_____

Name _____

Date _____ Class _____

HOLT
BIOSOURCES
LAB PROGRAM
LABORATORY TECHNIQUES

C11 *Modeling Meiosis*

Skills
- modeling a biological process

Objectives
- *Model* the stages of meiosis in an animal cell.
- *Demonstrate* genetic recombination.
- *Relate* the events of meiosis to the formation of haploid gametes.

Materials

PREPARATION NOTES
Time required:
50 minutes

- 4 pieces of string (1 m long)
- 4 pieces of string (40 cm long)
- 4 strips (2 cm × 6 cm) of paper, 1 each of light blue, dark blue, light green, and dark green
- scissors
- 8 paper clips
- metric ruler
- transparent tape
- 8 pieces of string (10 cm long)

Purpose

You are a student teacher in a high-school science class. Your supervising teacher has asked you to tutor a group of students who are having difficulty learning the process of meiosis. You decide to use paper models to show the students what happens during meiosis I and II.

Background

Preparation Tips
- Colored index cards can be used for the paper strips.
- Materials for this exploration can be made by the teacher or a science aide.
- Provide a container for the paper clips.
- Adhesive putty can be substituted for paper clips.

Disposal
- Remind students to place all paper scraps in the trash can.

The body cells of plants and animals are diploid. A **diploid** (*2n*) cell has two sets of chromosomes in its nucleus. A cell with only one set of chromosomes in its nucleus is termed **haploid** (*n*). **Gametes,** egg and sperm, are examples of haploid cells. When gametes fuse at fertilization, a diploid **zygote** is formed. The zygote contains one set of chromosomes from each parent.

Meiosis is a process that produces haploid (*n*) cells, such as gametes, from diploid (*2n*) cells. Before meiosis begins, DNA replication occurs. Following replication, each chromosome consists of two **chromatids** that are joined by a **centromere.** Meiosis involves two successive divisions of the nucleus. The first of these two divisions is called **meiosis I.** During meiosis I, the **homologous chromosomes** (chromosomes that carry the same genes and that are similar in size and shape) come together, or pair up, and then separate. The nuclei that result from meiosis I contain only one set of chromosomes, or one chromosome from each pair of homologous chromosomes. Therefore, meiosis I is also known as reduction division because each of the resulting nuclei contains half the number of chromosomes of the original cell. The second division of the nucleus is called **meiosis II.** During meiosis II, the chromatids separate, forming four haploid nuclei.

During meiosis I, the chromatids of a **homologue** (member of a pair of homologous chromosomes) may exchange parts. This exchange of segments between chromatids is called **crossing over.** Crossing over, as well as the fusion of two gametes during sexual reproduction, is a type of **genetic recombination,** which is the regrouping of genes into new combinations.

Procedure

PROCEDURAL NOTES

Safety Precautions
• Discuss all safety symbols and caution statements with students.

Procedural Tips
• Instruct students to read carefully as they follow the steps.

1. Using a 1 m piece of string, make a circle on your desk to represent the cell membrane of a cell. Using a 40 cm piece of string, make another circle inside the cell to represent the nuclear membrane.

2. Fold each of four strips of paper (one light blue, one dark blue, one light green, and one dark green) in half, lengthwise. Then place one strip of each color inside the nucleus to represent a chromosome before replication. The light and dark strips of the *same* color represent homologous chromosomes. The light-colored strips represent chromosomes from one parent, while the dark-colored strips represent chromosomes from the other parent.

3. ◆ **Interphase** To represent DNA replication, unfold each paper strip and cut each in half, lengthwise. The two pieces that result from cutting each homologous strip represent the chromatids. Attach the two identical chromatid strips at the center with a paper clip so an X is formed. Each paper clip represents a centromere.

   ♦ What process did you model when you cut the paper strips in half?

   DNA replication

   ♦ What is the function of the centromere?

   It holds the chromatids together after replication.

Meiosis I

4. **Prophase I** Remove the nuclear membrane. Place the blue chromosomes side by side. Do the same for the green chromosomes. Simulate crossing over by measuring and cutting a 2 cm tip from a light blue strip. Do the same with a dark blue strip. Tape the light blue tip to the dark blue strip, and tape the dark blue tip to the light blue strip. Do the same for the green strips.

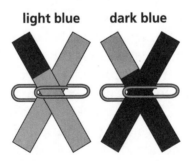

light blue   dark blue

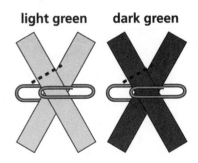

light green   dark green

   ♦ What is the purpose of placing the light and dark strips of the same color side by side?

   The two pairs are placed side by side to simulate the pairing of homologous chromosomes, which

   enables the exchange of genetic information in a process called crossing over.

5. **Metaphase I** Place four 10 cm strings inside the cell, so that two strings extend from one side into the center of the cell and two strings extend from the opposite side into the center of the cell. The strings represent spindle fibers. Attach a string to the centromere of each chromosome (chromatid pair) with tape. Move the chromosomes to the center of the cell. *Note: Make sure that the strings attached to similar colors come from opposite sides of the cell.*

6. **Anaphase I** Gather the loose ends of the strings on both sides of the cell, and pull the strings in opposite directions.

7. **Telophase I** Untape the string from each centromere. Place a 40 cm piece of string around each group of chromatids, forming two nuclei. Place a 1 m piece of string around each cell, forming two cell membranes.

    ◆ How many chromosomes are in each cell? Describe what each part represents.

    There are two chromosomes in each cell. Each consists of two chromatids, represented by the paper

    strips. They are attached by a centromere, represented by the paper clip.

Meiosis II

8. **Prophase II** Remove the strings that represent the nuclear and cell membranes in both cells. Attach a 10 cm piece of string to each chromatid.

    ◆ What must happen to the centromeres before the chromatids can separate?

    The centromeres must divide, or duplicate.

9. **Metaphase II** Move the chromosomes to the center of each cell. Make sure that the strings attached to the two strips in each chromosome come from opposite sides of the cell.

10. **Anaphase II** Gather the strings on both sides of each cell, and pull the strings in opposite directions, separating the paper strips. *Note: Only one strip in each pair will have a paper clip attached.*

11. **Telophase II** Untape the strings. Remove the strings and paper clips. Each strip of paper now represents a chromosome. Place a 40 cm piece of string around each group of chromosomes, forming four nuclei. Place a 1 m string around each cell, forming four cell membranes.

    ◆ How many cells did you make? How many chromosomes are in each cell? Are the cells haploid or diploid in number?

    There are four cells with two chromosomes in each cell. They are haploid.

12. Dispose of your materials according to the directions from your teacher.

13. Clean up your work area before leaving the lab.

Analysis

14. What is the diploid chromosome number of the original cell in your model? How many homologous pairs does this represent?

The diploid number of the model cell is four. This represents two pairs of

homologous chromosomes.

15. If a cell with a diploid number of 6 chromosomes undergoes meiosis, what will the cell look like after Telophase I? Draw it in the space below.

Conclusions

16. Give two reasons why meiosis is important in sexual reproduction.

Meiosis reduces the chromosome number by half. Without meiosis, each succeeding generation

would have twice the number of chromosomes. During fertilization, the diploid number is restored.

Meiosis also allows for new genetic combinations, providing variation within a species.

17. How might crossing over affect the rate of evolution?

If a chromosome failed to pair with its homologue, one cell might have extra genes and another might be missing genes. It could be lethal to the developing organism. If the organism did survive, it would probably be abnormal.

Crossing over increases the rate of genetic recombination and provides an additional source of

genetic variation. Increased genetic variation may provide a species with the ability to evolve and

adapt more quickly to its environment.

Extensions

18. Use your model to show what would happen if homologous chromosomes did not pair in Prophase I. Predict the outcome of this occurrence.

19. Distinguish between *oogenesis* and *spermatogenesis*. Creatively represent each process by using three-dimensional models.

Name _____

Date _____ Class _____

HOLT
BIOSOURCES
LAB PROGRAM
EXPERIMENTAL DESIGN

C12 | *Analyzing Corn Genetics*

Prerequisites   • **Laboratory Techniques C11: Modeling Meiosis** on pages 49–52

Review   • parental generation, F_1, F_2, phenotype, genotype, homozygous, and heterozygous
- using an F_2 ratio to find the F_1 generation genotype and finally the parental genotype
- use of the Punnett square

Storm Lake, Iowa

December 13, 1997

Caitlin Noonan
Research and Development
BioLogical Resources, Inc.
101 Jonas Salk Dr.
Oakwood, MO 65432-1101

Dear Ms. Noonan,

Our company packages and sells hybrid corn seeds to local farmers. The plants that produce our seeds are grown in small lots by many different farmers. Prior to marketing the seeds, samples from each lot are sent to a laboratory that determines what percentage of seeds in each lot germinate. Last year, the lab reported that each lot had a 95 percent germination rate (95 of every 100 seeds begin to develop and grow). This was the rate that we guaranteed on each package.

Recently we were notified that several farmers are filing a lawsuit against us. Apparently, last spring's seeds produced a crop yield far below the expected rate. When I received notice of the lawsuit, I tried unsuccessfully to contact the laboratory that determines our germination rates. I learned later that the company went out of business shortly after our tests were completed. I have been unable to contact anyone from the laboratory to ask questions about the data they provided. We will compensate the farmers for their losses, but we also want to protect our good reputation. Please help us identify the problem. Enclosed you will find seeds from last spring's lots A, B, and C. Please grow the seeds, determine the germination rate for each lot, and alert us of any abnormalities.

Sincerely,

Arthur Clements

Arthur Clements
President
Maize Seed Incorporated

BioLogical Resources, Inc. Oakwood, MO 65432-1101

M E M O R A N D U M

To: Team Leader, Genetics Dept.

From: Caitlin Noonan, Director of Research and Development

Have your teams get started on this problem right away. Determine the germination rate for each lot of Mr. Clements's corn seeds. Each group should test about 10 seeds from each of lots A, B, and C. Then compile the results from all research groups.

This morning, I had a phone conversation with Mr. Clements. He told me that some farmers who planted the seeds in question reported a greatly reduced corn yield, while others did not. This reduced yield could be the result of poor germination or a genetic problem. Grow the corn for about 2 or 3 weeks, determine the germination rate, and note any changes in the sample crop. Please be thorough and efficient in your procedures.

Proposal Checklist

Before you start your work, you must submit a proposal for my approval. **Your proposal must include the following:**

_____ • the **question** you seek to answer

_____ • the **procedure** you will use

_____ • two detailed **data tables** for recording observations

_____ • a complete, itemized list of proposed **materials** and **costs** (including use of facilities, labor, and amounts needed)

Proposal Approval: _____
(Supervisor's signature)

Report Procedures

When you finish your analysis, prepare a report in the form of a business letter to Mr. Clements. **Your report must include the following:**

_____ • a paragraph describing the **procedure** you followed to find the germination rate

_____ • complete group and class **data tables** showing the number of seeds planted, the number of normal and abnormal seeds that germinated, the number of normal and abnormal plants alive at 4 weeks, total germination rate, and total survival rate

_____ • your **conclusions** about the germination rate

_____ • a detailed **invoice** showing all materials, labor, and the total amount due

Safety Precautions

• Wash your hands before leaving the laboratory.

Disposal Methods

• Dispose of waste materials according to instructions from your teacher.

• Upon completion of the lab, corn seedlings can be thrown into a trash can.

• Place used potting soil in the container provided.

• Wash reusable materials such as glassware and lab utensils, and return them to the supply area.

FILE: Maize Seed Incorporated

MATERIALS AND COSTS (Select only what you will need. No refunds.)

I. Facilities and Equipment Use

Item	Rate	Number	Total
facilities	$480.00/day	_____	_____
personal protective equipment	$10.00/day	_____	_____
compound microscope	$30.00/day	_____	_____
microscope slide with coverslip	$2.00/day	_____	_____
plant tray	$5.00/day	_____	_____
thermometer	$5.00/day	_____	_____
scissors	$1.00/day	_____	_____
ruler	$1.00/day	_____	_____

II. Labor and Consumables

Item	Rate	Number	Total
labor	$40.00/hour	_____	_____
potting soil	$1.00/kg	_____	_____
seeds	provided	_____	_____
paper towels	$0.10 each	_____	_____
zippered plastic bag	$0.10 each	_____	_____
fertilizer	$2.00/kg	_____	_____
plant stakes	$2.00/stake	_____	_____
water	no charge		

Fines

OSHA safety violation	$2,000.00/incident	_____	_____

Subtotal	_____
Profit Margin	_____
Total Amount Due	_____

Name _____

Date _____ Class _____

HOLT
BIOSOURCES
LAB PROGRAM
LABORATORY TECHNIQUES

C13 *Preparing Tissue for Karyotyping*

Skills

PREPARATION NOTES

- making and staining a slide from a suspension of fixed human cells
- using a compound light microscope

Objectives

Time Required: two 50-minute periods

- *Prepare* slides from a suspension of fixed human cells arrested at metaphase.
- *Examine* the banding pattern of the chromosomes under a microscope.
- *Interpret* karyotypes of various human chromosome spreads.

Materials

Materials
• Materials for this lab activity can be purchased from WARD'S. See the *Master Materials List* for ordering instructions.

- safety goggles
- lab apron
- alcohol swab
- microscope slide
- bucket of ice
- pencil
- plastic Pasteur pipet
- pipet bulb

- microtube with fixed human cells
- gloves
- plastic staining jars (5)
- NaCl solution
- trypsin solution
- watch or clock with second hand
- Giemsa staining solution (buffered)
- compound microscope

Purpose

You are a lab technician for a genetic counselor at a local hospital. A young married couple expecting their first child wants to know the sex of their unborn baby. To determine the baby's sex, you prepare and stain cells extracted from the fluid surrounding the baby and arrested during mitosis, and then make slides so that a karyotype for the baby can be prepared.

Background

Preparation Tips
• To concentrate the cell suspension, either spin each microtube of cells at 1200 rpm (899 g) for 10 minutes in a standard table top centrifuge or let the cells settle to the bottom of the tube in a refrigerator overnight. Do not shake the settled cells. Carefully withdraw and discard excess fluid down to the black line, and then resuspend the cells in the remaining fixative just before the lab by gently tapping on the side of the tube.
• Set up one or two staining stations for all students to use.

Chromosomes are found in the nucleus of every cell. They are coiled packages of genes that provide the codes used to make proteins. In humans, the normal number of chromosomes is 46, arranged in 23 pairs. These chromosomes contain about 3 billion coding base pairs of DNA and 70,000 to more than 100,000 genes. Each person has 22 pairs of "identical" chromosomes that control almost all the body's characteristics. The twenty-third pair are the **sex chromosomes** (X and Y). Females have two **X chromosomes,** and males have one X chromosome and one Y chromosome. The **Y chromosome** is small and has only a small number of functional genes. Its major function is to turn the developing fetus into a male. In the absence of a Y chromosome, the fetus will develop into a female.

Genetic disorders can occur when an individual has too many or too few chromosomes. Sometimes a person with a genetic disorder has the correct number of chromosomes, but one or more of the chromosomes is abnormal. **Karyotypes** are pictures of paired human chromosomes arranged by size from largest to smallest. They are used in identifying chromosome abnormalities.

The first step in producing a karyotype is to take cells—from either the amniotic fluid of a fetus or the blood of a child. The cells are placed in a special culture that promotes their growth. When the cells begin to divide, a chemical called

- If an oven or incubator is available, heat it to 55° to 60°C. If not, prepare an alternative warm (40°C) surface.
- The following solutions come in a kit with premeasured amounts:
a. To prepare fresh 0.15 M NaCl solution, add the packet contents to 1 L of distilled water.
b. To prepare fresh trypsin solution, add 5 mL of distilled water to a vial of powdered trypsin.
c. To prepare phosphate buffer solution, add the packet contents to 1 L of distilled water.
d. To make fresh staining solution, add 3 mL of the supplied Giemsa stain to 47 mL of phosphate buffer solution.
e. Add 2 mL of Bacto-trypsin to 48 mL of 0.15 M NaCl. The trypsin differentially removes the protein from segments of the chromosomes so that banding patterns appear when the stain is applied.
- Always wear PPE when preparing and disposing of solutions.

Procedure

- Set up a row of plastic staining jars in the following order: NaCl rinse, trypsin, NaCl rinse, NaCl rinse, Giemsa stain.
- Prepare a separate container for the disposal of broken glass.

Disposal

Dilute solutions in a ratio of 1 part of solution to 20 parts of water, and pour down the drain.

colchicine is added to the culture to stop cell division during metaphase. Nuclei captured at this stage of cell division are called **metaphases.** Next, the cells are placed in a solution that ruptures the membranes, freeing the chromosomes and killing the cells. The chromosomes are stained, revealing dark and light bands in the chromosomes, and then photographed. The photograph is enlarged and cut into pieces containing one chromosome each. These pieces are then arranged on a sheet of paper in numbered homologous pairs, according to shape, size, and staining bands. An example of a prepared human karyotype is shown below.

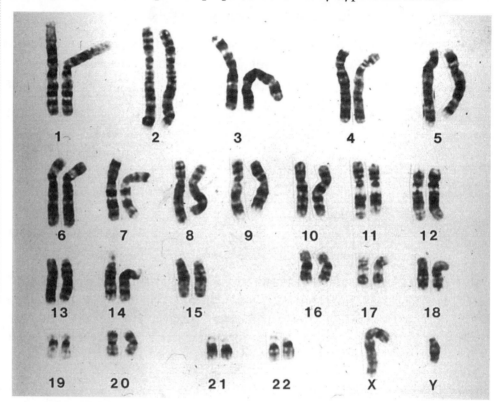

Part 1—Making a Slide

1. Put on safety goggles and a lab apron.

2. Thoroughly clean a glass microscope slide with a swab or tissue soaked with alcohol. Place the cleaned slide in a bucket of ice until you are ready to use it. **CAUTION: Glassware is fragile. Notify your teacher promptly of any broken glass or cuts. Do not clean up broken glass or spills unless your teacher tells you to do so.**

3. Remove the cleaned slide from the ice, but do not dry it. Place one end of the slide on a paper towel. Put a pencil under the other end of the slide as shown in the diagram on the next page. Hold a plastic Pasteur pipet (with pipet bulb) about 15 cm or less above the slide. Place one drop of the suspension of fixed human cells on the center of the slide. *Note: When using a Pasteur pipet to make the drops, take up just enough in the pipet to make the slide. Fill just the narrow part of the pipet; do not allow cells to go up into the large diameter section of the pipet.* Allow the excess liquid to drain off the slide onto the paper towel. Allow the slide to dry.

PROCEDURAL NOTES

Safety Precautions

• Discuss all safety symbols and caution statements with students.

• Review rules for carrying and using the compound microscope.

Procedural Tips

• Depending on the number of students per group, have students prepare and stain two or three slides for observation. After staining and initial observations under the microscope, have all students observe the slides that show the best results.

• Note that the cell suspensions contain a mixture of nuclei. The metaphases exist as swollen nuclear membranes containing the chromosomes. *Note: Any rough handling may cause these nuclei to burst and therefore be lost.* The metaphases also tend to be sticky and get caught and lost on surfaces such as the sides of tubes and pipets.

• Keep the cell suspension as cold as possible, without freezing.

• To ensure the best banding, make sure that the cell suspensions are handled as little as possible, the slides are clean and cold before beginning, the specimen slides are aged, the staining solutions are fresh, and the slides are rinsed between staining steps.

• If no banding can be seen, have students keep the slide in the trypsin solution for a slightly longer time. If the chromosomes are too dark to see the bands, have them use less Giemsa staining time. If the nuclei and metaphases are not

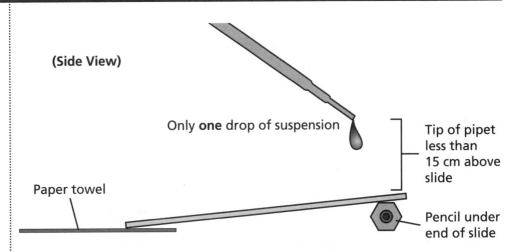

(Side View)

Only **one** drop of suspension

Tip of pipet less than 15 cm above slide

Paper towel

Pencil under end of slide

4. Let the slides age by keeping them for two or three weeks in a covered place at room temperature or by placing them overnight in a 60°C oven. Aging the slides this way results in better banding of the chromosomes.

5. Dispose of your materials according to the directions from your teacher.

6. Clean up your work area and wash your hands before leaving the lab.

Part 2—Staining Chromosomes

7. Put on safety goggles, gloves, and a lab apron.

8. To stain the chromosomes, place your slide in staining jars in the order shown in the diagram below. First, dip the slide momentarily in NaCl solution. **CAUTION: If you get a chemical on your skin or clothing, wash it off at the sink while calling to your teacher. If you get a chemical in your eyes, promptly flush it out at the eyewash station while calling to your teacher.** Then put the slide in trypsin solution for 15 seconds to 2 minutes to stain the chromosomes. Gently swish the slide momentarily in each of two jars of NaCl solution. Then put the slide in Giemsa solution for 30 to 60 seconds, swishing the slide every few seconds.

Order of staining jars (from left to right)

50 mL of 0.15 M NaCl	2 mL of Trypsin in 48 mL of NaCl	50 mL of 0.15 M NaCl	50 mL of 0.15 M NaCl	3 mL of Giemsa stain in 47 mL of buffer
Dip	**15 sec to 2 min**	**Rinse well**	**Rinse well**	**30–60 sec**

9. Rinse the slide *gently* in running water until all excess stain runs off. Dry the slide by waving it around in the air. **CAUTION: Do not allow the slide to hit other students or objects in the room.**

dark enough to see at all, have them use slightly longer Giemsa staining times. If the trypsin time is too long, the chromosomes become digested and appear faint and swollen.

• If you want to keep students' slides, have them add a drop of Piccolyte II mounting medium and a coverslip to their slides.

• Students who do Extension step 16 should discover that common numerical anomalies include Down syndrome (an

Analysis

extra 21), trisomy 18 (an extra 18), and trisomy 13 (an extra 13). Survival for trisomy 13 and trisomy 18 is usually only weeks or months, due to the presence of numerous life-threatening abnormalities. Down syndrome offspring have certain physical characteristics and mental retardation and can usually survive into adulthood.

Conclusions

• Students who do Extension step 17 may have researched the clinical abnormalities in Klinefelter (XXY) syndrome and XYY syndrome, which are often so mild that the

Extensions

diagnosis is made only when infertility is evident (XXY) or is never even suspected (XYY). Females with XXX syndrome usually have clinically significant developmental anomalies, including development delay.

---

10. Examine your slides with a compound light microscope. Scan the slide under low power until you see nuclei in metaphase. Look for a metaphase that has well-spread chromosomes that are not all twisted and clumped. *Note: These nuclei may be hard to see.* Once you have a metaphase in focus, switch to high power and examine the banding for detail. Examine the chromosomes, and note the differences in size and banding patterns. Try to count the total number of chromosomes in the metaphase. If there are 46, then the metaphase is complete and no chromosomes are missing. (Sometimes the metaphases burst too widely, and chromosomes float off to other areas of the slide.) Use the picture of the human karyotype shown on page 58 as a guide to identify individual chromosomes you can observe in your slide.

11. Dispose of your materials according to the directions from your teacher.

12. Clean up your work area and wash your hands before leaving the lab.

13. Study the picture of the karyotype shown on page 58. Is the individual shown in the karyotype a male or female? Explain your answer.

The individual is a male because both an X and a Y chromosome are present.

14. Does the individual shown in the karyotype have any genetic abnormalities? Explain your answer.

The individual does not appear to have any genetic abnormalities. All 23 pairs of chromosomes are present and appear to be complete and "normal."

15. Is the individual whose chromosomes you observed in your prepared slides a male or a female? Explain your answer.

Answers will vary. Students may be able to observe and identify the X and Y chromosomes. If they can observe both an X and a Y chromosome, the individual is a male. If they can observe two X chromosomes, the individual is a female.

16. Genetic abnormalities often occur in chromosomes 13, 18, and 21. Do library or on-line research to find out what syndromes these abnormalities cause.

17. Do library or on-line research to find out what syndromes occur due to the presence or absence of extra X or Y chromosomes in a human.

C14 *Karyotyping*

Skills
- constructing a karyotype
- analyzing a karyotype

Objectives

PREPARATION NOTES

Time Required:
50 minutes

- *Identify* pairs of homologous chromosomes by their length, centromere position, and banding pattern.
- *Determine* the sex of an individual from a karyotype.
- *Predict* whether an individual will be normal or have Down syndrome.

Materials

Materials

Materials for this lab activity can be purchased from WARD'S. See the *Master Materials List* for ordering instructions.

- chromosome spread
- scissors
- metric ruler
- WARD's Human Karyotyping Form
- transparent tape

Purpose

As a medical lab technician, one of your jobs is to assist with prenatal testing. Currently, you are working on the case of Mr. and Mrs. Smith. Mrs. Smith is pregnant, and her doctor has recommended amniocentesis, which is a type of prenatal testing. You have been given photomicrographs of the chromosomes in the unborn baby's cells, which were obtained through amniocentesis. Your job is to complete and analyze a karyotype of these cells to determine the sex of the Smiths' baby and whether the baby is normal or has Down syndrome.

Background

Additional Background

You may want to discuss genetic diseases such as Turner's syndrome and Klinefelter's syndrome. When discussing nondisjunction, you may want to introduce the following terms:

• deletion—a portion of a chromosome is lost

• duplication—the deletion becomes incorporated into its homologue so that the segment appears twice on the same chromosome

Humans have 46 chromosomes in every diploid (*2n*) body cell. The chromosomes of a diploid cell occur in **homologous pairs,** which are pairs of chromosomes that are similar in size, shape, and the position of their centromere. In humans, 22 homologous pairs of chromosomes are called **autosomes.** The twenty-third pair, which determines the individual's sex, make up the **sex chromosomes.** Females have only one type of sex chromosome, which is called an **X chromosome.** Males have two types of sex chromosomes, an X chromosome and a much smaller **Y chromosome.** The diagram at the top of the next page shows each of the 22 types of autosomes and the 2 types of sex chromosomes.

A **karyotype** is a diagram that shows a cell's chromosomes arranged in order from largest to smallest. A karyotype is made from a photomicrograph (photo taken through a microscope) of the chromosomes from a cell in metaphase. The photographic images of the chromosomes are cut out and arranged in homologous pairs by their size and shape. The karyotype can be analyzed to determine the sex of the individual and whether there are any chromosomal abnormalities. For example, the karyotype of a female shows two X chromosomes, and the karyotype of a male shows an X chromosome and a Y chromosome.

• translocation—the deleted portion is transferred to a chromosome other than its homologue

Preparation Tips

• If a karyotype form is not available, students can make their own. Be sure to have them label groups A–G and place the centromeres on the line. They also need to number them accordingly.

Disposal

• Remind students to place all paper scraps in the trash can.

PROCEDURAL NOTES

Safety Precautions

• Discuss all safety symbols and caution statements with students.

Procedural Tips

• Instruct students to cut out the chromosomes in a rectangular shape. It is a good idea to place the cut-out chromosomes face up in one pile and the scrap paper in another pile.

Procedure

• When taping down the chromosomes, students will find it easier to use short pieces of tape and to tape down one or two pairs at a time. Students are likely to make a mistake if they try to tape down an entire row at one time.

• When evaluating the karyotype, check that chromosome pairs are accurate and that each pair is centered on the line by their centromeres.

Human Chromosomes

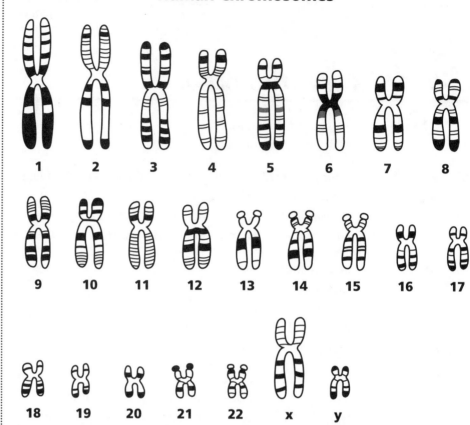

Chromosomal abnormalities often result from **nondisjunction,** the failure of chromosomes to separate properly during meiosis. Nondisjunction results in cells that have too many or too few chromosomes. **Trisomy** is an abnormality in which a cell has an extra chromosome, or section of a chromosome. This means that the cell contains 47 chromosomes instead of 46. **Down syndrome,** or **Trisomy 21,** is a chromosomal abnormality that results from having an extra number 21 chromosome.

1. Carefully cut each chromosome from the chromosome spread. Be sure to leave a slight margin around each chromosome.

2. Arrange the chromosomes in homologous pairs. The members of each pair will be the same length and will have the centromere in the same location. Use the ruler to measure the length of the chromosome and the position of the centromere. Arrange the pairs according to their length, from largest to smallest. The banding patterns of the chromosomes may also help you to pair up the homologous chromosomes.

3. Tape each homologous pair to a Human Karyotyping Form, positioning the centromeres on the lines. Place the pairs in order, with the longest pair at position 1, the shortest pair at position 22, and the sex chromosomes at position 23.

4. The diagram you have made is a karyotype. Analyze the karyotype to determine the sex of the individual and whether or not the karyotype is normal.

Analysis

5. 🜚 Dispose of your materials according to the directions from your teacher.

6. Examine your karyotype. Is the baby male or female? Will the baby have Down syndrome? How do you know?

The baby is a normal male. The sex is determined by examining the sex chromosomes. There is an X

and a Y. The autosomes are all paired and there are no extras, so the karyotype is normal.

7. The Y chromosome closely resembles many of the other chromosomes. What did you have to do to determine that it was the Y chromosome?

The length of the chromosome and position of the centromere were measured. Banding patterns

were compared, and it was determined that there was only one chromosome present and not a

pair. This helped to distinguish the Y chromosome from others of similar size and shape.

8. If the karyotype you constructed were for a female with Down syndrome, what chromosome changes would be evident?

There would be two X chromosomes and an extra 21st chromosome. The total chromosome

number would increase to 47.

Conclusions

9. If your job were to inform the Smiths of their test results, what would you say?

Answers will vary. Students should cite evidence to support their conclusion.

10. Why are karyotypes important tools for geneticists?

Karyotypes supply information related to genetic problems and the sex of an individual.

Extensions

11. *Genetic counselors* work with couples who are concerned that the father or mother may have a harmful gene that could be passed to their child. The counselors study the family history and perform blood tests to detect harmful genes. They then predict if there is a chance that a parent could contribute a harmful gene to a child. Parents can then make an informed decision about having children. Find out about the training and skills required to become a genetic counselor.

12. Research genetic diseases. Collect data about the incidences of genetic diseases by age group and number of cases. Present your findings with graphs and pictures.

Chromosome Spread

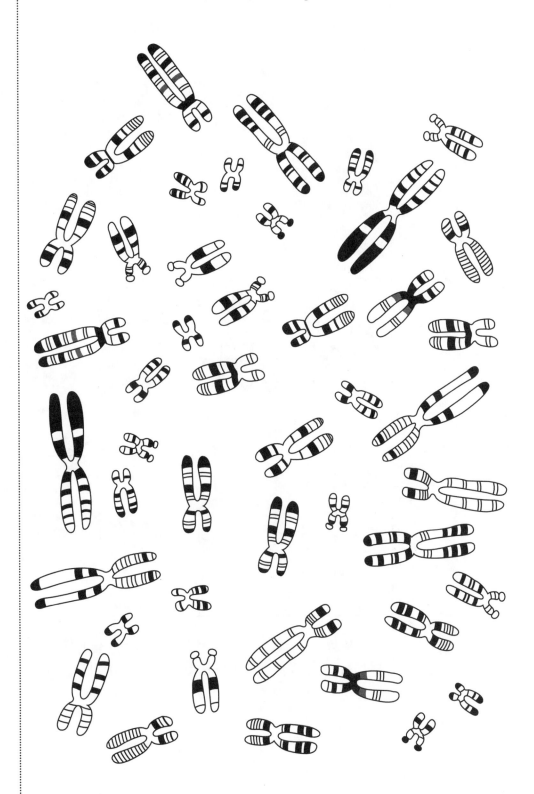

Human Karyotypes Answer Key

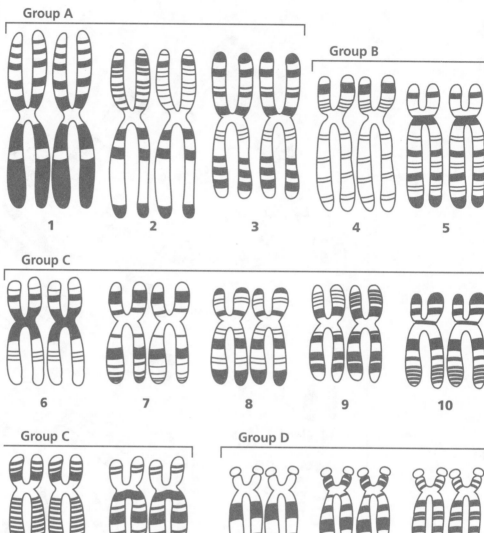

Group A
1   2   3

Group B
4   5

Group C
6   7   8   9   10

Group C
11   12

Group D
13   14   15

Group E
16   17   18

Group F
19   20

Group G
21   22

Sex Chromosomes
x   y

Name _____

Date _____ Class _____

HOLT
BIOSOURCES
LAB PROGRAM
EXPERIMENTAL DESIGN

C15 | *Karyotyping—Genetic Disorders*

Prerequisites • **Laboratory Techniques C14: Karyotyping** on pages 61–66

Review • the structure and appearance of chromosomes during mitosis

CENTRAL OREGON UNIVERSITY HOSPITAL

Bend, Oregon

February 12, 1998

Caitlin Noonan
Research and Development
BioLogical Resources, Inc.
101 Jonas Salk Dr.
Oakwood, MO 65432-1101

Dear Ms. Noonan,

Central Oregon University Hospital is conducting a study on the effectiveness of the hospital's prenatal screening methods. Our goal is to improve our success rate in determining who should be a candidate for amniocentesis or chorionic villi sampling. These tests are complicated and costly, and we want to avoid conducting unnecessary tests.

We have given our screening test to, and have taken chorionic villi samples from, all of the subjects in our test group. During the study, we will be comparing the results of the karyotypes for each test subject with the results of their screening tests. By comparing the list of candidates with the actual test results, we can determine the success rate of our standard screening process.

In order to conduct an effective double-blind study, we have decided to have our karyotypes analyzed out-of-house. I understand that your company has experience with this type of consulting. The hospital will provide you with photomicrographs of the chromosomes from each test subject. I'm sure you understand the importance of providing timely and accurate results to the patients. I appreciate your close attention to this matter.

Sincerely,

Teresa Flores

Teresa Flores, M.D.
Central Oregon University Hospital

BioLogical Resources, Inc. Oakwood, MO 65432-1101

M E M O R A N D U M

To: Team Leader, Medical Testing Dept.

From: Caitlin Noonan, Director of Research and Development

Have your teams analyze the photographs described in the attached letter and complete a karyotype for each subject. Additionally, please include a brief explanation of the test results and a summary of the characteristics of any genetic disorder, so that the hospital can include this information in their patient reports.

Please keep in mind that in addition to providing the laboratory with data, we will also be providing many sets of parents with information about the health of their child. Accuracy and speed are essential for this project. Divide these sets of chromosomes, and spread the work out among your research groups. Attached are the actual chromosome photomicrographs, which were delivered today by a special courier. Use our standard karyotyping form. To ensure privacy, each set of chromosomes has been given a letter for identification.

Proposal Checklist

Before you start your work, you must submit a proposal for my approval. **Your proposal must include the following:**

_____ • the **question** you seek to answer

_____ • the **procedure** you will use

_____ • a detailed **data table** for recording observations

_____ • a complete, itemized list of proposed **materials** and **costs** (including use of facilities, labor, and amounts needed)

Proposal Approval: _____
(Supervisor's signature)

Report Procedures

When you finish your analysis, prepare a report in the form of a business letter to Dr. Flores.
Your report must include the following:

_____ • a paragraph describing the **procedure** you followed to perform the karyotypes

_____ • a complete class **data table** showing the result, condition, and abnormality for each smear

_____ • your **conclusions** about the karyotype, and (if you did find a genetic disorder) an explanation of the disorder, the physical expression of the syndrome, and how this disorder may have occurred during meiosis

_____ • a detailed **invoice** showing all materials, labor, and the total amount due

Amniocentesis tests are generally given to women over 35 years of age, and these tests usually show a normal male or female child. I have listed below some genetic complications that may show up in a karyotype.

Name of Abnormality	Chromosome Affected	Brief Description of Abnormality
Down syndrome, or Trisomy 21	#21	47 chromosomes, mental retardation with specific characteristic features, may have heart defects and respiratory problems
Edwards' syndrome, or Trisomy 18	#18	47 chromosomes, severe mental retardation, very characteristic malformations of the skull, pelvis, and feet, among others. Die in early infancy
Patau syndrome, or Trisomy 13	#13	47 chromosomes, abnormal brain function that is very severe, many facial malformations, usually die in early infancy
Turner's syndrome	Single X in female (X0)	45 chromosomes, in females only, missing an X chromosome, do not develop secondary sex characteristics, are infertile
Klinefelter's syndrome	Extra X in male (XXY)	47 chromosomes, in males only, tall, sterile, small testicles, otherwise appear normal
XYY syndrome	Extra Y in male (XYY)	47 chromosomes, in males only, low mental ability, otherwise normal appearance
Triple X syndrome	Extra X in female (XXX)	47 chromosomes, sterility sometimes occurs, normal mental ability

Safety Precautions

• Handle sharp objects carefully.

• Wash your hands before leaving the laboratory.

Disposal Methods

• Dispose of waste materials according to instructions from your teacher.

• Place paper scraps and chromosome cutouts in the separate containers provided.

FILE: Central Oregon University Hospital

MATERIALS AND COSTS (Select only what you will need. No refunds.)

I. Facilities and Equipment Use

Item	Rate	Number	Total
facilities	$480.00/day	_____	_____
personal protective equipment	$10.00/day	_____	_____
compound microscope	$30.00/day	_____	_____
microscope slide with coverslip	$2.00/day	_____	_____
scissors	$1.00/day	_____	_____
ruler	$1.00/day	_____	_____

II. Labor and Consumables

Item	Rate	Number	Total
labor	$40.00/hour	_____	_____
chromosome photomicrographs	provided	_____	_____
tape	$0.25/m	_____	_____
karyotyping form	$10.00 each	_____	_____

Fines

OSHA safety violation	$2,000.00/incident	_____	_____

Subtotal	_____
Profit Margin	_____
Total Amount Due	_____

C16 DNA Whodunit

Skills

- simulating DNA fingerprinting
- collecting and comparing data from DNA fingerprinting

Objectives

- *Use* pop beads to model DNA fingerprinting.
- *Describe* the processes of DNA extraction, gel electrophoresis, and autoradiography as they relate to DNA analysis.
- *Solve* a murder by analyzing the DNA fingerprints of suspects and DNA samples collected at the scene of a crime.

PREPARATION NOTES

Time Required: one 50-minute period

Materials

Materials

Materials for this lab activity can be purchased from WARD'S. See the *Master Materials List* for ordering instructions.

- red beads (phosphate) (60)
- five-hole white beads (deoxyribose) (60)
- orange beads (thymine) (15)
- yellow beads (adenine) (15)
- blue beads (cytosine) (15)
- green beads (guanine) (15)
- plastic connectors (hydrogen bonds) (30)
- restriction enzyme card Jan I
- restriction enzyme card Ward II

- paper gel electrophoresis lane
- clear plastic alkali card (cDNA probes)
- white beads (deoxyribose) (6)
- pink beads (radioactive phosphate) (6)
- orange beads (thymine) (2)
- green beads (guanine) (2)
- blue beads (cytosine) (2)

Purpose

You are a police investigator trying to solve a murder case in your town. Blood stains of two different types were found at the murder scene. You suspect one of the blood stains belongs to the murderer. You have read about a technique called DNA fingerprinting, which can be used to identify a criminal from minute amounts of DNA left at the scene of a crime in the form of blood, semen, hair, or saliva. The process of making a DNA fingerprint is detailed and complicated. You know about a model that has been designed to explain this process. The district attorney would like to be able to use the model in court to explain the process to a jury. He has asked you to test the model to see how it works.

Background

Each person has a DNA profile that is as unique as his or her fingerprints. With the exception of identical twins, no two people have the same DNA sequence. A technique called **DNA fingerprinting** can be used to compare the DNA of different individuals. The end result is a photograph with a pattern of dark bands that reflect the composition of a DNA molecule. Because these band patterns are unique to each person, they are often called **DNA fingerprints.** DNA fingerprinting works because of the small differences in each individual's genetic makeup.

PROCEDURAL NOTES

Procedural Tips

• Have the class choose partners the day before the lab to save valuable time.

The process of DNA fingerprinting involves three steps. First, known and unknown samples of DNA are obtained. The samples are then digested, or cut into small fragments, by the same restriction enzyme. A **restriction enzyme** is an enzyme that recognizes and binds with a specific short base sequence of DNA. The restriction enzyme then cuts the DNA at specific sites within the sequence. The

- Make photocopies of the "Suspect/Victim DNA Samples" sheet. Cut apart the "Suspect" and "Victim" strips. Assign one strip to each group in a class. Be sure each class has the strips for all the suspects, the victim, and the murderer.

- Decide which of the suspects will be the murderer for each class. Use one of the photocopies to make strips labeled "Murderer's DNA." Cut out the strips of the five suspects' DNA and the victim's DNA. Let's say you have chosen Suspect 3 as the murderer. Select the Suspect 3 DNA strip, and cut out or blacken out the suspect's number so that the murderer's

Procedure

identity is a surprise to the class. Change the identity of the murderer from class to class to keep students guessing. Keep the identity of the murderer secret—let the teams do the work of solving the crime!

- Give the murderer's DNA sample strip to your most capable group of students. Procedural errors in this sample will destroy your class results. Give the victim's DNA strip to another capable team. Distribute the suspect DNA strips to the remaining teams. Hopefully, each suspect DNA strip will be analyzed by more than one team.

- Arrange the beads and other materials in a location that is easily accessible to students. A front desk or table is ideal.

resulting fragments are called **restriction fragment length polymorphisms— RFLPs** for short.

The next step in DNA fingerprinting is to separate the RFLPs by size. This is done with a technique called **gel electrophoresis.** The DNA is placed on a jelly-like slab called a gel, and the gel is exposed to an electrical current. DNA has a negative electrical charge, so the RFLPs are attracted to the positive pole when an electric current is applied. Smaller units travel farther up the gel than longer ones. The length of a given DNA fragment can be determined by comparing its mobility on the gel with that of a DNA marker, which is a sample of DNA that has been digested by restriction enzymes into fragments of a known size.

Using a technique called Southern blotting, the separated fragments are transferred to special nitrocellulose paper and labeled with a radioactive **cDNA probe** so that they can be seen. A cDNA probe is a relatively short (usually about 8 to 20 nucleotides long) piece of single-stranded DNA that finds and sticks to a complementary sequence in one or more of the RFLPs. At this point in the actual lab procedure, a piece of unexposed X-ray film is laid out on top of the paper. The radioactive probes expose the film wherever they bind to the RFLPs. The film is developed and dark bands mark the lengths of the RFLPs. This process is called **autoradiography.** The resulting pattern is unique for each individual.

1. You will test the DNA from a blood sample of one of the five suspects, the victim, or the murderer in a hypothetical murder case. At the end of the testing, DNA fingerprints from all five suspects, the victim, and the murderer's blood samples will be gathered from all of the teams and will then be compared to determine which of the suspects is the murderer.

2. Carefully assemble the DNA you were assigned with pop beads, using the DNA strip given to your group as a blueprint. *Be sure to assemble the beads in the precise pattern indicated, or your results will be incorrect!* The assembled bead chain represents your subject's DNA.

3. Place the DNA "molecule" you have just assembled on your work area so that the 5' end is on the top left side, as shown below. Be sure that the three orange beads (thymine) are in the following position on your work area:

5'   3'

TTT, etc.............................G

AAA, etc.............................C

3'   5'

Note: From this point on, it is important to keep the beads in this orientation. Never allow the chain to be turned upside down or rotated. The 5' TTT end should always be on the top left of the molecule. If your chain is accidentally turned upside down, refer to your Suspect/Victim DNA strip to obtain the correct orientation.

4. Next, use the model restriction enzymes Jan I and Ward II to chop up the DNA. Look at your two Restriction Enzyme Cards; they look like the cards in the diagram below. These "enzymes" will make cuts in the DNA in the manner indicated by the dotted lines.

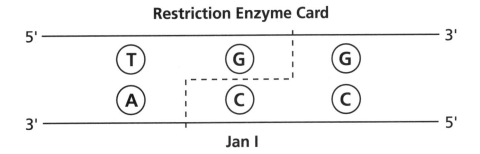

Restriction Enzyme Card

Jan I

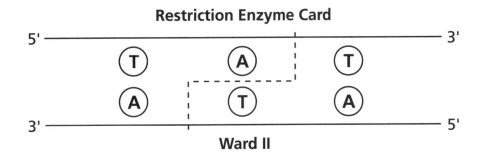

Restriction Enzyme Card

Ward II

Begin with Restriction Enzyme Card Jan I. Place it on top of the left side of the DNA chain so that its label is right side up. Move the card along the surface of the DNA until you match the precise sequence shown on the card. Stop and break the beads apart in the manner indicated by the dotted lines. Move the enzyme card until you reach the right end of the DNA. It is a good idea to double-check to ensure you have made all the possible cuts.

5. Repeat the procedure on the remaining DNA fragments using the Restriction Enzyme Card Ward II. *Note: Be sure to keep the DNA fragments in the orientation described above (5' orange thymine beads on the top left) throughout this exercise.* In reality, the RFLPs created in steps 4 and 5 might be thousands of base pairs long!

6. Now simulate separating the RFLPs by gel electrophoresis. Taking care to retain the proper 5' to 3' orientation, place the RFLPs at the negative pole of the gel electrophoresis lane shown on the next page. Remember, DNA has a negative electrical charge, so the RFLPs are attracted to the positive end of the gel lane when an electric current is applied. Simulate this by sliding your RFLPs along the lane. Shorter fragments are lighter and move farther than longer fragments. To determine the final position of the RFLPs, count the number of nucleotides on the longest side of each fragment. Place each measured RFLP next to its corresponding length marked on the gel lane.

• You may want to have students redraw the gel lane on this page at a larger scale on a separate sheet of paper. Suggest a scale of 1 in. to 1 cm.

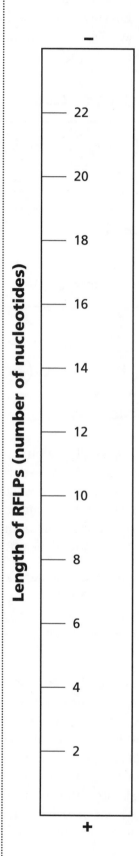

Length of RFLPs (number of nucleotides)

7. Observe the work of other teams in the class.

◆ Are the RFLPs of the other DNA samples the same length as yours? Why?

Answers will vary, but students should indicate that the lengths are different. They are different

because the DNA of each individual is unique.

8. Pick up an alkali card like the one shown below. Treatment with alkali causes a DNA molecule to unzip and become single-stranded. Pass this card over each of the RFLPs. Break the hydrogen bonds that hold the bases (adenine, cytosine, guanine, and thymine) together to produce single-stranded RFLPs.

Alkali Card

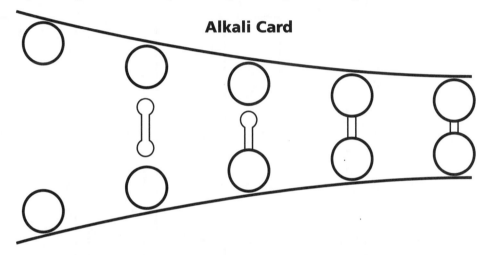

9. Using six white beads (D, or deoxyribose), six pink beads (P, or radioactive phosphate), two orange (T, or thymine), two green (G, or guanine), and two blue (C, or cytosine) beads, construct two model cDNA probe molecules like those shown below.

$$
\begin{array}{ccc}
\text{T} & \text{G} & \text{C} \\
| & | & | \\
\end{array}
$$

3' D–P–D–P–D–P 5'

Probe

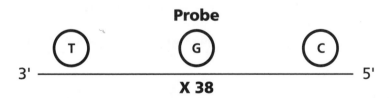

X 38

The model probe X38 will bind with any DNA that has the sequence 5' A C G 3'. Search the top halves (5' to 3') of your RFLP fragments for such a sequence. When you find a 5' A C G 3' sequence, attach the beads of the probe to the RFLP sequence. *Note: Remember, no two people have identical DNA. None of the five suspects' DNA is identical. Everyone in the class is using the same cDNA probes. The result is that the probes that bind to some of your RFLPs will not necessarily bind to the same RFLPs of others.*

10. Again, observe the work of other teams in the room. Carefully check the lengths of the RFLPs for each DNA sample. Also, check to see if the probes are attached to the correct sequence of the RFLPs. After you have done this, return to your seat.

11. In the seven gel electrophoresis lanes below, sketch dark bands at the correct positions in the gel lane reserved for your sample. Get up once again, and visit other teams. Obtain the banding patterns for each of the DNA samples. *Note: In the actual lab procedure, a single autoradiograph would contain the banding patterns for all seven lanes. The X ray is placed against a light background and the banding patterns are compared.*

Victim's Blood	Suspect 1	Suspect 2	Suspect 3	Suspect 4	Suspect 5	Murderer's Blood
22	22	22	22	22	22	22
20	20	20	20	20	20	20
18	18	18	18	18	18	18
16	16	16	16	16	16	16
14	14	14	14	14	14	14
12	12	12	12	12	12	12
10	10	10	10	10	10	10
8	8	8	8	8	8	8
6	6	6	6	6	6	6
4	4	4	4	4	4	4
2	2	2	2	2	2	2

Analysis

12. Each of the following played an important role in DNA fingerprinting. Describe the function of each.

a. restriction enzyme

The restriction enzyme has a specific shape that allows it to recognize and bind with specific short-

base sequences of DNA and then to cut the DNA at a specific site within the sequence.

b. gel electrophoresis

Gel electrophoresis separates the RFLPs by size.

c. alkali

Treatment with alkali causes the DNA molecule to unzip and become single-stranded.

d. cDNA probe

The radioactive cDNA probe fits and sticks to a specific complementary sequence in one or more

RFLPs.

e. autoradiograph

An autoradiograph is the exposed X-ray film that compares the banding patterns of numerous gel

electrophoresis lanes.

Conclusions   **13.** Which of the suspects is the real murderer? Explain your answer.

The identity of the murderer will vary. The suspect is identified as the murderer because the

patterns in his or her DNA exactly match those of the murderer's blood.

14. Imagine that you are on a jury and that DNA fingerprinting evidence is introduced. Explain how you would regard such evidence.

Answers will vary, but students should discuss the validity of the test.

Extensions   **15.** Look through newspapers and news magazines to find articles about actual court cases in which DNA fingerprinting was used to determine the innocence or guilt of a suspect in a crime. Share the articles with your classmates.

16. Do library research or search the Internet to explain how the DNA fingerprinting procedure could be used by scientists who study fossils.

Teacher's Guide to Suspect/Victim DNA and RFLP's

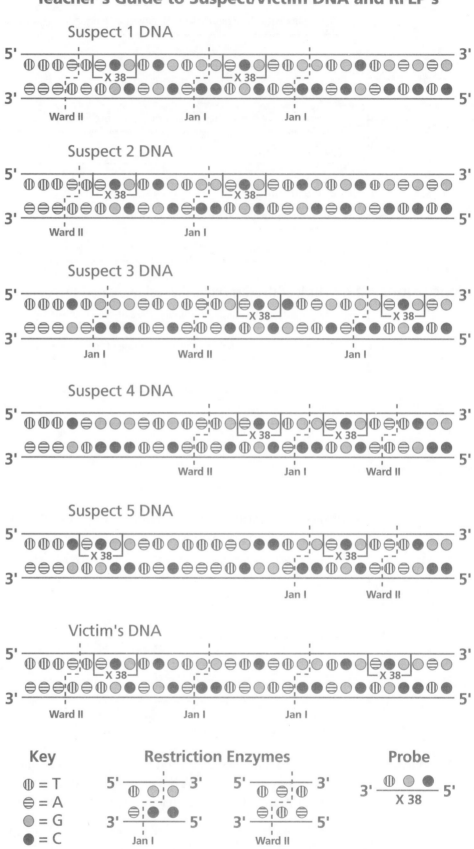

Suspect 1 DNA

Suspect 2 DNA

Suspect 3 DNA

Suspect 4 DNA

Suspect 5 DNA

Victim's DNA

Key

Restriction Enzymes

Probe

C17 Analyzing Blood Serum to Determine Evolutionary Relationships

Skills

- preparing a serial dilution of simulated blood serum
- collecting and analyzing data
- comparing and contrasting amounts of agglutination in simulated blood serum

Objectives

PREPARATION NOTES

Time required: one 50-minute period

- *Prepare* a serial dilution of simulated cow, frog, and human "blood serum."
- *Analyze* the amount of agglutination that occurs when simulated human antiserum is added to simulated cow and frog serums.
- *Relate* the amount of agglutination to the evolutionary relationships of these organisms.

Materials

Materials

Materials for this lab activity can be purchased from WARD'S. See the *Master Materials List* for ordering instructions.

- safety goggles
- gloves
- lab apron
- testing well trays (3)
- wax pencil
- plastic dropping pipets (7)
- simulated human serum (1 vial)

- beaker of distilled water
- "human antiserum" (1 vial)
- stirring sticks (24)
- stopwatch or clock with second hand
- simulated cow serum (1 vial)
- simulated frog serum (1 vial)

Purpose

You are a biochemist researching the evolutionary relationship of various animals. You have chosen to study three animals: frog, cow, and human. You are aware that other scientists have studied this relationship by comparing proteins such as the blood protein hemoglobin and cytochrome c, a protein found in mitochondria. You decide to use an alternative method, one that uses the immune response in animal blood.

Background

Preparation Tip

Provide separate containers for the disposal of solutions, broken trays and stirrers, and broken glass.

Disposal

Simulated serums and antiserum can be washed down the drain with copious amounts of water.

An important technique used in determining evolutionary relationships among organisms is biochemical comparisons. Organisms that are not similar in physical appearance may show "hidden" biochemical similarities that may provide insights into their evolutionary past.

One technique used to study the biochemical relationships among organisms is the **Nutall agglutination technique.** This technique takes advantage of the production of **antibodies,** defensive proteins in the blood that bind to foreign materials in the bloodstream. Any substance in the blood, foreign or otherwise, that causes the production of antibodies is an **antigen.** Once antibodies for a specific antigen are produced, they remain in the bloodstream as **antiserum,** the noncellular part of the blood that contains antibodies for a specific antigen. If the antigen is ever reintroduced into the bloodstream, these antibodies will cause cells in the blood to **agglutinate,** or clump together, making the foreign material harmless.

The Nutall agglutination technique takes advantage of this reaction. When a rabbit is injected with human serum, the rabbit will make human antiserum. This human antiserum can subsequently be mixed with the blood **serum,** the part of the blood that does not include cells, of a variety of animals. The greater the amount of agglutination, the closer the evolutionary relationship of the organism to humans.

Procedure

PROCEDURAL NOTES

Safety Precautions

• Under no circumstances are students to test any blood other than the simulated blood that you provide.

• Discuss all safety symbols and caution statements with students.

• Have students wear safety goggles, lab aprons, and gloves.

• Explain to students that even though they are working with simulated blood, all laboratory tests that involve blood require gloves.

Procedural Tips

• Discuss with students what a serial dilution is and why it is used in this lab.

• Discuss with students how to recognize a positive reaction.

• Discuss with students the concept of genetic distance and evolution.

1. Put on safety goggles, gloves, and a lab apron.

2. With a wax pencil, label one testing well tray "human serum." Number the wells 1 through 8. Repeat this procedure for the other two trays, labeling the second tray "cow serum" and the third tray "frog serum."

3. Use a clean dropping pipet to add drops of simulated human serum to the tray labeled "human serum." Use another clean dropping pipet to add drops of water to each well. Use the table below to determine the correct number of drops of serum and water for each well.

Serial Dilution Amounts

Well number	Drops serum	Drops water
1	8	0
2	7	1
3	6	2
4	5	3
5	4	4
6	3	5
7	2	6
8	1	7

4. Use a clean dropping pipet to add 8 drops of human antiserum to each well. Use a clean plastic stirrer to stir the contents of each well. *Note: Be sure to use a clean stirrer to stir each well to avoid contamination from one well to another.*

5. Observe the wells for 2 minutes, occasionally stirring each well. Remember to use the correct stirrer for each well to avoid contamination.

6. After 2 minutes, record your results in the data table on the next page using the following key. *Note: Ignore any reaction that may occur after 2 minutes.*

    +++ Heavy agglutination (fast and clear reaction)

    ++ Medium agglutination (precipitation is definite but not immediate)

    + Slight agglutination (weak reaction, may take longer amount of time)

    — No reaction

Agglutination Response

Organism	1	2	3	4	5	6	7	8
human	+++	+++	+++	+++	+++	+++	++	+
cow	+++	+++	+++	++	—	—	—	—
frog	+ or —	—	—	—	—	—	—	—

7. Repeat steps 3 through 6 with the tray labeled "cow serum" and simulated cow serum.

8. Repeat steps 3 through 6 with the tray labeled "frog serum" and simulated frog serum.

9. Dispose of your materials according to the directions from your teacher.

10. Clean up your work area and wash your hands before leaving the lab.

Analysis

11. What kind of simulated blood serum showed the greatest amount of agglutination?

human

12. What type of simulated blood serum showed the least amount of agglutination?

frog

13. Why was a serial dilution made? How did the dilutions help you interpret the data?

The serial dilution was made to help interpret the results. When there is a large amount of

agglutination, it is difficult to interpret results. Diluting the serum allows the amount of

agglutination to be viewed more easily.

14. When you made the serial dilutions of each type of serum, why did you increase the number of drops of water as you decreased the number of drops of serum?

Water was added in increasing amounts as the serum decreased to provide a uniform amount of

liquid in each well. This allowed for uniform comparisons of the wells.

15. What does an agglutination reaction demonstrate?

The antibodies are attacking the foreign material present in the serum. The more clumping, the

greater the reaction or the closer the evolutionary relationship.

Conclusions

16. Based on agglutination results, which organism is more closely related to humans? Explain your answer.

Cow—there was a greater amount of agglutination in the cow serum than in the frog serum.

Extensions

17. A biologist was walking around the Mojave Desert. He discovered a new animal that seems to have many of the structural characteristics found in a lizard, a tortoise, and a snake. Scientists are stumped about how to place this new organism in the evolutionary scale. What further tests can be done that will help solve this problem?

18. Find out how scientists use amino acid sequence in proteins such as hemoglobin and cytochrome c to determine the evolutionary relationship between organisms.

LAB PROGRAM

EXPERIMENTAL DESIGN

C18 Analyzing Blood Serum—Evolution of Primates

Prerequisites	• **Laboratory Techniques C17: Analyzing Blood Serum to Determine Evolutionary Relationships** on pages 79–82
Review	• the Nutall agglutination technique
	• how to make an evolutionary tree

CONCORD UNIVERSITY
New York, New York

June 2, 1998

Caitlin Noonan
Research and Development Division
BioLogical Resources, Inc.
101 Jonas Salk Dr.
Oakwood, MO 65432-1101

Dear Ms. Noonan,

I am an anthropologist who has been studying primates in Africa. Last month, while I was following an adolescent male gorilla into the mountains, I discovered what appears to be a previously unknown primate. The animals have opposable thumbs, grasping hands, and a forward placement of the eyes. I have even observed the animals using rocks and sticks as tools. I have measured the skulls of these primates and the measurements fall in a range between chimpanzee and human skull measurements. I believe that this new primate may be the closest living relative to humans.

This may be the find of the century. I want to be completely sure of my accuracy before I make any announcements. I heard that your research group has experience with the Nutall agglutination technique. I would like you to perform this test on the animal's blood serum and determine its evolutionary relationship to humans.

I have included some of the animal's blood serum for your tests. As I am sure you are aware, this is an extremely confidential matter. Please use the utmost discretion. I am looking forward to your assessment.

Sincerely,

Jonatha Hill

Jonatha Hill
Professor of Anthropology
Concord University

BioLogical Resources, Inc. Oakwood, MO 65432-1101

M E M O R A N D U M

To: Team Leader, Medical Testing Dept.

From: Caitlin Noonan, Director of Research and Development

When you read the attached letter, I am sure you were as excited as I was to be asked to participate in this research. I want your team to start on the appropriate tests for Dr. Hill right away. Our procedures will be under strict scrutiny, and it is possible that many institutes will want to test our findings. Be sure to write down your procedural steps clearly and thoroughly.

Test samples of human blood serum, chimpanzee blood serum, and blood serum from the unknown primate. Test each serum sample against human antiserum to determine the amount of agglutination. A serial dilution should be used to ensure an accurate analysis. Please include an evolutionary tree with your test results.

Proposal Checklist

Before you start your work, you must submit a proposal for my approval. **Your proposal must include the following:**

_____ • the **question** you seek to answer

_____ • the **procedure** you will use

_____ • a detailed **data table** for recording results

_____ • a complete, itemized list of proposed **materials** and **costs** (including use of facilities, labor, and amounts needed)

Proposal Approval: _____
(Supervisor's signature)

Report Procedures

When you finish your analysis, prepare a report in the form of a business letter to Dr. Hill. **Your report must include the following:**

_____ • a paragraph describing the **procedure** you followed to test the serum samples for agglutination

_____ • a complete **data table** showing agglutination results for the human, chimpanzee, and unknown primate

_____ • your **conclusions** about the evolutionary relationship of the three primates

_____ • a detailed **invoice** showing all materials, labor, and the total amount due

Safety Precautions

• ◇ ◇ ◆ Wear safety goggles, disposable gloves, and a lab apron.

• ◆ Glassware is fragile. Notify your teacher promptly of any broken glass or cuts. Do not clean up broken glass or spills unless your teacher tells you to do so.

• ◆ Wash your hands before leaving the laboratory.

• Under no circumstances are you to test any blood other than the simulated blood provided by your teacher.

Disposal Methods

• ◆ Dispose of waste materials according to instructions from your teacher.

• Place broken glass, serum samples, and antiserum in the separate containers provided.

• Place used stirring sticks in a trash can.

• Wash reusable materials such as glassware and lab utensils, and return them to the supply area.

FILE: Concord University

MATERIALS AND COSTS (Select only what you will need. No refunds.)

I. Facilities and Equipment Use

Item	Rate	Number	Total
facilities	$480.00/day	_____	_____
personal protective equipment	$10.00/day	_____	_____
compound microscope	$30.00/day	_____	_____
microscope slide with coverslip	$2.00/day	_____	_____
testing well tray	$8.00/day	_____	_____
50 mL beaker	$5.00/day	_____	_____
watch or clock	$10.00/day	_____	_____

II. Labor and Consumables

Item	Rate	Number	Total
labor	$40.00/hour	_____	_____
vial of human antiserum	$20.00 each	_____	_____
vial of human serum	$20.00 each	_____	_____
vial of chimpanzee serum	$20.00 each	_____	_____
vial of serum from unknown primate	provided	_____	_____
vial of cow serum	$20.00 each	_____	_____
distilled water	$0.10/mL	_____	_____
plastic dropping pipets	$2.00 each	_____	_____
stirring sticks	$1.00 each	_____	_____
wax pencil	$2.00 each	_____	_____

Fines

OSHA safety violation	$2,000.00/incident	_____	_____

Subtotal	_____
Profit Margin	_____
Total Amount Due	_____

C19 Analyzing Amino-Acid Sequences to Determine Evolutionary Relationships

Skills
- comparing and contrasting sequences of amino acids
- interpreting data

Objectives

PREPARATION NOTES

Time Required: 1 class period

- *Identify* the differences in the amino-acid sequences of the cytochrome c and hemoglobin molecules of several species.
- *Infer* the evolutionary relationships among several species by comparing amino-acid sequences of the same protein in different organisms.

Materials
- paper
- pencil

Purpose

No outside preparation is required for this lab.

You are a zoologist who specializes in the classification of vertebrates according to their evolutionary relationships. In your research, you examine the amino-acid sequences of particular protein molecules found in vertebrates to determine the degree of biochemical similarity between vertebrate species. Today you will compare portions of human cytochrome c and hemoglobin molecules with the same portions of those molecules in other vertebrate species. Your goals are to determine the differences in the amino-acid sequences of the molecules and to deduce the evolutionary relationships among the species.

Background

The biochemical comparison of proteins is a technique used to determine evolutionary relationships among organisms. Proteins consist of chains of amino acids. The sequence, or order, of the amino acids in a protein determines the type and nature of the protein. In turn, the sequence of amino acids in a protein is determined by the sequence of nucleotides in a gene. A change in the DNA nucleotide sequence (mutation) of a gene that codes for a protein may result in a change in the amino-acid sequence of the protein.

Biochemical evidence of evolution compares favorably with structural evidence of evolution. Even organisms that appear to have few physical similarities may have similar sequences of amino acids in their proteins and be closely related through evolution. Researchers believe that the greater the similarity in the amino-acid sequences of two organisms, the more closely related they are in an evolutionary sense. Conversely, the greater the time that organisms have been diverging from a common ancestor, the greater the differences that can be expected in the amino-acid sequences of their proteins.

Two proteins are commonly studied in attempting to deduce evolutionary relationships from differences in amino-acid sequences. One is cytochrome c, and the other is hemoglobin. **Cytochrome c** is a protein used in cellular respiration and found in the mitochondria of many organisms. **Hemoglobin** is the oxygen-carrying molecule found in red blood cells.

Procedure

Part 1—Cytochrome c

1. A cytochrome c molecule consists of a chain of 104 amino acids. The chart below shows the amino-acid sequence in corresponding parts of the cytochrome c molecules of nine vertebrates. The numbers along the side of the chart refer to the position of these sequences in the chain. The letters identify the specific amino acids in the chain.

Cytochrome c Amino-Acid Sequences

AA #	Horse	Chicken	Tuna	Frog	Human	Shark	Turtle	Monkey	Rabbit
42	Q	Q	Q	Q	Q	Q	Q	Q	Q
43	A	A	A	A	A	A	A	A	A
44	P	E	E	A	P	Q	E	P	V
46	F	F	Y	F	Y	F	F	Y	F
47	T	S	S	S	S	S	S	S	S
49	T	T	T	T	T	T	T	T	T
50	D	D	D	D	A	D	E	A	D
53	K	K	K	K	K	K	K	K	K
54	N	N	S	N	N	S	N	N	N
55	K	K	K	K	K	K	K	K	K
56	G	G	G	G	G	G	G	G	G
57	I	I	I	I	I	I	I	I	I
58	T	T	V	T	I	T	T	T	T
60	K	G	N	G	G	Q	G	G	G
61	E	E	N	E	E	Q	E	E	E
62	E	D	D	D	D	E	E	D	D
63	T	T	T	T	T	T	T	T	T
64	L	L	L	L	L	L	L	L	L
65	M	M	M	M	M	R	M	M	M
66	E	E	E	E	E	I	E	E	E
100	K	D	S	S	K	K	D	K	K
101	A	A	A	A	A	T	A	A	A
102	T	T	T	C	T	A	T	T	T
103	N	S	S	S	N	A	S	N	N
104	E	K	—	K	E	S	K	E	E

2. On a piece of scratch paper, write the name of each vertebrate in the chart on the previous page. Compare the amino-acid sequence of human cytochrome c with that of each of the other eight vertebrates. For each vertebrate's sequence, count the number of amino acids that differ from those in the human sequence. Write the number of differences in the amino-acid sequences under the vertebrate's name. When you have completed your comparisons, transfer your data to the data table below. As you do, list the eight vertebrates in order from fewest differences to most differences.

Cytochrome c Amino-Acid Sequence Differences Between Humans and Other Vertebrate Species

Species	Number of differences from human cytochrome c
monkey	1
rabbit	4
horse	6
chicken	7
turtle	8
frog	8
tuna	9
shark	14

◆ According to this line of evidence, which organism is most closely related to humans? Which is least closely related to humans?

The chimpanzee is the most closely related to humans. The shark is the least closely related.

◆ Frog and turtle cytochrome c molecules have the same number of differences from human cytochrome c. Which vertebrate, frog or turtle, would you put higher on the list? Explain.

Answers may vary. Because reptiles are more advanced than amphibians, students should list the

turtle ahead of the frog.

Part 2—Hemoglobin

3. Look at the amino-acid sequences shown below. These sequences are portions of the hemoglobin molecules of five organisms. The portion of the chains shown are from amino acid number 87 to amino acid number 116 in a sequence of 146 amino acids.

Hemoglobin Amino-Acid Sequences

AA #	Human	Chimpanzee	Gorilla	Monkey	Horse
87	THR	THR	THR	GLN	ALA
88	LEU	LEU	LEU	LEU	LEU
89	SER	SER	SER	SER	SER
90	GLU	GLU	GLU	GLU	GLU
91	LEU	LEU	LEU	LEU	LEU
92	HIS	HIS	HIS	HIS	HIS
93	CYS	CYS	CYS	CYS	CYS
94	ASP	ASP	ASP	ASP	ASP
95	LYS	LYS	LYS	LYS	LYS
96	LEU	LEU	LEU	LEU	LEU
97	HIS	HIS	HIS	HIS	HIS
98	VAL	VAL	VAL	VAL	VAL
99	ASP	ASP	ASP	ASP	ASP
100	PRO	PRO	PRO	PRO	PRO
101	GLU	GLU	GLU	GLU	GLU
102	ASN	ASN	ASN	ASN	ASN
103	PHE	PHE	PHE	PHE	PHE
104	ARG	ARG	LYS	LYS	ARG
105	LEU	LEU	LEU	LEU	LEU
106	LEU	LEU	LEU	LEU	LEU
107	GLY	GLY	GLY	GLY	GLY
108	ASN	ASN	ASN	ASN	ASN
109	VAL	VAL	VAL	VAL	VAL
110	LEU	LEU	LEU	LEU	LEU
111	VAL	VAL	VAL	VAL	ALA
112	CYS	CYS	CYS	CYS	LEU
113	VAL	VAL	VAL	VAL	VAL
114	LEU	LEU	LEU	LEU	VAL
115	ALA	ALA	ALA	ALA	ALA
116	HIS	HIS	HIS	HIS	ARG

4. Compare the amino-acid sequence of human hemoglobin molecules with that of each of the other four vertebrates. For each vertebrate's sequence, count the number of amino acids that differ from the human sequence and list them in the table below. Be sure to list the animal species in descending order according to their degree of evolutionary closeness to humans.

Hemoglobin Amino-Acid Sequence Similarities Between Humans and Other Vertebrate Species

Species	Number of differences from human hemoglobin
chimpanzee	0
gorilla	1
monkey	2
horse	5

♦ In the study of hemoglobin, which vertebrate is most closely related to humans? Least closely related?

The chimpanzee is the most closely related; the horse is the least closely related.

Analysis

5. What are some methods biologists use to determine evolutionary relationships?

Methods include biochemical comparison, examination of the fossil record, and evaluation of

existing and vestigial structures in modern organisms.

6. Why can it be said that proteins behave like molecular clocks?

Proteins can be said to behave like clocks because they change gradually over time due to

mutations. The number of differences in amino-acid sequences might be considered a measure of

the passage of time.

7. When the portions of the gorilla and human hemoglobin molecules were compared, there was only one difference in the amino-acid sequence. What could have been responsible for this change?

a mutation in the DNA of one species or the other after the two lines diverged

8. If the amino-acid sequences are similar in gorillas and humans, will the nucleotide sequence of their DNA also be similar? Why or why not?

Yes, because the amino-acid sequence of a protein is encoded by the nucleotide sequence of DNA

Conclusions

9. Examine the data table you completed in step 2 of Part 1. The values listed for the chicken and the horse differ by only one. Can you deduce from this that the chicken and the horse are closely related to each other? Why or why not?

No. The data in the table resulted from comparing chickens and horses with humans, not with

each other.

10. How is biochemical comparison different from other methods of determining evolutionary relationships?

Biochemical comparison is based directly or indirectly on the nucleotide sequence of DNA, which

determines the structure and appearance of an organism. Biochemical comparison enables

biologists to collect numerical data that shows the degree of similarity between the DNA of

different organisms. It can also be used to infer the passage of time. Other methods rely on more

indirect comparisons, such as comparisons of structure and appearance, or on determinations of

age based on relative or absolute dating of fossils.

Extensions

11. Do library or on-line research to discover what other types of molecules can be used to determine the evolutionary relationships among organisms based on biochemical comparisons.

12. Do research to find out how biologists determine the amino-acid sequence of a protein molecule.

C20 *Observing Animal Behavior*

Skills
- observing animal behavior
- analyzing animal behavior
- preparing summaries and questions for a research proposal

Objectives
- *Conduct* a preliminary study for a research proposal.
- *Organize* data and research in preparation for writing a grant application.

Materials
- none

Purpose

PREPARATION NOTES

Time Required: 3 class periods

Materials
None required

Additional Background
None

You are an intern at a biological research and consulting firm. In response to growing concern over the extinction of animal species, the company has recently decided to undertake a major behavioral study of an endangered animal. The study will require applying for a grant from one of several national organizations. The company has narrowed down its research topic to one of five endangered animal species: the cheetah, the orangutan, the Indian elephant, the red wolf, and the California condor. Your job is to choose one of these species, complete a preliminary observational study of the animal's behavioral patterns, and make three suggestions about a preliminary study.

Part 1—Conducting an Animal-Behavior Study

Background

Preparation Tips
- You may wish to make this a homework assignment.
- You may wish to extend this project by having students conduct a preliminary observation at a zoo. If classroom animals are

The response of an organism to its environment is called **behavior.** Two categories of behavior are innate and learned. **Innate behavior** is instinctive, or unlearned, such as a dog turning around in a circle before it lays down. A reflex action, like an involuntary blink, is also innate behavior. **Learned behavior** is voluntary and conditioned, such as leaving the classroom when the bell rings.

Behavioral patterns describe sets of behaviors repeated by an organism. These include habitat preference, locomotion, sensory behaviors, protection, activity level, feeding habits, territoriality, communication, and sleeping habits.

Procedure

used, arrange the setting so students will be able to see the animal without obstructions.
- You may wish to assign student groups to a particular animal ahead of time.
- Arrange class time for library research.

1. Choose the endangered animal that you wish to study.

2. Do some background research on the animal you have chosen. Find out about the status of the species and about why it has become endangered. Use the following list of behavioral patterns as a guide for your research. If possible, research each of these topics. Record your findings in the data table that follows.

   a. **Habitat preference** What is the animal's natural habitat? Is the animal territorial? Is it solitary, or does it live in a group?

   b. **Locomotion** How does the animal move?

   c. **Sensory behaviors** What senses does the animal rely on when exploring its environment? Does it have a preference for cold or warmth, light or dark, etc.?

Disposal

PROCEDURAL NOTES

Safety Precautions

• If you chose to have students conduct an actual observation, caution students not to touch or approach any animal in the wild.

Procedural Tips

• If students are to observe animals in the wild, require students to obtain parental permission before going to their observation site.

• Below is a list of wildlife organizations and addresses. Let students choose from this list to complete step 17 in *Extensions*.

1. Center for Marine Conservation
1725 DeSales Street, NW, Suite 500
Washington, DC 20036

2. Defenders of Wildlife
1101 14th St., NW, Suite 1400
Washington, DC 20005

3. Ducks Unlimited
One Waterfowl Way
Memphis, TN 38120-2351

4. Friends of the Earth
1025 Vermont Ave., NW, Suite 300
Washington, DC 20005

5. National Audubon Society
700 Broadway
New York, NY 10003

6. National Fish and Wildlife Foundation
1120 Connecticut Ave., NW, Suite 900
Washington, DC 20036

7. National Wildlife Federation
1400 16th Street, NW
Washington, DC 20036

d. Protection How does the animal protect itself from danger? What special adaptations does the animal have for protection?

e. Level of activity How active is the animal? How often does it sleep? (If observed in a zoo, the amount of activity may be misleading.)

f. Feeding habits What does the animal eat? How often does it eat? If the animals eat in a group, is there a hierarchy for the order in which the animals eat?

g. Communication What noises does the animal make, and when does it make them?

h. Sleeping habits Is the animal nocturnal or diurnal?

Table 1: Animal Behavior

Behavior	Description
Habitat preference	
Locomotion	
Sensory behaviors	
Protection	
Level of activity	
Feeding habits	
Communication	
Sleeping habits	

8. Natural Parks and Conservation Association
1776 Massachusetts Avenue, NW, #200 Washington, DC 20036

9. Sierra Club Foundation
220 Sansome Street, Suite 1100 San Francisco, CA 94104

10. Student Conservation Association
PO Box 550 Charlestown, NH 03603

11. The Nature Conservatory
1815 North Lynn Street Arlington, VA 22209

12. The Wilderness Society
900 17th Street, NW Washington, DC 20006

13. Wildlife Conservation Society
185th Street and Southern Boulevard Bronx, NY 10460-1099

14. World Wildlife Fund
1250 24th Street, NW Washington, DC 20037

3. Review the animal behavioral patterns described on pages 93 and 94.

◆ Which of these behavioral patterns seem most relevant to the endangered status of the species?

Answers will vary, but students should make educated guesses based on which behaviors are most

vital to the species and which behaviors are most easily influenced by human activity.

4. Complete a preliminary observational study of the animal you have chosen. Observe the animal for one hour. Every two minutes, note the behavior (or most pronounced behavior if the animal is exhibiting more than one) the animal is engaged in. At the end of the hour, you should have a list of 30 observations of behavior, some of which repeat. Record the behaviors and the number of times they were exhibited in the data table below. Then calculate the percentage of total behaviors represented by each type of behavior, using the following formula.

$$\frac{\text{Number of times behavior is exhibited}}{\text{Total number of behaviors}} \times 100 = \text{Percentage of total behaviors}$$

Table 2: Behaviors

Behavior	Number of times exhibited	Percentage of total behaviors
eating	5	16.7%
sleeping	11	36.7%
vocalizing	1	3.3%
stretching	3	10.0%
grooming	4	13.3%
resting	6	20.0%

Part 2—Beginning the Grant-Application Process

Background

A **grant application** is both a research proposal and a request for funding. The application normally addresses a specific problem and makes a case for its relevance. The grant application generally includes several sections. A **cover sheet** lists the research project title, the organization requesting the grant, the amount requested, and an address. An **abstract** defines the objective and the hypothesis, and summarizes the need for research. A **description** outlines research that has already been done and explains how the proposed research would be conducted. An **evaluation** section explains how the success of the research will be evaluated. A **budget** lists the projected expenses of the project. Finally, a staff **vitae** gives a brief background about the people working on the project. When completed, the grant application is sent to an organization, which will evaluate the project's merits and choose to either fund or reject it.

5. Prepare to write a grant application by identifying three possible research topics.

♦ Based on your research, what are three questions about animal behavior that may be relevant to the endangered status of the species you chose to study?

Answers will vary. Students' questions may focus on animal behaviors that could easily be

influenced by human activity.

6. Restate each of these questions in terms of a research objective (a statement that summarizes what you intend to find out or study).

Answers will vary.

7. For each objective, write a hypothesis that specifically addresses the problem.

Answers will vary, but the hypothesis should be phrased as an expectation that can be disproven or

supported.

8. Decide which problem is the best one to pursue. Write the abstract for a grant application related to this problem.

Answers will vary. The abstract should give a background of the animal and its endangered status,

raise a question about a behavioral pattern, and clearly state an objective for study. Students

should provide arguments about the relevance of the study, using the preliminary observations and

background research to support these arguments.

_____

_____

_____

_____

_____

_____

_____

_____

_____

_____

_____

_____

Analysis

9. What animal did you observe? Name three behaviors that the animal exhibits frequently, and give the percentage of total observed behaviors that each behavior accounts for.

Answers will vary.

_____

_____

10. What type of behaviors did you suggest for further study and why?

Answers will vary. Students may base their suggestions on the frequency of a behavior or on its

importance to the survival of the species.

_____

_____

11. Characterize the behaviors chosen for further study as innate or learned, and explain your reasoning.

Answers will vary, but students should be able to support their answers. Some behaviors may be

difficult to classify.

12. Why is it necessary to conduct background research before writing a research proposal?

Background research provides information that allows researchers to focus their research and to

develop appropriate and effective procedures. Background research also helps prevent mistakes

that could undermine the study, waste time and materials, or harm the environment.

Conclusions

13. Based on your preliminary research and observational study, explain how your proposed study is relevant to the endangered status of the species.

Answers will vary. Students may explain, for example, how human activity has interfered with

animal behavior.

14. List two reasons why the study of animal behavior is important.

Answers will vary. If behavioral traits are known, scientists can use the information to develop

techniques to protect and increase the populations of endangered species and to rehabilitate sick

or injured animals.

15. What would make your proposal successful? Why?

Answers will vary, but students should address such components as the relevance and feasibility of

the study and the thoroughness of their background research.

Extensions

16. *Ethologists* play an important role in wildlife management. They are specialists who study animal behavior and how behavioral patterns develop in a given habitat. They study all aspects of animal behavior and work closely with the animals they study. Find out about the training and skills needed to become an ethologist and what opportunities exist in the field of wildlife management.

17. Write to a wildlife organization. In your letter, request information about the goals of the organization and how the goals are achieved. Make a classroom display that outlines this information.

Name _____

Date _____ Class _____

HOLT
BioSources
LAB PROGRAM
EXPERIMENTAL DESIGN

C21 Observing Animal Behavior—Grant Application

Prerequisites • **Laboratory Techniques C20: Observing Animal Behavior** on pages 93–98

Review • innate and learned behavior
• types of behavioral patterns
• components of a grant application

BioLogical Resources, Inc. Oakwood, MO 65432-1101

M E M O R A N D U M

To: Lee Kwan, Director of Macrobiology
From: Sandra Berkham, President *SB*

You are well aware that we will soon be launching into a behavioral study on an endangered species. As we have discussed previously, it has been my goal to get our student interns involved in the planning and execution of a major study. I know how busy your research teams are right now, and I felt that much of the preliminary research could be handled by interns.

At my request, each student intern has done some background research and preliminary observations of an endangered species of his or her choice. The interns have also completed an abstract describing a question or problem they would like to see addressed in a more extensive study. Please have the appropriate team or teams evaluate these abstracts and use them to design a research project.

I recently received a listing of several grants that are available for the research of animal behavior. I would like each research team in your department to apply for one of these grants. You will need to advise your team members about the procedures they should follow and their choice of grants. Before that happens, I would like to meet with you to discuss this subject further. At that time, I will give you the list of available grants and the application requirements.

Please call to let me know when you are available to meet with me.

BioLogical Resources, Inc. Oakwood, MO 65432-1101

M E M O R A N D U M

To: Team Leader, Behavioral Studies Dept.

From: Lee Kwan, Director of Macrobioloby

Attached is a copy of a memo I received from the president of the company. I have discussed the project with her, and we are both very excited about this new direction for our company.

Please prepare a grant application using the information provided to you by your intern. Prior to writing the grant application, you will need to evaluate the abstracts presented to you by the interns and design an experiment to address one of the proposed questions. I have included a list of available grants given to me by Ms. Berkham. Please consider your intern's recommendations, and choose the grant for which you will apply.

Also attached is the Grant Application Requirements form, which should help you organize your proposal. Use the intern's abstract in your application. You may adapt it if you need to. When completing the vitae, please include the intern's educational background with your own, since he or she will be working with you on this project if it is funded.

You are limited to one grant application, so choose wisely. Make sure that it fits your grant topic and financial needs. The available grants are listed below:

Aquatic Life — $500,000.00
This group grants money for studies on all types of aquatic organisms.

United States Department of the Interior—$1,000,000.00
This group grants money for the study of animals within the United States.

Avian Society of America—$50,000.00
This group provides money for the study of birds.

USA Parks and Recreation—$100,000.00
This grant is given for the advanced study of plants and animals within parks in cities, counties and states.

National Science Foundation—$2,000,000.00
This group grants money for almost any type of scientific study.

Wildlife America—$200,000.00
This group allots grant money for the study of wildlife both in captivity and in natural environments.

United Zoos of America Fund—$150,000.00
This group grants money for the study of animals in zoos, marine parks, and wild-animal parks.

Proposal Checklist

Before you start your work, you must submit a proposal for my approval. **Your proposal must include the following:**

_____ • the **question** you seek to answer in the form of an abstract

_____ • the **procedure** you will follow to complete your grant application

Proposal Approval: _____

(Supervisor's signature)

Report Procedures

When you finish your grant application, prepare a report in the form of a memo to Ms. Berkham, and attach it to the complete grant application. **Your report must include the following:**

_____ • a paragraph describing your **evaluation** of the abstract presented by the intern

_____ • a complete **grant application**

_____ • your **predictions** about the success of the study

_____ • a detailed **invoice** showing all labor (at $40.00 per hour) and the total cost (for tax records)

Grant Application Requirements

I. Grant Cover Sheet

The grant cover sheet should include the name of the grant you are applying for and the dollar amount of the grant. Each person involved in the grant should complete a personal data summary similar to the one below:

Name _____

Affiliation _____

Work Address _____

Home Address _____

Signature _____

II. Proposal Abstract

Briefly summarize the research question and the reason there is a need for this question to be answered. Include a summary of your preliminary research and conclusions. This should not exceed 100 words.

III. Description of Procedure

Write a brief description of the goals of your grant research and the activities you will be doing to answer the research question. This should not exceed 250 words.

IV. Evaluation

Write a brief description of how you will evaluate the success of your research. In other words, specify the measures in your study, such as data analysis and control methods, that ensure the production of accurate results. This should not exceed 100 words.

V. Budget

Prepare a listing of estimated expenses for your proposed project. Be sure to list all equipment, supplies, and labor. This should not exceed 1 page.

VI. Project Staff Vitae

Limit this to one page per group member. The following information should be included in each vitae: name, address, education, work experience (including field experience), publications, other grants received, awards received, and any other pertinent information.

HOLT
BioSources
LAB PROGRAM
LABORATORY TECHNIQUES

C22 *Examining Owl Pellets*

Skills

- observing the skeletal structure of organisms
- using a dichotomous key
- dissecting an owl pellet and separating its contents

PREPARATION NOTES

Time Required: one 50-minute period. To do a thorough job, however, this activity may require two 50-minute periods, one to dissect the pellets and another to identify and analyze the findings.

Objectives

- *Dissect* an owl pellet.
- *Identify* the species eaten by an owl from the remains found in an owl pellet.
- *Construct* a food web containing an owl from a list of species found in owl pellets.

Materials

- gloves
- lab apron
- northwestern barn-owl pellet
- dissecting tray
- dissecting needle
- forceps
- white paper
- metric ruler
- tape or glue

Purpose

Materials

Materials for this lab activity can be purchased from WARD'S. See the *Master Materials List* for ordering instructions.

You are a field biologist for the state division of wildlife. While doing a study to determine the cause of a recent decrease in the vole population, you also noticed a decrease in the population of barn owls. You determine that the vole population is down because of the recent drought conditions in the area. You suspect that the decrease in owl population might be due to a decrease in the owl's food supply—namely voles. Your job is to dissect owl pellets collected from the area to determine whether the owls are eating voles.

Background

Additional Background

The pellets used in this investigation were produced by barn owls *(Tyto alba).* Barn owl pellets have been chosen because this owl feeds primarily on small mammals and birds, which are swallowed whole. The heads of long-billed birds and large rats are often removed, but most pellets contain the entire skeletons of birds and small rodents on which the owls have fed.

Owls are **raptors,** or birds of prey. They catch their prey, including small birds and rodents, and swallow them whole. Enzymatic juices in the owl's digestive system break down the body tissues of the prey but leave the bony materials, hair, and feathers undigested. Depending on the prey eaten, the undigested portions may include beaks, claws, scales, or insect exoskeletons. This type of material has little nutritional value and must be eliminated from the owl's body.

Since owls do not have teeth for grinding and cannot pass whole bones and claws through their digestive tract safely, these indigestible materials form a **bolus,** or lump, called a **pellet.** The pellet is composed of fur, feathers, bones, and other undigested parts of the consumed prey. Pellets begin forming within the digestive tract of the owl as soon as the prey is swallowed. The pellets are then coughed up, or regurgitated, and the owl begins feeding once more.

Scientists take advantage of this adaptation by collecting these pellets. Owl pellets are dried and either fumigated (treated with chemicals) or sterilized so that their contents can be examined safely. Since owls are not selective feeders, the pellets can be used to estimate the diversity of available prey. The contents are also a direct indicator of what an owl has fed on—information that is crucial for species management and protection.

There are many genera of prey that occur in the northwest. The Analysis of Owl Prey data table contains nine mammalian prey types that account for 96–100% of the prey your students will find. Any other prey will be composed of birds, bats, insects, and small reptiles. These are occasional and too diverse to address in detail.

The contents of an owl pellet can be identified with a **dichotomous key,** a tool used to identify an object or an organism. A dichotomous key has a series of statements or questions that compare contrasting characteristics among a group of items or organisms. For example, assume that you want to identify a common U.S. coin using a dichotomous key. The key might read as shown below. Compare the first pair of statements and determine which one best fits the coin you are trying to identify. After you pick one of the paired statements, you will be directed to another paired statement until you reach an answer.

Dichotomous Key of Coins

1.	coin edge smooth	go to 2
	coin edge grooved	go to 3
2.	coin copper in color	penny
	coin silver in color	nickel
3.	picture of Roosevelt on front	dime
	picture of Washington on front	quarter

Procedure

Preparation Tip

Owl pellets may be found in roosting and feeding areas, old farm buildings, woodlands, and parks. It is easier to buy pellets that are already dried and sterilized or fumigated. Purchased pellets are accompanied by information and identification guides. If you collect your own pellets, dry them and fumigate them in polyethylene bags with naphthalene to destroy insect eggs. Soaking the pellets in water for about two hours before the lab begins softens the mucilage and makes dissection easier.

Part 1—Dissecting an Owl Pellet

1. Put on gloves and a lab apron. **CAUTION: Do not touch your hands to your face or mouth during this investigation.** Place an owl pellet in a dissecting tray. Remove the pellet from the aluminum foil casing. Use a dissecting needle and forceps to carefully break apart the owl pellet. Remove the fur and feathers from the bones. *Note: Be careful to avoid damaging small bones while you are pulling the pellet apart.* As the bones are uncovered, use forceps to carefully place them on a sheet of paper. Take care to remove all skulls and bones from the fur mass. You will identify the animals the owl has eaten mainly by the skulls, mandibles (jaws), and teeth, so be especially careful when dissecting the pellet. Use the following diagrams of the bird and mammal skeletons to help you distinguish among types of bones.

Sometimes, undigested beetles and pill bugs are found in owl pellets. They are small animals that find the expelled pellets and use them as a food source and nursery for their eggs and larvae. Therefore, these organisms should not be included as owl prey.

♦ In what way might the formation of owl pellets increase an owl's chances of survival in an ecosystem?

Owl-pellet formation allows owls to be less selective of the kinds of prey they can feed on. Also,

owls can ingest prey whole. This ability speeds up feeding and decreases exposure to other

predators and scavengers.

**PROCEDURAL
NOTES**

Safety Precautions
• Discuss all safety sym-
bols and caution state-
ments with students.
• Remind students to
wash their hands after
handling owl pellets.

Procedural Tips
• Reassure students
that, while it may
sound unpleasant to
dissect something owls
cough up, it is really
nothing more than
examining animals'
bones and tissues
"cleaned" by stomach
enzymes.
• Pass out owl pellets
to the students for dis-
section. Students must
first remove the pellets
from the aluminum
foil. Next, have stu-
dents label a clean
sheet of paper with
their name. Then have
them use a dissecting
needle to loosen the
hair of the owl pellet.
As bones are uncov-
ered, they should be
carefully removed and
placed on the sheet of
paper. If you prefer,
students can label a
small sheet of paper to
hold the bones of each
prey item that they
extract. After all bones
have been removed,
identification of the
skulls can take place.

Generalized Bird Skeleton   **Generalized Mammal Skeleton**

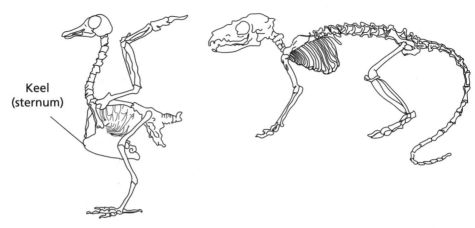

Keel
(sternum)

2. Assemble similar bone parts to see how many prey types are represented
   in the pellet. Count the number of skulls to determine how many prey were
   in the pellet. Refer to the diagrams of the skulls shown below to aid you in
   identifying types of skulls.

**Skull
Comparisons**

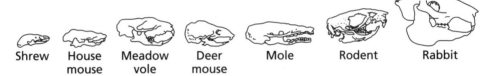

Shrew   House   Meadow   Deer   Mole   Rodent   Rabbit
         mouse   vole   mouse

◆ What are some ways you can predict which species of animals you might
   find in an owl pellet?

Answers will vary. Students can base their predictions on the other nocturnal organisms found in

the owl's ecosystem.

◆ How can you distinguish between vertebrate and invertebrate material?
   between birds and mammals?

Vertebrate and invertebrate material can be distinguished by the presence of vertebrae and chitin,

respectively. Birds and mammals can be distinguished by their skeletons and the presence of fur or

feathers.

3. Reassemble the skeletons by laying them out on a piece of paper with
   appendages arranged and spread out to the side. Glue or tape the assembled
   skeletons to the paper.

• Egg cartons or petri dishes may be helpful storage containers when students are separating their bones into like piles.

• Suggest that students first sort bones by shape. Then have them sort the bones by size after the relationship among different kinds of bones has been determined. When students identify prey by size, be sure to tell them that some animals may be immature, so size can be deceiving.

• Another method of identifying raptor prey is through comparison to materials that have already been identified. Most biologists keep a set of identified skulls along with hair and feather samples. These can be very useful when identifying prey remains. This process can speed up the identification of large numbers of similar items by eliminating the need for a key once all of the common items have been identified.

• Students can compare the skeletons of the prey with the skeletons of humans.

♦ How many skeletons were you able to assemble from your owl pellet?

Answers will vary.

_____

_____

Part 2—Using a Dichotomous Key to Identify Prey

4. Use the diagrams below of the side and top views of a skull to become familiar with terminology used in a dichotomous key of small mammal skulls. A gap between the incisors (front teeth) and cheek teeth (molars and premolars) is called a *diastema*. The *infraorbital canal* refers to an opening just below the eye socket. The *zygomatic arch* is a ridge of bone located on the side of the skull making up part of the cheekbone.

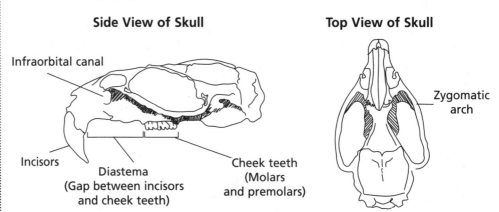

Side View of Skull

Infraorbital canal

Incisors

Diastema
(Gap between incisors
and cheek teeth)

Cheek teeth
(Molars
and premolars)

Top View of Skull

Zygomatic
arch

5. Use the diagrams below to aid you in determining the length of the mandible, or lower jaw. The bottom view of the skull is actually a view of the roof of the mouth. When determining the posterior edge of the palate, look for the position of the *posterior palatine foramina*. The palate ends beyond the posterior palatine foramina. The skull on the left shows the posterior edge of the palate ending even with the last cheek teeth. The right-hand skull shows the palate ending beyond the last cheek teeth.

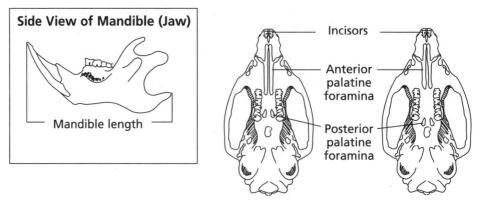

Side View of Mandible (Jaw)

Mandible length

Bottom View of Skulls

Incisors

Anterior
palatine
foramina

Posterior
palatine
foramina

Anterior and posterior palatine foramina location will
vary from species to species. Use these as examples only.

- Students should construct a food web in the space provided. The food web should contain a barn owl at the highest trophic level and grass and seeds at the lowest. The intermediate organisms should include the prey found in your students' owl pellet investigations. You may wish to include other potential prey as suggested by students. You may also wish to do this activity as a class on an overhead transparency or a chalkboard.

- Class input should be used to fill in the food web. Everything listed in the Analysis of Owl Prey data table is eaten by barn owls. Most prey listed will eat plant material. Birds, bats, shrews, and some rats will feed on insects.

6. To determine cheek teeth types or incisor types, refer to the diagrams below.

Cheek Teeth Types

| Acute (bottom view) | Acute (side view) | Unicusp (bottom view) | Unicusp (side view) | Lobed s-shaped (bottom view) | Lobed s-shaped (side view) |

Incisor Types

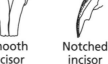

| Grooved incisor (front view) | Smooth incisor (front view) | Notched incisor (front view) | Unnotched incisor (front view) |

7. Use the dichotomous key below and the diagrams above and on the preceding page to identify the skulls of small mammals found in your owl pellet. Record the number of each type of skull found in the Analysis of Owl Prey data table on the next page.

Dichotomous Key

Skulls of Small Mammals Found in Northwestern Barn-owl Pellets

No gap (diastema) between incisors and cheek teeth. . . . . Order: Insectivora
Gap (diastema) between incisors and cheek teeth. . . . . Order: Rodentia

Order Insectivora (moles and shrews)

Zygomatic arch complete; skull flat and broad . . . . *Scapanus* (mole)
Zygomatic arch not complete; skull not flat and broad . . . . *Sorex* (shrew)

Order Rodentia (rats, voles, and mice)

1. Infraorbital canal not present . . . . Go to 2
   Infraorbital canal present. . . . . Go to 3
2. Upper incisors distinctly grooved . . . . *Perognathus* (mouse)
   Upper incisors not distinctly grooved . . . . *Thomomys* (pocket gopher)
3. Skull flat and broad; cheek teeth acutely angled and
   may appear as one continuous tooth. . . . . *Microtus* (vole)
   Skull generally rounded; cheek teeth lobed or rounded
   and easily distinguished individually. . . . . Go to 4
4. Upper incisors distinctly grooved. . . . . *Reithrodontomys* (harvest mouse)
   Upper incisors not distinctly grooved . . . . Go to 5
5. Posterior edge of palate ending even with or
   only slightly beyond last cheek teeth; cheek
   teeth not capped with enamel . . . . *Peromyscus* (deer mouse)
   Posterior edge of palate ending beyond last cheek teeth;
   cheek teeth capped with enamel . . . . Go to 6
6. Upper incisors notched, mandible length less than 16 mm . . . *Mus* (house mouse)
   Upper incisors not notched, mandible length greater than 18 mm. . . . *Rattus* (rat)

Analysis of Owl Prey

Prey	Occurrence	Number found in sample	Prey biomass (in grams)	Total number in all samples	Total biomass
Pocket Gopher *Thomomys*	+++		150 g		
Rat *Rattus*	+		150 g		
Vole *Microtus*	+		40 g		
Mice *Peromyscus* *Mus* *Reithrodontomys* *Perognathus*	++ +++ ++ ++	Answers will vary.	40 g 18 g 12 g 25 g	Answers will vary.	Answers will vary.
Mole *Scapanus*	+		55 g		
Shrew *Sorex*	++		4 g		
Other Prey bats birds insects	++ ++ ++		7 g 15 g 1 g		
Cumulative total biomass					

Key: +++ = common, ++ = occasional, + = rare

Part 3—Analyzing the Diet of a Barn Owl

8. As you have seen from the dichotomous key on the previous page, there are many genera of prey that occur in the northwestern region of the United States. Any other animal remains you find in your pellet will be from a bird, bat, or insect. List them as "other prey" in the data table above.

   ◆ How many different species of animals did you find in the owl pellet?

   Answers will vary.

9. Your teacher will gather class totals for each type of prey. Record these in the *Total number in all samples* column in your data table.

10. To calculate the total biomass, multiply the numbers in the *Total number in all samples* column by the numbers listed in the *Prey biomass* column. For example:

    If five pocket gophers *(Thomomys)* were recorded as the *Total number in all samples,* and from the chart the prey mass is 150 g, then

    $$5 \times 150 \text{ g} = 750 \text{ g}$$

Compute totals in this way for all species found, and enter your answers in the column headed *Total biomass*. Calculate *Cumulative total biomass* by summing all data in the *Total biomass* column.

11. It is now possible to get a good idea what each prey species contributes to the diet of the northwestern barn owl population that the owl pellets represent. To calculate percentage (in biomass) of owl diet, divide each number in the *Total biomass* column by the *Cumulative total biomass* and multiply your answer by 100 to give a percentage. For example, assuming a cumulative total biomass of 5,000 g and using the pocket gopher example:

$$\frac{750 \text{ g}}{5,000 \text{ g}} \times 100 = 15\%$$

You can conclude that the pocket gopher contributes, on the average, about 15 percent of the biomass of a northwestern barn owl's diet.

◆ What was the total biomass from all species found in your class's owl pellets?

Answers will vary.

Part 4—Constructing a Food Web

12. In the space below, construct a food web with the owl at the highest trophic level. *Note: Be sure to include producers (green plants) and decomposers in your food web.* The intermediate organisms should include the prey found in your owl pellets in class.

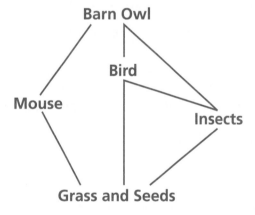

◆ How many different trophic levels are represented in your food web?

Answers will vary. Students should include three or four trophic levels in their drawings.

13. Dispose of your materials according to the directions from your teacher.

14. Clean up your work area and wash your hands before leaving the lab.

Analysis

15. If an owl needs 120 g of food per day, how many *Sorex* would it need to capture? How many *Microtus*?

30 *Sorex* or 3 *Microtus*

16. Were the remains of any voles found in the owl pellets?

Answers will vary.

Conclusions

17. How would a sudden decrease in the shrew population affect the barn-owl population?

If owls could no longer capture shrews, they would not be seriously affected. Shrews occasionally

occur in large numbers in raptor pellets but seldom contribute significantly to a raptor's diet

because of their low biomass.

18. How would a sudden decrease in the vole population affect the barn-owl population?

A crash in voles might have a detrimental effect on barn owls. Voles usually account for the

majority of barn-owl prey in the Northwest both in numbers and in biomass.

19. Assume an owl eats 100 1-gram insects and 1 100-gram rat. Did the insects or rat contribute more to the owl's diet? How does foraging time affect this outcome?

The rat contributed more because the owl used less time and energy to capture the same amount

of biomass.

20. Is quantity or quality of prey more important?

Answers will vary. An abundant supply of a low-quality prey may be as good as high-quality prey

that is scarce.

Extensions

21. Predators such as owls are subjected to poisons that have been concentrated through the food chain. Some species of owls are considered endangered, threatened by pesticides and heavy metals that contaminate the foods eaten by their prey. Do library research, and write a report on this topic. Explain how a grain crop contaminated with a compound containing lead can affect the owl population.

22. *Wildlife biologists* study and help manage wildlife populations in state and federal parks and preserves. They conduct research on predator-prey relationships, interactions among species, and effects of pollution and human intervention on population changes. Find out about the training and skills required to become a wildlife biologist.

Name _____

Date _____ Class _____

HOLT
BIOSOURCES
LAB PROGRAM
EXPERIMENTAL DESIGN

C23 | *Examining Owl Pellets—NW vs. SE*

Prerequisites · **Laboratory Techniques C22: Examining Owl Pellets** on pages 103–110
Review · procedure for dissecting owl pellets
· food webs
· predator/prey relationships

 National Park Service

Washington, D.C.

October 14, 1997

Rosalinda Gonzales
Environmental Studies Division
BioLogical Resources, Inc.
101 Jonas Salk Dr.
Oakwood, MO 65432-1101

Dear Ms. Gonzales,

In the last few years, the effects of increased tourism in several national parks have caused us some concern. Because rats eat the food left at campsites, an increase in tourism can lead to an increase in the rat population. We are afraid that a dramatic increase in the rat population would have a negative effect on park ecosystems. We have considered poisoning some of the rats, but we are concerned that the poison might be harmful to the barn owl population.

The barn owl preys on rats and is prevalent in many of our national parks. We are hoping that the barn owl population will control the rat population naturally, but we need more information before we decide how to approach the situation.

I want to hire your firm to find out how much of the owl's diet is composed of rats. We will send you samples of owl pellets from each of two parks. Please dissect them, analyze the contents, and let us know what you find.

Sincerely,

Phillip Chaney

Phillip Chaney
Director of Biological Research
National Park Service

BioLogical Resources, Inc. Oakwood, MO 65432-1101

M E M O R A N D U M

To: Team Leader, Ecology Dept.

From: Rosalinda Gonzales, Director of Environmental Studies

I am sure that I do not need to tell you that a contract with the National Park Service is a terrific opportunity. I am confident that your department can handle this task.

I called Mr. Chaney to clarify some points in his letter. He told me that the pellets he has sent came from a park in the Northwest and a park in the Southeast. You will be dissecting and examining the pellets of the Northwestern and Southeastern barn owls.

Please dissect these pellets and examine the contents. You will be provided with a prey identification key and illustrations. Identify the different types of prey, and record the number of each type found in each pellet. Upon completing your examination, pool the data from all dissections, and determine the percentage of biomass that rats occupy in the owl's diet. Finally, compare the diets of the Northwestern and Southeastern barn owls.

Proposal Checklist

Before you start your work, you must submit a proposal for my approval. **Your proposal must include the following:**

_____ • the **question** you seek to answer

_____ • the **procedure** you will use

_____ • a detailed **data table** for recording observations and calculations

_____ • a complete, itemized list of proposed **materials** and **costs** (including use of facilities, labor, and amounts needed)

Proposal Approval: _____

(Supervisor's signature)

Report Procedures

When you finish your analysis, prepare a report in the form of a business letter to Mr. Chaney **Your report must include the following:**

_____ • a paragraph describing the **procedure** you followed to examine, sort, identify, and analyze the contents of the Northwestern and Southwestern barn owl pellets

_____ • a complete class **data table** showing the number of specimen found, prey weight, prey biomass, and percentage of total biomass for each species of prey

_____ • your **conclusions** about the differences in the diets of the Northwestern and Southeastern barn owl

_____ • a detailed **invoice** showing all materials, labor, and the total amount due

Safety Precautions

• ◇ ◆ Wear disposable gloves and a lab apron.

• ◆ Handle sharp objects carefully.

• ◆ Wash your hands before leaving the laboratory.

Disposal Methods

• ◆ Dispose of waste materials according to instructions from your teacher.

• Place dissected owl pellets in a trash can.

• Place mammal skeletons and aluminum foil wrappers in the separate containers provided.

• Clean your desk with disinfectant before leaving the laboratory.

• Wash reusable materials such as glassware and lab utensils, and return them to the supply area.

FILE: National Park Service

MATERIALS AND COSTS (Select only what you will need. No refunds.)

I. Facilities and Equipment Use

Item	Rate	Number	Total
facilities	$480.00/day	_____	_____
personal protective equipment	$10.00/day	_____	_____
compound microscope	$30.00/day	_____	_____
microscope slide with coverslip	$2.00/day	_____	_____
scissors	$1.00/day	_____	_____
dissecting tray	$5.00/day	_____	_____
forceps	$5.00/day	_____	_____
dissecting needle	$3.00/day	_____	_____
light source	$15.00/day	_____	_____

II. Labor and Consumables

Item	Rate	Number	Total
labor	$40.00/hour	_____	_____
Northwestern barn owl pellet	provided	_____	_____
Southeastern barn owl pellet	provided	_____	_____
dichotomous key	$10.00 each	_____	_____
skull characteristics sheet	$10.00 each	_____	_____
paper	$0.10/sheet	_____	_____

Fines

OSHA safety violation	$2,000.00/incident	_____	_____

Subtotal	_____
Profit Margin	_____
Total Amount Due	_____

HOLT
BIOSOURCES
LAB PROGRAM
LABORATORY TECHNIQUES

c24 *Mapping Biotic Factors in the Environment*

Skills
• counting populations of organisms in the field

Objectives
• *Study* populations using the quadrat sampling method.
• *Identify* the dominant plant type in an area.
• *Estimate* tree height.
• *Construct,* on paper, a nature walkway.

PREPARATION NOTES
Time Required: 2 class periods

Materials

Materials
Materials for this lab activity can be purchased from WARD'S. See the *Master Materials List* for ordering instructions.

• wooden stakes (4)
• 15 m piece of string
• meter stick
• hammer
• field guides to plants and animals
• paper

• pencil
• protractor with string
• clipboard
• calculator
• 18 × 24 in. piece of construction paper
• large sheet of white paper

Purpose

You are a landscape architect who has been commissioned to design a natural walkway on undeveloped land next to a school. Your goal is to develop a walkway that will enable visitors to view the native plants and animals that inhabit the area without disturbing the environment. To help accomplish your goal, you need to complete a population study of all the organisms that inhabit this area. With that knowledge, you can begin construction on the nature walkway.

Background

Preparation Tips
• For this activity, select a vegetated area that is representative of the natural vegetation in your area.
• The 2 m × 5 m sample quadrats should be a convenient size for most areas, but you may wish to decrease the quadrat size where the vegetation is particularly heavy.

An **ecosystem** consists of all of the living and nonliving things in an area. **Biotic factors,** or the living things in the ecosystem, include plants, animals, and microorganisms. Together, the many different types of living things in an area make up a **community.** All of the members of a single species within a community make up a **population.** For example, in a forest community, there may be a population of oak trees, a population of sparrows, and a population of ferns. Ecologists refer to all of the plants in an area as a **stand.**

To best describe the biotic factors in an ecosystem, you should study each organism. Sometimes an area is too large, or the organisms are too numerous, to study each organism within the area. In such cases, the area is divided into **quadrats,** which are smaller areas that represent samples of the entire area. All the individuals in a quadrat are studied (counted, measured, massed, etc.). Quadrats are randomly designated. This is called the **random quadrats method of sampling.** The data collected from each quadrat is used to draw conclusions about the community of organisms in the entire area. For example, if five 10 m² quadrats surveyed had an average of 16 dandelions and 2 grasshoppers per quadrat, then you could estimate that 160 dandelions and 20 grasshoppers live in a 100 m² area. The population of each quadrat must be known, the area of each quadrat must be calculated, and the sample quadrats must be representative of the entire area for the random quadrats method of sampling to be accurate.

Procedure

• This lab is designed for groups of 4 to 8 students. It is not necessary to have five quadrats. Use any number that is convenient for your class size, but remember that too few quadrats may bias your data averages. The use of multiple quadrats will illustrate how sample areas can differ within a stand and how data can be collected differently by different investigators.

• Wooden stakes can be obtained at a hardware store or a lumber yard. Be sure the stakes are sturdy enough to be pounded into the ground.

• *The Peterson Field Guide* series, *Audubon Field Guides*, and *Golden Guides* are excellent field guides for identification of trees, shrubs, birds, insects, and mammals. The mammal guides contain pictures of animal tracks.

• Take time to familiarize students with the field guides. Suggest that students look through the table of contents and browse through each guide they will use to see how they are organized.

• Prepare the protractor apparatus using tape and a piece of string approximately 30 cm long. Tape the string to the center of the straight edge on the protractor. Be sure that the string moves freely about its axis.

• Remind students to dress appropriately for fieldwork. Wear comfortable clothing and shoes with closed toes.

1. In the area indicated by your teacher, select a site to analyze. Using a meter stick, four wooden stakes, a hammer, and string, mark off a quadrat 2 m wide and 5 m long. **CAUTION: Be careful not to touch or brush against any poisonous plants. Do not touch or approach any animal in the wild.**

2. Using plant field guides, identify each type of plant in your group's quadrat. Make counts of each of the following types of plants: grasses, forbs (nonwoody plants other than grasses), vines, shrubs, and trees. *NOTE: If there are very large numbers of small plants, such as grasses or forbs, you can estimate the total number for the quadrat by counting plants in a 1 m × 1 m area and multiplying by the area of the quadrat.* Record your data in the data table below.

Individual Data: Plant Types in a Quadrat

Plant type	Total number of plants	Measurements of plant heights	Average height
grasses			
forbs			
vines	Answers	will	vary.
shrubs			
trees			

3. Select a few plants of each plant type, and measure their heights using a meter stick. Then calculate the average height of each plant type. Record your measurements and calculations in the data table above. To estimate tree height, hold a protractor with the curved side downward, as seen in the diagram below. Sight along the straight edge of the protractor to view the top of the tree. With your lab partner watching the protractor, move back and forth until the string is hanging from the protractor at a 45° angle. Add your height and your distance from the tree to obtain the estimated tree height.

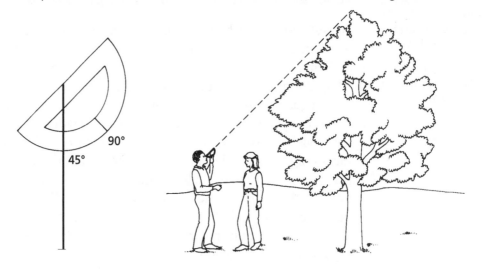

• Provide students with map pencils, markers, or crayons for making their sketches.

PROCEDURAL NOTES

Safety Precautions
• Before going into the field, discuss some of the more dangerous organisms students might find and how they may avoid them. Show them pictures of poison ivy and other plants to avoid.
• This is an outdoor activity. Be sure to communicate with administrators and obtain permission from parents if necessary.

Procedural Tips
• The most important aspect of this activity is to show the students that there are methods by which communities can be studied and described.
• It may take the entire class period to perform the field study.
• Remind students not to pick leaves and plants. If they do, they are altering and destroying part of the ecosystem.
• It may not be necessary to measure the height of each plant in the quadrat. Depending on your area, you can tell students how many to measure to obtain an average. If there are no trees in your area of study, you can estimate the height of a tree that may be elsewhere on your campus.

4. Add the total numbers for each plant type to determine the total number of plants in your quadrat.

◆ How many plants did you count in your quadrat?

Answers will vary.

5. Calculate the percentage of plants in your quadrat that constitute the dominant plant type. *NOTE: The dominant plant type is the most numerous.* To calculate the percentage, use the following formula:

(number of dominant plant type ÷ total number of plants) × 100
= percentage of dominant plant type

For example:

(300 grasses ÷ 500 plants) × 100 = 60% grasses

Record your results in the first row of the table below. Share your data with all lab groups. Add each group's data to the table below.

Percentage of Dominant Plant Types

Group	Dominant plant type	Number of dominant type	Total number of plants	Percentage of dominant plant
		Answers will vary.		

6. Compare the vegetation of your quadrat with that of the stand. Record your observations in the table on the next page.

◆ What is the difference between a stand and a quadrat?

A stand refers to a group of plants growing within a continuous area. A quadrat is a sampling area

within a stand.

• The designs of the walkways and the ideas for signs the students create could be implemented as a class or school project to create an outdoor classroom on campus.

Appearance of Sampled Area

Description	Quadrat	Stand
Dominant plant		
Tallest plants (average height)	Answers will vary.	
Percentage of ground shaded or covered by vegetation		
Percent of ground without vegetation		
Distribution of plants (even/uneven)		

♦ What is the general description of the stand?

Answers will vary.

_____

_____

_____

7. As you study and count the plants in your quadrat, observe the presence of animals. Look for insects that live on plants, flying insects, and birds. Note any bird and mammal tracks and any fecal material. Look for small-mammal burrows. *NOTE: Mobile animals cannot be accurately counted in an ecosystem in a short period of time.* Use animal field guides to help you identify animal tracks. Record your observations of animals in the table on the next page.

♦ What animals or indications of animals did you observe?

Answers will vary.

_____

_____

♦ Why do you think the animals you saw or found traces of actually live in the quadrat?

Answers will vary.

_____

_____

♦ Select an animal species and tell why you think it is dominant in this stand.

Answers will vary.

_____

_____

Animal Descriptions

Animal	Description
	Answers will vary.

8. Return your materials according to the directions from your teacher.

9. Leave the stand showing little impact from your visit. Upon returning, wash your hands before leaving the lab.

10. Work as a class to write a description of the plant community studied. Write the description below.

Answers will vary.

Analysis 11. List the dominant plant type and its percentage in decreasing order for each quadrat studied. Choose the dominant plant for the stand, and support your selection.

Answers will vary. The individual plant types will vary within any given area.

12. Environmental factors that stabilize a population's size and keep species from reaching their biotic potential are called limiting factors. What limiting factors could exist to limit the growth rate of one of the plants or animals in your quadrat?

Limiting factors include the availability of space, food, and nesting materials. Soil nutrients could

influence a plant's ability to grow. The availability of plants would influence the insect population.

Conclusions

13. Based on your observations, would you describe the stand as well established? Cite evidence for your answer.

Answers will vary, but students should use information from their data tables to help them answer

the question.

Sketches will vary.

14. Make a sketch of your quadrat on a piece of construction paper, using the symbols below to represent plant types. Represent plant numbers by using estimates. For example, if you counted 300 grasses, and 2 trees, sketch in most of your quadrat with grass symbols, and sketch in 2 tree symbols. Write an average height on one of each of the plant type symbols. Design and draw a walkway throughout your quadrat to view the plants and animals.

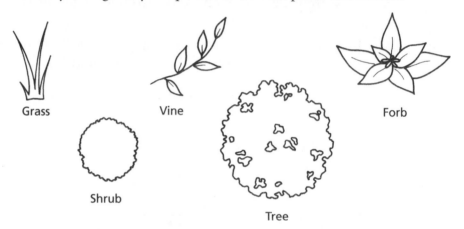

Grass   Vine   Forb

Shrub

Tree

15. Obtain a large sheet of white paper to represent the area of the stand. Transfer your quadrat sketch to its approximate location on the paper. As a class, sketch in plant types in the blank areas to represent the stand not sampled. Connect your walkway throughout the stand. Create three sayings that could be painted on signs to help visitors notice specific vegetation or animal life as they tour the stand.

Answers will vary. Some sample signs might be: "Cardinal Corner," "Pine Tree Grove," and "Native

Vegetation."

Extensions

• Do not allow students to conduct experiments they design until you have approved their procedures.

16. *Landscape architects* design and plan outdoor areas with respect to the needs of the humans and other organisms in the environment. They study the site, mapping features such as plants, trees, and slope of the land. Landscape architects work with many people, such as engineers, real estate agents, and zoning experts, when working on a project. Find out the training and skills needed to become a landscape architect.

17. Design an experiment to find out how space availability, temperature, and soil nutrients affect a plant population.

HOLT

BIOSOURCES
LAB PROGRAM
LABORATORY TECHNIQUES

C25 Assessing Abiotic Factors in the Environment

Skills

- measuring wind speed and temperature
- calculating moisture content in soil
- testing soil for pH, nitrates, phosphates, and potassium

Objectives

- *Identify* the abiotic factors in an environment.
- *Determine* the best location to place a flower garden.

Materials

PREPARATION NOTES

Time Required:
2 class periods

Materials
Materials for this lab activity can be purchased from WARD's. See the *Master Materials List* for ordering instructions.

- meter stick
- anemometer
- thermometer
- stakes (9)
- string (32 m)
- soil thermometer
- garden trowel
- 1 gallon sealable plastic bag
- metal can with both ends removed

- filter paper
- rubber band
- balance
- permanent marker
- hot pad
- soil-test kit
- safety goggles
- gloves
- lab apron

Purpose

You are the grounds maintenance supervisor for your local school district. The biology club at one of the schools in the district wants to plant a perennial flower garden on their campus. Your job is to help the club identify the best location for the garden. To do this, you and your grounds-maintenance crew must assess the abiotic factors of the environment at various locations around the campus.

Background

Preparation Tips
- Prepare a map of the school campus showing the locations of several 5 × 5 meter plots, or quadrats. The number of quadrats may vary depending on class size and the number of students per group.
- Divide the students into groups and assign four groups to each location. Assign each group to an area of their plot.

Disposal
- Have students return the soil sample to the area where they collected it.
- Soil-test chemicals can be poured down the drain with water.

The nonliving parts of an ecosystem are called **abiotic factors.** They include light, wind, temperature, water, and mineral nutrients. Abiotic factors have many influences on organisms such as where they may live, rate of growth, reproductive potential, and genetic expression. Environmental factors that stabilize population size and keep species from reaching their potential are called **limiting factors;** they include availability of space, food, water, and nesting materials.

Organisms are affected by the intensity of the sun, hours of daylight, and the different seasons. Wind affects temperature, the rate of evaporation of moisture from the soil, and plant growth. Wind speed can be measured by an instrument called an **anemometer.** Both air and soil temperatures are important abiotic factors. Soil moisture also has a major influence on plant growth.

Soil contains inorganic mineral nutrients such as nitrates, phosphorus, and potassium. The presence of these minerals can be detected with a soil-test kit. The **soil pH,** or acidity of soil, often determines whether necessary mineral nutrients are available to plants. Soil pH can also be determined with a soil-test kit.

Procedure

PROCEDURAL NOTES

Safety Precautions
• Before going into the field, discuss some of the more dangerous organisms students might find and how they may avoid them. Show them pictures of poison ivy and other plants to avoid.
• Remind students to dress appropriately for the lab activity. Students should wear shoes with closed toes.
• Read and discuss with students all safety information related to the soil-test kit.

Procedural Tips
• Model the procedure for using an anemometer and both types of thermometers.
• Remind students not to force a soil thermometer into the soil.
• Model the procedures for using a soil-test kit.
• Remind students to leave the environment as they found it and to fill in the holes when returning the soil.
• Research the best perennials for your particular area. Provide students with photographs and descriptions of these plants for use in making recommendations for the placement of the perennial flower garden.

1. Practice using an anemometer and a thermometer by taking readings at five locations on campus. Record the anemometer readings in mph. At each location, take two temperature readings, one at 4–5 cm above the ground, and one at 1 m above the ground, using a meter stick to approximate the distance. Record your results in the table below.

Table 1 Sample Measurements

Measurement	Location 1	Location 2	Location 3	Location 4	Location 5
Wind speed (mph)					
Temperature at a height of 4–5 cm (°C)			Answers will vary.		
Temperature at a height of 1 m (°C)					

2. Work with the other groups assigned to the same location. Using stakes and string, mark off a 5 m × 5 m plot, or quadrat, and divide it into four areas as seen in the diagram below.

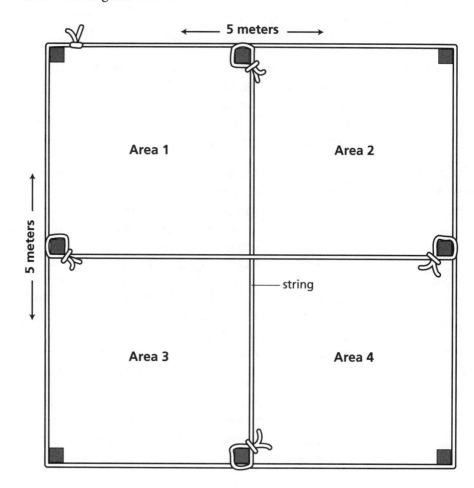

3. Using an anemometer, record the wind speed in your group's area of the quadrat. Share your measurement with the other three groups working in your quadrat. Record all measurements in the table below.

Table 2   Wind Speed, Temperature, and Shade Data

Measurement	Area 1	Area 2	Area 3	Area 4
Wind speed (mph)				
Temperature at a height of 4–5 cm (°C)		Answers will vary.		
Temperature at a height of 1 m (°C)				
Soil temperature at a depth of ____ cm (°C)				
Amount of shade				

4. Using a thermometer, measure the air temperature in your group's area of the quadrat. Take the first reading 4–5 cm above the ground. Take the second reading 1 m above the ground. Insert a soil thermometer about 6 cm into the ground to measure soil temperature. Allow the thermometer to remain in the soil until the reading is constant. *NOTE: Do not force the soil thermometer into the soil.* Record all temperature measurements for your quadrat in the table above. Note the depth at which you measured the soil temperature.

◆ Why do you take temperature readings at three different heights?

It allows you to compare temperatures and helps you analyze how temperature differences affect

the organisms that live there.

5. Observe your area for the amount of shade it receives, and record the approximate amount in the table above using descriptions such as: none, less than half, about half, more than half, or all. Share your observations with the other groups working in the same quadrat.

◆ Why is it important to note the amount of shade?

Temperature readings will vary depending on the amount of shade in an area. Soil temperature is

cooler in shade than in sun.

6. Collect a soil sample from your group's area, place it in a sealable plastic bag, and take it to the lab.

7. Using a piece of filter paper and a rubber band, cover one end of a metal can that has both ends removed, as shown in the diagram below.

8. Using a balance, measure 100 g of soil from your sample, and place it in the container you made in step 7. Measure and record the mass of the container and soil in the appropriate column for your area in the table below. Using a permanent marker, label your container with the initials of each member of your group and the number of the area where the soil was collected. Place your container in a drying oven set at 100°C for 24 hours.

Table 3 Soil Moisture Data

Measurements	Area 1	Area 2	Area 3	Area 4
Mass before drying				
Mass after drying		Answers will vary.		
Mass of soil moisture				
Percent of moisture in the soil				

9. Using a hot pad, carefully remove your container from the oven, and let it cool. Then, using a balance, find the mass of the container with the soil, and record it in the table above.

10. Calculate the mass of the soil moisture in your sample by using the following formula:

mass of container with soil before drying − mass of container with soil after drying = mass of soil moisture

Record this number in the table above.

11. Calculate the percentage of moisture in your soil by using the following formula:

(mass of soil moisture ÷ mass of soil before drying) × 100 = percentage of moisture in soil

Record this number in the table above.

12. Complete the table above by entering data obtained by other groups for the remaining three areas.

13.   Put on safety goggles, gloves, and a lab apron. Obtain a soil-test kit, and carefully read the safety rules included with the kit. Following the directions in the soil-test kit, test your dried soil for pH, nitrate content, phosphate content, and potassium content. Record your results in the table below. Complete this table by entering data obtained by other groups for the remaining three areas.

Table 4   Soil-Test Kit Data

Test	Area 1	Area 2	Area 3	Area 4
pH				
Nitrate		Answers will vary.		
Phosphate				
Potassium				

14. When all measurements and tests have been completed, share the data on the abiotic factors in your quadrat with the other groups in your class.

15. Dispose of your materials according to the directions from your teacher.

16. Clean up your work area and wash your hands before leaving the lab.

Analysis

17. Account for any differences in wind speed that were noted among the four areas in your quadrat and among the quadrats at other locations.

Students should observe that wind speed is related to location. If they are behind a building, the

wind speed might be slower.

18. Compare the soil moisture in the four areas of your quadrat and in the quadrats at other locations. Explain how any difference might relate to the amount of light, soil temperature, wind speed, and plant distribution.

Answers will vary. There may be a difference in soil moisture, depending on the amount of wind

and shade in the area tested. Shady areas will have more soil moisture, less light, a lower soil

temperature, and possibly more plant variety. Wind speed is affected by obstructions, such as tall

plants, buildings, or fences. Wind will reduce soil moisture.

19. How did the results of your soil test compare with those of the other areas of your quadrat? the quadrats at other locations? Account for any differences.

Answers will vary.

20. Was there any variation of plant type between sunny areas and shady areas? If so, offer an explanation.

Answers will vary. Students should notice that some plant species require direct sunlight, while

others thrive in a shady environment.

Conclusions

21. Based on your data, which abiotic factors are most likely to be limiting factors for plant growth in areas sampled?

Answers will vary.

22. Recommend the best location to place a perennial flower garden.

Answers will vary. Ideally, perennial flower gardens should be placed in areas that are protected

from the wind and where the soil (pH, texture, and fertility) and available light are suited to the

plants that are to be grown.

Extensions

23. *Soil scientists* study all aspects of the soil, from its origin to its composition and distribution. A soil scientist may test the effects of fertilizer on soil quality or assess how a field drains. One job a soil scientist might do is develop practices to increase soil productivity. Chemical analyses of soil and examining soil for bacterial and mineral content are other jobs that a soil scientist might perform. Find out the training and skills needed to become a soil scientist.

24. Select a plant that you like and determine what is necessary for it to thrive in any given region. List the abiotic factors that influence the success of your plant. Present your findings in a poster.

C26 Assessing and Mapping Factors in the Environment

Prerequisites
- Laboratory Techniques C24: Mapping Biotic Factors in the Environment on pages 115–120
- Laboratory Techniques C25: Assessing Abiotic Factors in the Environment on pages 121–126

Review
- biotic and abiotic factors

Chemical Wonders, Inc.

Atlanta, Georgia

April 4, 1998

Rosalinda Gonzales
Environmental Studies Division
BioLogical Resources, Inc.
101 Jonas Salk Dr.
Oakwood, MO 65432-1101

Dear Ms. Gonzales,

I am the president of a growing company that is donating a large plot of land to the city for the development of a natural reserve. The reserve will offer citizens a chance to learn about and enjoy the ecosystems in their community. It will serve as a living museum, with hiking trails, informative signs describing local plant and animal species, and a visitor's center.

To further help the city in its efforts, our company has offered to fund a study of the ecosystem on the reserve. I have recently received a memo from the city council about what should be included in this study. The study should include the following components: 1) a list of plants, insects, and other animals found on the property, 2) a list of abiotic factors of the environment, 3) a map illustrating the distribution of plant and animal populations, 4) recommendations about appropriate locations for hiking trails, and 5) signs illustrating the dominant plants and animals.

I am looking forward to your report. We are willing to pay your normal fee. Do not hesitate to call me if you have any questions.

Sincerely,

Mark Grayson

Mark Grayson
President, Chemical Wonders, Inc.

BioLogical Resources, Inc. Oakwood, MO 65432-1101

M E M O R A N D U M

To: Team Leader, Ecology Dept.

From: Rosalinda Gonzales, Director of Environmental Studies

We have our work cut out for us on this project, so I suggest we get started immediately. We will have a meeting soon to discuss our plans. Please prepare a list of questions and concerns that you would like to have clarified at this meeting.

Have each of your teams analyze a different plot of land, each plot measuring 5 m × 5 m. Groups should use stakes , strings, and flags to mark off the area they will be studying. When preparing your proposed procedure, refer to the requirements listed in Mr. Grayson's letter.

I recently spoke to Mr. Grayson, and he informed me of a few details. He also requested that you identify the dominant plant and animal species. He asked that your list of abiotic factors include air temperature, wind speed, soil moisture content, soil temperature, soil pH, soil nitrate content, soil potassium content, and soil phosphate content. Finally, he asked that on your map you use symbols to indicate differences in elevation and that you include a key to define any symbols used.

Proposal Checklist
Before you start your work, you must submit a proposal for my approval. **Your proposal must include the following:**

_____ • the **question** you seek to answer

_____ • the **procedure** you will use

_____ • a detailed **data table** for recording observations and measurements

_____ • a complete, itemized list of proposed **materials** and **costs** (including use of facilities, labor, and amounts needed)

Proposal Approval: _____
(Supervisor's signature)

Report Procedures

When you finish your analysis, prepare a report in the form of a business letter to Mr. Grayson. **Your report must include the following:**

_____ • a paragraph describing the **procedure** you followed to investigate the ecosystem

_____ • a complete **data table** showing all observations of abiotic and biotic factors

_____ • a **map** illustrating land characteristics and the distribution of plant and animal species

_____ • a **model sign** illustrating dominant plant and animal species

_____ • your **conclusions** about the dominant species and abiotic factors in the ecosystem and recommendations for hiking trails

_____ • a detailed **invoice** showing all materials, labor, and the total amount due

Safety Precautions

• When using soil test kits, always wear safety goggles, a lab apron, and disposable gloves. Be sure to follow all instructions provided by your teacher.

• Do not touch or approach any animal in the wild.

• Do not touch any plant in the wild unless otherwise instructed by your teacher.

• Handle sharp objects carefully.

• Wear oven mitts when handling hot objects.

• Wash your hands upon leaving the field or the laboratory.

• Do not jam a thermometer into the soil to determine soil temperature. It will break. Dig out a little hole using a stick or trowel.

• Never go alone beyond where you can be seen or heard. Always travel with a partner.

Disposal Methods

• Dispose of waste materials according to instructions from your teacher.

• Place string, stakes, flags, and protractors in the separate containers provided.

• Place used soil samples in a trash can or return them to the field.

• Wash reusable materials such as glassware and lab utensils, and return them to the supply area.

FILE: Chemical Wonders, Inc.

MATERIALS AND COSTS (Select only what you will need. No refunds.)

I. Facilities and Equipment Use

Item	Rate	Number	Total
facilities	$480.00/day		
personal protective equipment	$10.00/day		
scissors	$1.00/day		
protractor	$1.00/day		
anemometer	$20.00/day		
meter stick	$1.00/day		
thermometer	$5.00/day		
garden trowel	$2.00/day		
drying oven	$30.00/day		
field guides	$20.00/day		
balance	$10.00/day		
hot plate	$15.00/day		

II. Labor and Consumables

Item	Rate	Number	Total
labor	$40.00/hour		
soil-test kit for pH	$10.00 each		
soil-test kit for nitrogen	$10.00 each		
soil-test kit for phosphorous	$10.00 each		
soil-test kit for potassium	$10.00 each		
500 mL container	$3.00 each		
metal can with both ends cut out	$2.00 each		
filter paper	$0.10/sheet		
rubber bands	$0.10 each		
stakes	$0.50 each		
colored flags	$0.20 each		
string	$0.20/m		
zippered plastic bags	$0.20 each		
nontoxic permanent marker	$2.00 each		

Fines

OSHA safety violation	$2,000.00/incident		

Subtotal	
Profit Margin	
Total Amount Due	

Name _____

Date _____ Class _____

C27 *Studying an Algal Bloom*

Skills
- using a compound light microscope
- making a wet mount
- collecting and analyzing data
- constructing a bar graph (histogram)

PREPARATION NOTES

Objectives

Time Required: two 50-minute periods, then 15 minutes every other day for five to seven days

- *Investigate* how algae are affected by common pollutants.
- *Determine* how common pollutants contribute to environmental problems.
- *Analyze* the effect of different concentrations of common pollutants on the growth of algae.

Materials

Materials
Materials for this lab activity can be purchased from WARD'S. See the *Master Materials List* for ordering instructions.

- safety goggles
- lab apron
- wax pencil
- plastic test containers (3)
- plastic graduated pipets (3)
- culture of green algae (30 mL)
- nitrate pollutant solution (30 mL)

- phosphate pollutant solution (30 mL)
- aged water (71 mL)
- sheet of white paper
- fluorescent lights
- glass microscope slides (3)
- coverslips (3)
- compound microscope

Purpose

You are a naturalist specializing in aquatic ecosystems for the state park service. During your annual visit, you realize that the state's largest lake is covered by a thick, green scum. Plants and fish in the lake are dying. You suspect that runoff containing pollutants such as fertilizers from a nearby housing development may be contributing to increased algae growth in the lake. To test your hypothesis, you decide to test the effects of two common pollutants, nitrate and phosphate, on the rate of algal growth.

Background

Additional Background
Lakes and ponds undergo a natural succession. When a lake is young and deep, there are fewer aquatic organisms because the nutrients they require are scarce. Over time, runoff into the lakes brings nutrients needed to sustain phytoplankton and other primary producers. This allows an increase in population at all levels of the food web and in the overall biomass of

An **algal bloom** is a sudden massive growth of algae. At first, you may think this rapid growth of algae is a good thing, but it is not. The pond is not in natural balance and is actually dying. It is undergoing a process of **eutrophication,** which means that bacteria are using up all the available dissolved oxygen in the water, eventually causing plants and animals to die. This process, which usually takes thousands of years by natural causes, is greatly accelerated by the addition of pollutants to the lake or pond.

Algae grow very quickly when they have a good supply of nutrients. Sources of nutrients for algae and plants, such as nitrates and phosphates, are beneficial in small amounts. These nutrients become pollutants, however, when they are present in excessive amounts. Instead of being toxins that kill the algae, these "pollutants" are actually rich sources of nutrients. **Nitrates** are salts or esters commonly found in fertilizers, and many laundry detergents contain **phosphate,** another salt or ester. Nitrates and phosphates are carried into ponds and lakes by seepage from household waste water and runoff from rain.

the lake. At this point the lake is becoming eutrophic—bacteria are consuming all the available dissolved oxygen in the water, and plants and animals are dying.

Ironically, because of human impact, runoff into lakes and ponds can have devastating effects. Drainage from agricultural lands transports nitrogen and

Procedure

phosphorus used for fertilization into the surrounding bodies of water. Under some conditions, fecal material from animals raised on farms also adds to the nitrate and phosphate totals.

Nitrogen and phosphorus stimulate a rapid growth of vegetation in lakes and ponds. Shallow areas quickly become choked with weeds, destroying their recreational value. As these weeds die, large amounts of organic debris accumulate on the lake bottoms, filling them in. Eutrophication, which normally takes thousands of years, is drastically accelerated.

Preparation Tips

• Pollutant solutions must be prepared one week in advance to allow the breakdown of detergent and fertilizer into phosphates (PO_4^{-3}) and nitrates (NO_3^-).

• Use powdered laundry detergent to make the pollutant containing phosphate. Make sure the label says that it contains phosphates and is biodegradable.

• Use either a liquid or a solid fertilizer (plant food) with nitrates to make the pollutant with nitrate.

Once a pond or lake has additional amounts of phosphates and nitrates, algae multiply rapidly. At first, the algae produce oxygen. But as they increase in number, they consume more oxygen at night through cellular respiration than they produce during the day through photosynthesis. The algae eventually form a thick green scum that blocks out sunlight to lower-level algae and submerged plants. Starved for light and crowded out by the algae, the plants and algae die and accumulate in shallow areas on the pond bottom. A rapid increase in the number of bacteria decompose the dead plants and algae, using up oxygen in the process. With the food web out of balance and the dissolved oxygen levels drastically reduced, aquatic organisms, including fish, begin to die.

Part 1—Creating an Algal Bloom

1. Put on safety goggles and a lab apron.

2. Use a wax pencil to label three test containers. Label one container "control." Label the second container "low" and the third container "high." Also label these containers "N" for nitrates, "P" for phosphates, or "NP" for both pollutants, depending on which pollutant your group is testing.

   ♦ How will you set up a control for your experiment?

   Set up one test container with aged water and algae but no pollutant, and incubate it just as you

   do the other cultures.

3. Use graduated pipets to add the correct amount of algae, pollutant, and aged water in each container according to the table below. *Note: Be sure to use a clean pipet for the algae, pollutant, and aged water.*

Table 1 Amount of Algae, Pollutant, and Water

Test container	Amount of algae culture	Amount of N, P, or NP pollutant	Amount of aged water
Control	5.0 mL	0 mL	25.0 mL
Low concentration	5.0 mL	1.0 mL	24.0 mL
High concentration	5.0 mL	3.0 mL	22.0 mL

4. Swirl each container gently but thoroughly after adding all ingredients. Cap all containers lightly. *Note: Do not tighten the caps so tightly that air cannot pass into and out of the container.*

   ♦ What do green algae need to grow?

   light, water, carbon dioxide, nutrients (e.g., nitrates and phosphates)

   ♦ Why is it important for oxygen to pass into and out of the container?

   The algae in the pond water need oxygen to grow and reproduce.

• You will need to mix equal parts nitrate pollutant with phosphate pollutant to make the combined "NP" solution.

• If time permits, have students prepare the stock solutions of pollutants. To prepare the pollutant solutions, add 12.5 grams (e.g., powdered detergent) or 12.5 mL (e.g., liquid fertilizer) and enough tap water to an Erlenmeyer flask to make 1 L. To break down the detergent and fertilizer so nutrients are available to the algae, add about 5 g dirt that is not sterile and that has had no insecticides or herbicides added to it. Stopper the flask with cotton, and leave it in a dark place for one week. Filter solutions before using them.

• Prepare a quantity of "aged" water by filling a container (a large flask or gallon jug) with tap water and allowing it to sit at least 24 hours to remove chlorine.

• Purchase *Chlorella, Closterium, Chlamydomonas,* or a combination of these three algae. Dilute these cultures with aged water.

Disposal

Wash all solutions down the drain with copious amounts of water.

PROCEDURAL NOTES

Safety Precautions
• Discuss all safety symbols and caution statements with students.

• Review the rules for carrying and using the compound microscope.

• Have students wear safety goggles and a lab apron. Allow students to remove safety goggles when viewing slides with a microscope.

5. Place all containers on a sheet of white paper near a light source for five to seven days. If possible, place the containers about 2 ft below fluorescent lights. If fluorescent lights are not available, place the containers by a window and cover them with translucent paper to keep them from overheating.

Part 2—Observing Algal Blooms

6. Make a visual inspection of each container after you have placed all containers in the light. Note the differences in color intensity and cloudiness between each container. Make additional observations every other day for five to seven days, filling in the data table below with a brief description of the appearance of each container. Circle N, P, or NP in the heading column indicating the pollutant for which your group is responsible.

Table 2 Effects of Pollutants on Algal Growth

Day	Control	Low N, P, or NP	High N, P, or NP
1			
3		Answers will vary.	
5			
7			

7. When your observations are complete (after five or seven days), use a wax pencil to label one glass slide "control," another "low" (and N, P, or NP depending on the pollutant you are testing), and a third "high" (and the pollutant you are testing). **CAUTION: Glassware is fragile. Notify your teacher promptly of any broken glass or cuts. Do not clean up broken glass or spills unless your teacher tells you to do so.**

8. Gently swirl each of the containers to mix. Make a wet-mount slide by placing one drop of water on a slide and covering with a coverslip. *Note: You will make a wet mount for each of your samples but wait until you are ready to view them before you make the other slides. They may dry out while you're looking at the first slide.*

9. Place the slide under a compound microscope, and observe it on low power. Count the number of algal cells present in each of four different viewing fields. Start with a viewing field in the top left corner, and work clockwise as shown in the diagram at right. Record your findings in the data table on the next page.

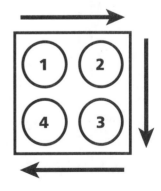

Procedural Tips

Divide students into groups of three to five. Some teams will test the effects of nitrogen pollution, others will test the effects of phosphate pollution, and still others will test the effects of the combined pollutants. Because of growth variations, you will want as many groups as possible.

Table 3 Number of Algal Cells (pollutant tested: _____)

Test container	Number of cells/field				Number of cells/ undiluted drop	Total number of algal cells
	1	2	3	4		
Control						
Low concentration					Answers will vary.	
High concentration						

10. Total the number of algal cells in all four fields. This will tell you approximately how many algal cells are in one undiluted drop (approximately 0.05 mL) of each sample. To estimate the total number of algal cells (the concentration of algae) in your 30 mL container, use the following equation. Record the information in the data table above.

total number of algal cells = total number of algal cells in four fields $\times$ 600

11. Your teacher will draw a table like the one below on the chalkboard. Work with the other groups in your class to complete the data table.

Table 4 Total Number of Algal Cells

Classroom group	Control	N pollutant		P pollutant		NP pollutant	
		Low	High	Low	High	Low	High
1							
2							
3							
4							
5							
6							
7							
8							
9							
10							
Class average							

Answers will vary, but the results should follow the basic formula: NP>N>P>C. Your students should find that if the control is designated x, then the low test will be approximately 5x and the high test will be approximately 10x. Of course, there can be a wide variance due to light conditions, temperature, and preexisting algae and minerals in any collected water, to name a few variables.

12. 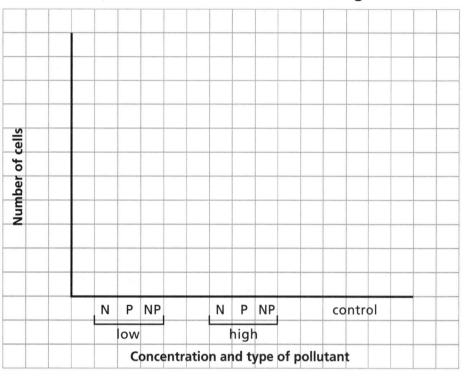 Dispose of your materials according to the directions from your teacher.

13. Clean up your work area and wash your hands before leaving the lab.

Analysis 14. On the grid below, make a bar graph (histogram) of the class results of the average number of algal cells versus concentration and type of pollutant ("N," "P," or "NP"). *Note: Use a different color for each nutrient.*

Effect of Pollutants on Number of Algal Cells

	N	P	NP		N	P	NP		control
		low				high			

Y-axis: Number of cells

X-axis: Concentration and type of pollutant

15. What effect did the addition of nitrates have on algal growth?

The increase of nitrates increased the number of algal cells.

16. What effect did the addition of phosphates have on algal growth?

The increase of phosphates increased the number of algal cells.

17. What effect did the addition of both pollutants together have on algal growth?

The increase of nitrates and phosphates increased the number of algal cells.

18. Which additive(s) had the greatest effect on algal growth?

The addition of both nitrates and phosphates had the greatest effect on algal growth, followed

first by nitrates and then by phosphates.

19. What effect did increasing the concentration of the pollutants have?

Increasing the concentration of the pollutants had a greater effect on (speeded up) algal growth.

20. Which pollutant or combination of pollutants had the greatest effect on algal growth at low concentration? At high concentration?

At both low and high concentrations, the combination of nitrates and phosphates had the greatest

effect on algal growth, followed first by nitrates and then by phosphates.

Conclusions

21. Why can excessive levels of nitrates and phosphates be harmful to a lake?

They can cause radically increased plant or algal growth that can lead to fish kills and accelerated

eutrophication.

22. Based on your results, is it more important to limit nitrate runoff or phosphate runoff? Explain your answer.

Both should be limited, even though the nitrates promote more algal growth than phosphates.

23. Do you think it would be easy to reverse the effects of eutrophication caused by pollution?

No. The source of the pollution would have to be discovered and stopped. The large quantity of

organic matter would have to be dredged or otherwise removed. Animal life that was destroyed

would have to be reintroduced.

24. What are three things you would suggest to the homeowners to decrease the amounts of nitrates and phosphates entering the lake?

Acceptable answers include sensible fertilization of home lawns and gardens (and farmland), the

use of phosphate-free detergents, peripheral sewer service around lakes, reforestation to minimize

runoff, and better water purification and treatment plants.

Extensions

• Caution students not to carry out their experiments until you have checked their procedures to make sure they are safe.

25. Identify some common pollutants that you believe might be harmful to algae. These substances might include motor oil, antifreeze, granular road salt, and household chlorine bleach. Design an experiment to determine at what concentration these substances might be toxic to algae.

26. Not all laundry detergents have phosphates. Many parts of the country have bans on the use of laundry detergents containing phosphates or require the use of detergents with reduced phosphate amounts. Find out if there are any restrictions in your area on the use of detergents containing phosphates. What are the restrictions? Why do these restrictions exist in your area? How are the restrictions enforced?

HOLT
BIOSOURCES
LAB PROGRAM
EXPERIMENTAL DESIGN

C28 Studying an Algal Bloom—Phosphate Pollution

Prerequisites • **Laboratory Techniques C27: Studying an Algal Bloom** on pages 131–136

Review • limiting factors
• role of minerals in plant growth
• the process and causes of eutrophication

**CITY OF
CRYSTAL LAKE
PLANNING COUNCIL**
Crystal Lake City, Montana

March 5, 1998

Rosalinda Gonzales
Environmental Studies Division
BioLogical Resources, Inc.
101 Jonas Salk Dr.
Oakwood, MO 65432-1101

Dear Ms. Gonzales,

The Crystal Lake City planning council is currently considering a request for a building permit from a company that makes a variety of phosphate compounds. The phosphates are used in commercial products, such as detergents and fertilizers. The planned location of this plant is on the shoreline of our lake. Many residents and local business owners are concerned about the possible effects of phosphate pollution. They fear that the quality of the lake's ecosystem may be compromised if the company is allowed to build. However, the new plant should bring a number of jobs to our community.

The council is under tremendous pressure from people on both sides of the issue. The company has owned the land for the past 50 years and is only now ready to build. It is not willing to sell the land and build elsewhere, and our community does not have enough money to purchase the land from the company.

Our major concern is the potential of a low-level phosphate leak that might go undetected. Before we make a decision, the council must know how a small phosphate leak would affect the lake. Please design and conduct an experiment that simulates a low-level phosphate leak.

Sincerely,

Sally Jacobs

Sally Jacobs
City Planner

M E M O R A N D U M

To: Team Leader, Ecology Dept.

From: Rosalinda Gonzales, Director of Environmental Studies

I want your research teams to complete a study of the effect of low levels of phosphate on Crystal Lake water. Ms. Jacobs has sent me water samples from the lake, which I will send to you.

Test the effects of low levels of phosphate on algal growth. I suggest using a grid slide when counting organisms under the microscope. Start by testing three mixtures of lake water and phosphate in the following ratios of phosphate to lake water: 1 to 29, 3 to 27, and 5 to 25.

In your conclusions, along with the results of your work, make sure that you discuss the effects of adding nutrients to an environment. Be sure to include an explanation of what happens when a limiting factor changes.

Proposal Checklist

Before you start your work, you must submit a proposal for my approval. **Your proposal must include the following:**

_____ • the **question** you seek to answer

_____ • the **procedure** you will use

_____ • a detailed **data table** for recording observations

_____ • a complete, itemized list of proposed **materials** and **costs** (including use of facilities, labor, and amounts needed)

Proposal Approval: _____
(Supervisor's signature)

Report Procedures

When you finish your analysis, prepare a report in the form of a business letter to Ms. Jacobs. **Your report must include the following:**

_____ • a paragraph describing the **procedure** you followed to test the effect of low levels of phosphate on algal growth in lake water

_____ • a complete **data table** showing algal growth over several days for different levels of phosphate

_____ • your **conclusions** about the effect of low levels of phosphate on lake water

_____ • a detailed **invoice** showing all materials, labor, and the total amount due

Safety Precautions

• 🥽 🦺 Wear safety goggles and a lab apron.

• 🧪 Glassware is fragile. Notify your teacher promptly of any broken glass or cuts. Do not clean up broken glass or spills unless your teacher tells you to do so.

• 🔌 Never use electrical equipment around water, nor with wet hands or clothing. Never use equipment with frayed cords.

• 🧼 Wash your hands before leaving the laboratory.

• If you are using a microscope with a mirror, do not use direct sunlight as a light source.

Disposal Methods

• ☣ Dispose of waste materials according to instructions from your teacher.

• Place disposable material in a trash can.

• Place broken glass, phosphate, pond water, and containers of phosphate and lake water mixtures in the separate containers provided.

• Wash reusable materials such as glassware and lab utensils, and return them to the supply area.

FILE: City of Crystal Lake Planning Council

MATERIALS AND COSTS (Select only what you will need. No refunds.)

I. Facilities and Equipment Use

Item	Rate	Number	Total
facilities	$480.00/day	_____	_____
personal protective equipment	$10.00/day	_____	_____
compound microscope	$30.00/day	_____	_____
microscope slide with coverslip	$2.00/day	_____	_____
grid slide with coverslip	$2.00/day	_____	_____
depression slide with coverslip	$2.00/day	_____	_____
100 mL graduated cylinder	$5.00/day	_____	_____
25 mL graduated cylinder	$5.00/day	_____	_____
10 mL graduated cylinder	$5.00/day	_____	_____
small jars or 50 mL beakers	$2.00/day	_____	_____
medicine dropper	$2.00/day	_____	_____
balance	$10.00/day	_____	_____
grow lamp	$15.00/day	_____	_____
hot plate	$15.00/day	_____	_____

II. Labor and Consumables

Item	Rate	Number	Total
labor	$40.00/hour	_____	_____
filter paper	$0.10/sheet	_____	_____
funnel	$5.00 each	_____	_____
phosphate solution	$0.50/mL	_____	_____
lake water	provided	_____	_____
adhesive labels	$0.10 each	_____	_____

Fines

Item	Rate	Number	Total
OSHA safety violation	$2,000.00/incident	_____	_____
		Subtotal	_____
		Profit Margin	_____
		Total Amount Due	_____

Name _____

Date _____ Class _____

HOLT
BIOSOURCES
LAB PROGRAM
EXPERIMENTAL DESIGN

C29 | *Classifying Mysterious Organisms*

Prerequisites   • **Laboratory Techniques C1: Using a Microscope** on pages 1–8

Review
- characteristics that modern taxonomists use to classify organisms
- major differences among groups of protists
- problems biologists have when trying to classify organisms

YOUNG ECOLOGISTS
DILLARD, GEORGIA 30687

March 17, 1998

Karl Smith
Microbiology Division
BioLogical Resources, Inc.
101 Jonas Salk Dr.
Oakwood, MO 65432-1101

Dear Mr. Smith,

The Young Ecologists is an organization run by college students committed to protecting our local environment. We were recently very excited to hear that the city had accepted our proposal for a local environmental center. We have received a donation of land and a grant to construct a covered pavilion that will be used as an information center and a facility for our meetings and workshops. We have been working very hard to learn about the park ecosytem. There is a pond on the property. We have been able to identify most of the plants and larger organisms in the pond, but we would also like to know what kinds of microorganisms live there.

We have sent a sample of pond water that contains four different kinds of microorganisms. We have identified three of these as *Volvox, Paramecium,* and *Amoeba*. We think the other organism is a protist, but in some ways it resembles an alga. Please confirm the classification of the three known protists and identify the unknown microorganism. Since we have limited resources, we would appreciate knowing in advance how much you will be charging. Thank you for your time.

Sincerely yours,

Raydelle Chapman

Raydelle Chapman
President
Young Ecologists

M E M O R A N D U M

To: Team Leader, Limnology Dept.

From: Karl Smith, Director of Microbiology

Attached is a request that I want you to handle with your usual care and professionalism. I was really inspired by the determination and ingenuity of this group of young people. Approach this job just as you would any other, with the exception of payment. The rest of the board and I have decided to provide our services free of charge.

I have recently talked to Ms. Chapman and she has told me that Young Ecologists is registered as a nonprofit organization. This will not affect your procedures or your paperwork. Please prepare your invoice with all normal fees and costs, as we will need these amounts for our files.

On the phone, Ms. Chapman told me that the nature center would be opening in three months. I assured her that we will move as quickly as possible on this project. I would like to provide these students with as much information as possible. Along with your report, please include additional information about each identified organism, including facts about such areas as nutrition and reproduction. Please also include a sketch of each organism you observe.

Proposal Checklist

Before you start your work, you must submit a proposal for my approval. **Your proposal must include the following:**

_____ • the **question** you seek to answer

_____ • the **procedure** you will use

_____ • a detailed **data table** for recording observations and research

_____ • a complete, itemized list of proposed **materials** and **costs** (including use of facilities, labor, and amounts needed)

Proposal Approval: _____

(Supervisor's signature)

Report Procedures

When you finish your analysis, prepare a report in the form of a business letter to Ms. Chapman. **Your report must include the following:**

_____ • a paragraph describing the **procedure** you followed to determine the identities of the unknown organisms

_____ • a complete **data table** showing observations, the identity, and additional information for each organism

_____ • your **conclusions** about the identities of the unknown organisms

_____ • labeled **sketches** of each organism observed

_____ • a detailed **invoice** showing all materials, labor, and the total amount due

Safety Precautions

• Wear safety goggles and a lab apron.

• Glassware is fragile. Notify your teacher promptly of any broken glass or cuts. Do not clean up broken glass or spills unless your teacher tells you to do so.

• Never use electrical equipment around water, nor with wet hands or clothing. Never use equipment with frayed cords.

• Wash your hands before leaving the laboratory.

• If you are using a microscope with a mirror, do not use direct sunlight as a light source. Using direct sunlight causes eye damage.

Disposal Methods

• Dispose of waste materials according to instructions from your teacher.

• Used paper towels can be thrown into a trash can.

• Place broken glass, pond water, and slowing agent in the separate containers provided.

• Wash reusable materials such as glassware and lab utensils, and return them to the supply area.

FILE: Young Ecologists

MATERIALS AND COSTS (Select only what you will need. No refunds.)

I. Facilities and Equipment Use

Item	Rate	Number	Total
facilities	$480.00/day	_____	_____
personal protective equipment	$10.00/day	_____	_____
compound microscope	$30.00/day	_____	_____
depression slide with coverslip	$2.00/day	_____	_____
eyedropper	$2.00/day	_____	_____
clock with second hand	$10.00/day	_____	_____
scissors	$1.00/day	_____	_____
ruler	$1.00/day	_____	_____

II. Labor and Consumables

Item	Rate	Number	Total
labor	$40.00/hour	_____	_____
pond water sample	provided	_____	_____
iodine solution	$20.00/bottle	_____	_____
slowing agent	$1.00/drop	_____	_____
wax pencil	$2.00 each	_____	_____
paper towels	$0.10/sheet	_____	_____

Fines

Item	Rate	Number	Total
OSHA safety violation	$2,000.00/incident	_____	_____
	Subtotal		_____
	Profit Margin		_____
	Total Amount Due		_____

Name _____

Date _____ Class _____

HOLT
BIOSOURCES
LAB PROGRAM
LABORATORY TECHNIQUES

C30 Screening for Resistance to Tobacco Mosaic Virus

Skills

PREPARATION NOTES

Time Required: one 50-minute period to set up experiment; 10–15 minutes every 5 days for 30 days to make observations

- designing a controlled experiment
- recognizing dependent and independent variables in an experiment
- setting up and controlling variables in an experiment
- collecting and evaluating data in an experiment

Objectives

- *Inoculate* healthy plant tissue with a pathogen.
- *Observe* plant leaves after they have been treated with tobacco mosaic virus.
- *Determine* the time necessary for tobacco mosaic virus to affect tomato plants.

Materials

Materials

Materials for this lab activity can be purchased from WARD'S. See the *Master Materials List* for ordering instructions.

- marking pencil
- tomato plants in pots (5)
- safety goggles
- gloves
- lab apron
- tobacco from several brands of cigarettes

- mortar and pestle
- 0.1 M dibasic potassium phosphate solution (10 mL)
- 100 mL beaker (2)
- sterile cotton swabs (5)
- 400 grit carborundum powder

Purpose

Preparation Tips
• Conduct this investigation away from greenhouses and gardens because of the possible spread of TMV.

You are a county horticulturist in your state. A local tomato grower who wishes to import a new variety of tomatoes from South America has come to you for advice. The grower believes this tomato variety will be able to resist infection by tobacco mosaic virus, or TMV, which seriously stunts plant growth. The farmer would like you to investigate this variety of tomato before he invests a lot of money in purchasing a large volume of seed to plant. Your job is to screen the tomato plants for resistance to TMV and report your findings to the grower.

Background

A **virus** is a biological particle composed of genetic material and protein. Outside of host cells, a virus is a lifeless particle with no control of its movements. Viruses are spread via wind, water, food, blood, or other bodily secretions. Many viruses attack plants, causing damage to crops that are valuable to humans. The **tobacco mosaic virus,** or **TMV,** infects tobacco and other plants such as tomatoes. TMV causes lesions on the leaves of the plant, which then become mottled yellow and eventually die. The tobacco mosaic virus eventually destroys the plant.

Procedure

1. Use a marking pencil to number five pots of tomato plants 1, 2, 3, 4, and 5. On each pot, also include the date, your class period, and your name.

2. Put on safety goggles, gloves and a lab apron.

3. Place pinches of tobacco from different brands of cigarettes into a mortar. Add 5 mL of potassium phosphate solution to the tobacco. Grind the mixture with a pestle until the tobacco is thoroughly ground.

• From a local nursery or greenhouse, purchase a variety of tomato plants that is *not* resistant to TMV. Or grow non-TMV-resistant seeds available from seed supply catalogs. Otherwise, students will see no results.

• In advance of this lab, prepare five 10 cm (4 in.) pots of tomato plants for each group. If you start with transplants, place one in each pot. If you start with seeds, plant two or three per pot and then thin to one. Allow the plants to grow to a height of 7–12 cm before conducting the experiment. *Note: These plants require plenty of light and should be maintained in a moist but well-drained soil.*

• Keep the infected and control plants in the same light and temperature conditions, but separate them to protect against infecting the control plants.

• To prepare a 0.1 M solution of dibasic potassium phosphate (K_2HPO_4), dissolve 17.4 g potassium hydrogen phosphate powder in enough distilled water to make 1 L of solution. This solution can also be purchased ready-made.

4. Pour the mixture into a beaker. This mixture will be used to infect the tomato plants with TMV. **CAUTION: Glassware is fragile. Notify your teacher promptly of any broken glass or cuts. Do not clean up broken glass or spills unless your teacher tells you to do so. CAUTION: Wash your hands and all laboratory equipment used in this step with soap and water.**

♦ What does this step assume about cigarette tobacco?

It assumes that cigarette tobacco is infected with TMV.

5. *Wash your hands with soap and water before handling any plants.* Moisten a cotton swab with distilled water. Apply the water to the uppermost leaves of the tomato plant in pot 1. Dispose of the swab as instructed by your teacher.

♦ Why is it important to wash your hands before handling any plants?

Washing reduces the chance of contaminating plants with TMV after preparing the tobacco mixture.

6. Moisten a clean cotton swab with potassium phosphate solution (without tobacco). Sprinkle a small amount of carborundum powder onto the moistened swab. Apply the mixture to the uppermost leaves of the tomato plant in pot 2. Move the swab several times over the surface of each leaf. Dispose of the swab as instructed by your teacher.

7. Moisten a clean cotton swab with the tobacco mixture, and sprinkle a small amount of carborundum powder onto the moistened swab. Apply the mixture to the uppermost leaves of the tomato plant in pot 3. Move the swab several times over the surface of each leaf. Dispose of the swab as instructed by your teacher.

♦ What is the purpose of the carborundum powder?

The carborundum powder abrades the leaf surface and increases the chance of viral infection

through the artificially created "wound."

8. Moisten a clean cotton swab with the tobacco mixture, and sprinkle a small amount of carborundum powder onto the moistened swab. Apply the mixture to only the lowest leaves of the tomato plant in pot 4. Move the swab several times over the surface of each leaf. Dispose of the swab as instructed by your teacher.

9. Repeat the procedure in step 8, applying the mixture to all the leaves of the tomato plant in pot 5.

10. Place the tomato plants in a location designated by your teacher.

11. Dispose of your materials according to the directions from your teacher.

12. Clean up your work area and wash your hands before leaving the lab.

Disposal

- After the investigation has been concluded, incinerate or discard infected plants in a manner appropriate to removal of pathogenic bacteria.
- Provide waste containers for cotton swabs and other materials possibly contaminated by TMV.
- Solid contaminated trash should be placed in a specially marked disposal bag and autoclaved at 121°C at 15 psi for 20 minutes. Place autoclaved bags in an outer sealed container.

PROCEDURAL NOTES

Safety Precautions

- Discuss all safety symbols and caution statements with students.
- Remind students that although they cannot be infected by TMV, they should thoroughly wash their hands and all laboratory equipment that has come into contact with TMV. Students should also conduct this investigation away from gardens and house plants because TMV spreads from plant to plant.

Analysis

Procedural Tips

- Cigarette tobacco is commonly infected with TMV. Washing the equipment and hands decreases the possibility of accidental spread of the virus. To avoid infecting all the plants, stress the importance of using sterile cotton swabs and using each swab only once.
- Encourage students to consider the possibility that older leaves may have a different

13. Observe the plants every 5 days for about 30 days. Watch for symptoms of TMV. Record your observations in the data table below. *Note: During the time you are observing the plants, make sure the plants are watered every two to three days. Do not let the plants dry out between watering.*

Effects of TMV on Tomato Plants

Day	Pot #1: distilled water	Pot #2: potassium phosphate solution and carborundum powder	Pot #3: tobacco mixture on uppermost leaves	Pot #4: tobacco mixture on lower leaves	Pot #5: tobacco mixture on all leaves
Day 5	no change	Students may see some brown spots on leaves due to	no change	no change	yellow spots on at least one leaf
Day 10	no change	mechanical damage by carborundum but no	yellow on upper leaves	some yellow spots on upper leaves	stunted growth
Day 15	no change	yellowing, wrinkling, or stunting	poor growth with plant very stunted	yellow spots on both newer (upper) and older (lower) leaves	poor growth with plant very stunted
Day 20	no change		yellow spread to older (lower) leaves	more yellow on new (upper) leaves	more yellow spots spread to older (lower) leaves
Day 25	no change		older leaves curling	both new (upper) and old (lower) leaves yellow and curled	older (lower) leaves yellow and curling
Day 30	no change		plant dying	new (upper) leaves are more infected than old (lower) leaves	plant dying

14. What are some indications that TMV has infected the plants in pots 3, 4, and 5?

Some indications are wilted leaves, yellow spots on the leaves, and stunted growth.

15. Which plant(s) is/are the control? What treatment does/do the control plant(s) receive?

Plants in pots 1 and 2 are the control plants. The leaves of the plant in pot 1 are smeared with

distilled water only. The leaves of the plant in pot 2 are abraded with carborundum powder on a

swab moistened with potassium phosphate solution without the tobacco added. The plants should

be kept in the same lighting and temperature environment as the experimental plants.

degree of susceptibility to TMV than younger leaves.

• Some variations of the basic experimental design for further investigation include the following:

a. Infect only one plant in a flat of plants and record the rate of infection.

b. Cut the infected leaves from a plant. Use the same cutting tool to cut a leaf from a healthy plant to determine if the virus can be spread by this means.

c. Inject the TMV virus solution directly into the stem of the plant. Compare the spread of infection with the original experiment.

16. What are the independent and dependent variables in your experiment?

The independent variable is the virus mixture application. The dependent variable is any observed

changes in the tobacco leaves at the application sites.

17. Compare the susceptibility of older leaves and younger leaves to the virus.

Answers should be supported by data. Generally, older leaves are less susceptible to the effects of

TMV; they show fewer signs of yellowing.

18. What were some of the possible sources of error in your experiment?

Errors could include insufficient grinding of tobacco leaves to release the virus, inappropriate

amounts of carborundum, uneven amounts or uneven application of the virus mixture, insufficient

sampling of each leaf type, and contamination due to carelessness.

19. Why is tobacco from different brands of cigarettes used?

It increases the chance of obtaining TMV.

20. What is the purpose of grinding the leaves in step 3?

It releases the virus from infected cells.

Conclusions

21. Using the information in your data table, explain why plant viruses could pose a problem to farmers.

Plant viruses such as TMV appeared to spread throughout the plant once the plant became

infected, and the plant eventually died. Exposure to viruses such as TMV could cause widespread

loss of crop plants and reduced yield.

22. What would you report to the tomato grower about the results of your experiment and the resistance of the tomato plants to TMV?

Answers will vary but should mention that the plants are not resistant to TMV and would not be a

good financial investment for the grower.

Extensions

23. Greenhouse operators generally do not allow smoking in their greenhouses. Aside from health and safety issues, why do you think this is so? Do research to find out and verify if viruses are or could be transmitted by cigarette smoke.

24. Some biotechnology companies use viruses to alter the genetic structure of other organisms. Research how this is done, and share your findings with the rest of the class.

C31 *Using Aseptic Technique*

Skills

- using a Bunsen burner
- using sterile technique during laboratory procedures
- making a bacterial streak
- making a bacterial spread

PREPARATION NOTES

Objectives

Time Required: one 50-minute period, 5–10 minutes daily for 5–7 days

- *Demonstrate* aseptic technique for handling bacteria in a laboratory setting.
- *Prepare* a bacterial streak petri dish and a bacterial lawn petri dish.
- *Determine* the effect of temperature on the growth of bacteria.

Materials

Materials

Materials for this lab activity can be purchased from WARD'S. See the *Master Materials List* for ordering instructions.

Preparation Tips

- Prepare several spray bottles of 10% household bleach solution, and label them "disinfectant solution. To

Purpose

prepare 10% household bleach solution, add 900 mL of water to 100 mL of household bleach.

- Prepare at least two stock cultures of *Micrococcus luteus* for each class to use in this lab.

Background

- To prepare stock culture petri dishes, always use aseptic technique. Label a prepared tryptic soy agar petri dish with the date and "*M. luteus* Stock Culture." Use a sterile inoculating loop to scoop one colony from the stock culture tube of *M. luteus*. Streak each petri dish. Place the inoculated petri dishes in a 37°C incuba-

- antibacterial soap
- safety goggles
- gloves
- lab apron
- disinfectant solution in spray bottle
- paper towels
- Bunsen burner
- igniter
- wax pencil
- plastic petri dishes with sterile agar (6)
- stock culture of *Micrococcus luteus*

- inoculating loop
- tape
- 24 hour culture of *M. luteus* bacteria in tryptic soy agar
- test-tube rack
- 10 mL sterile graduated pipet
- pipet bulb
- sterile swab applicators (3)
- incubator set at 37°C
- refrigerator
- masking tape

You are a microbiologist for a city water department. Recent tests have shown that bacterial levels are unusually high in the city's reservoirs. You wonder if the increased bacterial levels are due to the record-breaking high temperatures in the area over the last several weeks. The high temperatures caused a slight increase in water temperature in the reservoirs. To find out if temperature has an effect on bacterial growth, you practice aseptic, or sterile, technique to grow cultures of bacteria in the lab at different temperatures.

When working with bacteria in the laboratory, safety is a primary concern. Bacteria are found everywhere: on countertops, in the air, and on your skin. **Aseptic, or sterile, technique** is necessary to prevent the contamination of bacterial cultures you are working with and to avoid the further contamination of countertops, the air, and your skin. Aseptic technique is also used in the lab when working with other microorganisms such as viruses and fungi.

Aseptic technique is always used when transferring bacteria from a prepared culture to another growth medium. Bacteria can be transferred to a petri dish containing agar in one of two ways. The first method is a **streak,** which is used to produce single, isolated colonies of bacteria. In this technique, a sample from a bacterial culture is drawn across the surface of the agar with a sterile inoculating

tor for at least 24 hours prior to the start of the lab.

• To prepare a 24-hour bacteria culture in nutrient broth, use a sterile inoculating loop to inoculate the broth with a colony from the stock culture petri dish. Incubate the tube for 24 hours at 37°C. Shake the tube periodically to aerate. Broth should appear turbid after the incubation period.

• To prepare nutrient broth media, dissolve 5 g of peptone and 3 g of beef extract in 1 L of

Procedure

distilled water. Heat gently until all solids are dissolved. Fill each culture tube with 10 mL of solution, close each tube with a cotton plug, and sterilize at 120°C, 15 psi pressure, for 15 minutes.

• To prepare nutrient agar media, add 5 g of peptone, 3 g of beef extract, and 15 g of agar to 1 L of distilled water in an Erlenmeyer flask and mix well. Heat gently and stir until solids are dis- solved. Boil the mixture for one minute. Plug the flask with cotton and sterilize at 120°C, 15 psi pressure, for 15 minutes. While the solution is still hot, pour it into sterile petri dishes. *Note: Some- times sterile petri dishes are available from a local hospital.*

• Prepare separate containers for the dis- posal of petri dishes, other contaminated items, and broken glass.

Disposal

• Any bacterial cultures and contaminated materials should be

loop or applicator. Successive streaks are made from the original streak. During **incubation,** the time the bacteria are allowed to grow under specific conditions, each viable but invisible bacterial cell will give rise to an individual—and visi- ble—colony of bacteria containing millions of cells.

The second method is a **spread,** which involves pouring bacteria from a stock culture evenly over the entire surface of the agar. A bacterial spread produces a **bacterial lawn** in which the entire surface of the agar is covered with bacteria. This technique is usually used when a test is being conducted to determine if bacteria will or will not grow under a certain set of conditions.

It is important to plan ahead when working with bacterial cultures in the lab. You must carefully follow sterile technique. If working conditions are not sterile, your results could be incorrect due to contamination. Also, nonsterile conditions could be a risk to your health, especially if you are working with pathogenic bacteria.

Part 1—Sterilizing the Work Area

1. 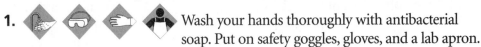 Wash your hands thoroughly with antibacterial soap. Put on safety goggles, gloves, and a lab apron.

2. Before you begin working with bacteria in the laboratory, you must first sterilize your work area. To do this, spread a disinfectant solution over the entire work area. Wipe the area clean with paper towels. Throw the paper towels into the laboratory trash can.

   ◆ Why is it important to sterilize your work area before you begin working with bacteria?

   It is important to sterilize the work area before beginning to prevent contamination of the

   microorganisms to be studied from microorganisms that may be present on the work area.

Part 2—Learning How to Use a Bunsen Burner

3. When using an inoculating loop to transfer bacteria, use a Bunsen burner to sterilize the loop. Compare the parts of a Bunsen burner with the diagram on the next page. Bunsen burners may differ slightly, but they all have the same basic parts. Use the following procedure to light the Bunsen burner.

   a. Examine the hose to make sure that it has no holes in it. If it has holes, get a new hose from your teacher.

   b. Fit one end of the hose securely over the gas outlet. If the hose does not fit tightly, get a new hose from your teacher. Fit the other end of the hose to the Bunsen burner.

   c. Partially close the air intake openings at the bottom of the barrel.

   d. Turn the gas on fully at the main outlet.

   e. Hold the igniter about 5 cm above the top of the barrel. Use the igniter to light the burner.

   f. Adjust the gas valve until the flame extends about 8 cm above the barrel.

   g. Adjust the air supply until the flame is steady and has a light blue inner core.

autoclaved for 15 minutes at 120°C and 15 psi. If an autoclave is not available, soak all materials in full strength household chlorine bleach for 30 to 60 minutes prior to disposal. Wash all lab surfaces with a disinfectant solution before and after handling bacterial cultures.

• Make sure all inoculating loops are sterilized before students leave the lab.

• Discard disposable gloves in a biohazard bag or specially designated plastic bag.

• Solid trash must be placed in a separate and specially marked disposal bag. Make sure no pipets protrude through the disposal bag. After autoclaving, disposal bags should be placed in sealed containers.

PROCEDURAL NOTES

Safety Precautions

• Discuss with students the need to treat all microbes as pathogenic, seal with tape all petri dishes containing bacterial cultures, and follow aseptic technique at all times.

• Have students wear safety goggles, laboratory aprons, and gloves.

• Discuss all safety symbols and caution statements with students.

• Never allow students to clean up bacteriological spills. Keep on hand a spill kit containing 400 mL of full strength household chlorine bleach, biohazard bags (autoclavable), forceps, and paper towels. In the event of a bacterial spill, cover the area with a layer of paper towels. Wet the paper

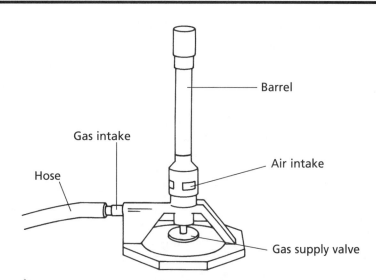

Barrel

Gas intake

Air intake

Hose

Gas supply valve

4. Practice lighting a Bunsen burner until you are familiar and comfortable with this procedure. **CAUTION: Keep combustibles away from flames. Do not light burners when others in the room are using alcohol.**

Part 3—Making a Bacterial Streak

5. Use a wax pencil to label the bottoms of three plastic petri dishes that contain agar. *Note: The agar in the petri dishes is sterile. Be careful not to open the dishes unless you are instructed to do so.* Label the first petri dish "cold streak," the second dish "room temperature streak," and the third dish "warm streak." Also, write the name of the bacteria, the date, your class period, and the initials of one or two members of your lab group.

6. Light a Bunsen burner, and sterilize an inoculating loop. Hold the tip of the inoculating loop in the flame of the Bunsen burner until the entire length of the loop glows red hot. Allow the loop to cool. *Note: Once the loop is sterile, do not set it down on the lab table.*

7. Open the top of a stock culture of *Micrococcus luteus* bacteria at a 45° angle, as shown in the diagram below. *Note: Never remove the top of the petri dish from the stock culture.* **CAUTION: Always practice aseptic technique when using bacteria in the lab. Never completely remove the cover from a petri dish. Never lay contaminated materials on the lab table.**

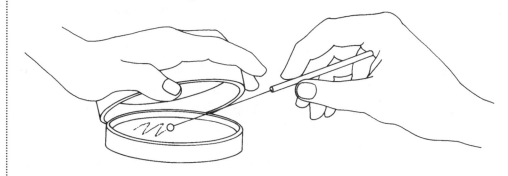

towels with the bleach solution; allow to stand for 15 to 20 minutes. Wearing gloves and using forceps, place the residue in the biohazard bag. If broken glass is present, use a brush and dustpan to collect material. Place all material in a suitably marked container.

• When students have completed this lab, autoclavable materials should be sterilized either in an autoclave or pressure cooker. All materials in contact with bacteria should be autoclaved.

Procedural Tips

• Explain to your students that even though they will be working with nonpathogenic *M. luteus,* it is important to maintain sterile, or aseptic, technique. Not only does this technique preserve the validity of an experiment, but the sterile technique protects against accidental contamination with pathogens.

• *Demonstrate aseptic technique before students try it.* Be sure students read and understand the procedure before beginning the lab. Never transfer liquid media by mouth; always use a pipet bulb. Wherever possible, use sterile cotton applicator sticks in place of inoculating loops and Bunsen burner flames for culture inoculation. Remember to use appropriate precautions when disposing of cotton applicator sticks. They should be autoclaved or sterilized before final disposal.

• "Normal" refrigeration is colder than water will ever be in

8. Make sure the tip of the inoculating loop is cool by testing it on the edge of the agar in the petri dish. *Note: If the agar melts, the loop is still too hot. Close the petri dish. Wait a few seconds, and test the loop again.* Find a *single* colony of bacteria that is isolated from the other colonies. Touch the edge of the inoculating loop to the bacterial colony to pick up some of the bacteria. *Note: Be careful to touch only the bacteria. Do not gouge the agar.* Close the lid of the bacterial stock culture.

♦ Why should you never set down inoculating loops and petri dish covers while you are performing transfers of bacteria?

For two reasons: you may contaminate the loop with organisms on the work bench, or you may

contaminate the work bench with organisms from inside the petri dish.

9. Open the top of the petri dish marked "cold streak" at a 45° angle. Remember: *Never open the petri dish completely.* Gently streak the bacteria back and forth along one edge of the agar, as shown in the diagram below. When you have completed this streak, close the dish and sterilize the inoculating loop in the flame of the Bunsen burner as described in step 6. Allow the loop to cool.

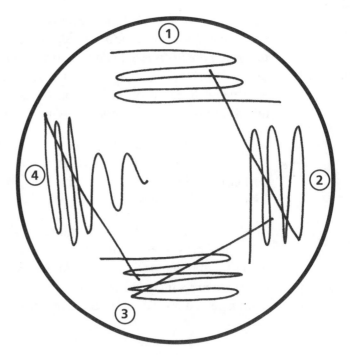

10. Rotate the petri dish 90 degrees in one direction. *Note: You may rotate the dish to either the left or right, but always rotate the dish in the same direction each time you make a new streak.* Once again open the top of the petri dish at a 45° angle. Touch the sterile loop to the end of the first streak. Be sure the loop touches some of the bacteria from the first streak. Without lifting the inoculating loop, make a second streak along another side of the petri dish, as shown in the diagram above. When the second streak is completed, close the lid of the petri dish and once again sterilize the inoculating loop.

most reservoirs. Room temperature is warmer than the water temperature during most of the year in most of the country. The warm temperature in this investigation will actually be closer to the temperature of a hot spring.

11. Again rotate the petri dish 90 degrees in the same direction as before. Place the cooled, sterile loop at the end of the second streak. Make a third streak along another side of the petri dish as shown in the diagram on the previous page. When the third streak is completed, close the petri dish and sterilize the inoculating loop.

12. Repeat the procedure one more time, making a fourth streak as shown in the diagram on the previous page. Close the petri dish and sterilize the inoculating loop in the flame of the Bunsen burner. When the fourth streak is completed, tape the petri dish closed and set it aside.

13. Repeat steps 6 through 12 for the petri dishes labeled "room temperature streak" and "warm streak."

14. Turn off the Bunsen burner at the main gas valve.

Part 4—Making a Bacterial Spread

15. Label the bottoms of three plastic petri dishes as described in step 5. This time label the first dish "cold spread," the second dish "room temperature spread," and the third dish "warm spread." Carefully set the labeled petri dishes aside.

16. Obtain a culture tube of *M. luteus* bacteria. Take the tube to your lab table and place it in a test-tube rack. *Note: Always hold and store a culture tube in an upright position to prevent bacteria from coming into contact with the tube's cap.*

17. 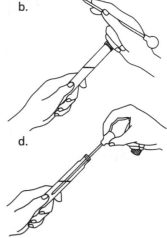 Light a Bunsen burner as described in step 3.

18. Place a pipet bulb on a sterile 10 mL pipet without touching the pipet's tip.

19. Hold the culture tube of bacteria in your left hand as shown in the diagram below. While still holding the pipet, quickly remove the cap from the culture tube, holding it with two fingers of your right or left hand, as shown below. *Note: Do not set the cap down on the lab table.* Pass the mouth of the tube through the flame of the Bunsen burner two or three times to sterilize it. Insert the pipet into the tube, and carefully withdraw 3 mL of broth containing bacteria. Once again pass the mouth of the tube through the flame of the Bunsen burner two or three times to sterilize. Replace the cap and return the tube to the test-tube rack. *Note: Do not set down the pipet containing the bacterial broth while you are returning the tube to the test-tube rack.* Turn off your Bunsen burner.

a.

b.

c.

d.

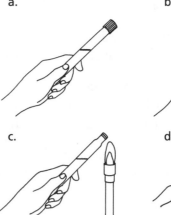

◆ Why is it important to sterilize the loop and the mouth of the culture tube both before and after removing the bacterial sample?

to prevent contamination of the stock culture by airborne spores or microbe-carrying dust particles

and to kill any microscopic drops that escape while transferring the culture

20. Open one petri dish at a 45° angle. *Note: Do not remove the lid completely.* Release 1 mL of the broth on the surface of the agar. *Note: Be very careful not to release more than 1 mL of the broth onto the surface of the agar.* Close the petri dish, and set it aside. Repeat this procedure for the remaining two dishes.

21. Place the pipet into the designated container provided by your teacher.

22. Open one petri dish at a 45°angle. Place a sterile swab applicator in the bacteria in the center of the dish. Carefully spread the bacteria evenly over the surface of the agar by turning the petri dish in circles until the bacterial broth completely covers the agar. Immediately close the lid of the petri dish. Dispose of the applicator in the designated container provided by your teacher. Tape the dish closed. Wait a few minutes for the broth to dry, and then turn the petri dish upside down, and set it aside.

23. Repeat step 22 for the remaining two petri dishes. *Note: Remember to use a new sterile swab applicator with each dish.*

Part 4—How Does Temperature Affect Bacterial Growth?

24. Sort the six petri dishes representing the two different transfer techniques into three groups: cold, room temperature, and warm. Stack the two "cold" dishes on top of each other with the bottom of one petri dish next to the top of the second dish. Use masking tape to tape the two dishes together. Repeat for the other two groups of petri dishes.

25. Place the "cold" dish in a refrigerator, and place the "warm" dish in an incubator set at 37°C. Place the "room temperature" dish in a drawer. Incubate the dishes for five to seven days. Observe your dishes daily for signs of growth.

26. Dispose of your materials according to the directions from your teacher.

27. Sterilize your work area as described in step 2. Wash your hands with antibacterial soap before leaving the lab.

28. After five to seven days, collect the six petri dishes. Remove the masking tape holding the pairs of dishes together. *Note: Never open sealed petri dishes.* Turn the petri dishes right side up, and observe the bacterial growth in each petri dish.

Analysis

29. Which temperature showed the most bacterial growth?

Student answers may vary but should indicate that more bacterial growth occurred at room

temperature.

30. Which temperature showed the least bacterial growth?

Student answers might vary but should indicate that the least bacterial growth occurred at lower

(cold) temperature.

31. What is the difference between a culture made by streaking a petri dish and one made by spreading a petri dish?

Streaking a petri dish gives individual colonies; spreading a petri dish gives a uniform growth of

bacteria all over the petri dish.

32. List the techniques you used in this lab that are part of aseptic technique.

washing hands with antibacterial soap before and after lab; wearing proper safety clothing; not

eating or having food or drinks in the lab; sterilizing lab tables before and after lab; sterilizing the

inoculating loop before and after each use; not opening any of the petri dishes more than 45°;

flaming the mouth of the culture tube before and after transfer; placing the items for disposal into

the correct waste containers

33. What was the purpose of sterilizing the work area before and after the lab?

Sterilizing the work area before the lab eliminates the possibility of contamination from bacteria

on the lab table. Sterilizing the work area after the lab ensures that no bacteria used in the lab

exercise will contaminate the work area.

34. What was the purpose of heating the inoculating loop before and after each use?

to stop bacterial contamination by sterilizing the loop

35. What was the purpose of storing the petri dishes upside down during incubation?

During incubation, water may condense. Condensation occurs on the lid when the dish is stored

right side up; the water can drop down on the bacteria. When the petri dish is stored upside down,

condensation occurs on the lid, where it cannot fall onto the bacteria.

Conclusions

36. Which of the two techniques learned in this investigation—making a bacterial spread or a bacterial streak—would you use if you wanted to study individual colonies of bacteria? Explain your answer.

You would use the bacterial streak technique because it allows for the isolation of a single bacterial

colony. The bacterial spread technique, in contrast, is used primarily to indicate bacterial growth. It

does not allow for the isolation of a single bacterial colony.

37. Why is it important not to have food or open containers in the laboratory during this lab?

Because bacteria are being used, the food and drink has a greater chance of contamination. It is

very unsafe for the health of the humans present.

38. Why should the cover of a petri dish never be opened completely during an experiment with bacteria? Why not after counting the colonies?

Bacteria exist everywhere: in the air, on our skin, in our breath. To open the petri dish for streaking,

spreading, or counting exposes the petri dish to this bacteria and us to the bacteria on the dish.

39. Why must the inoculating loop be cool before picking up a bacterial colony?

Extreme heat will kill the bacteria. If you pick up the colony with a very hot wire, it may kill the

bacteria before they are streaked. You, therefore, will get no growth on your agar.

40. When collecting a bacterial colony for streaking, why is it important to pick up only one colony?

Because a colony is a group of bacteria that originates from one cell, to pick up more than one

colony would add a variable to the experiment.

41. What effect did temperature have on the growth of the bacteria? What could the microbiologist conclude about the increase of bacteria in the city's reservoirs?

The warmer temperatures tended to produce more bacterial growth. Higher water temperatures

might have caused an increase in bacterial growth in the reservoirs.

• Students should recognize that aseptic techniques are important in daily living. Students should suggest that foods such as chicken be refrigerated until ready to cook,

Extensions

that countertops be wiped with antibacterial soap before and after exposure to food, that utensils and cutting boards be washed in hot water with antibacterial soap, and that hands be washed with antibacterial soap. Meat and eggs should be cooked thoroughly at a high temperature to kill microorganisms that might be present.

42. When working with food, such as chicken, in the kitchen, what techniques can you use to keep bacteria from spreading?

43. A *microbiologist* studies the growth, structure, and functions of microorganisms such as bacteria and protists. Because there are so many different kinds of microorganisms, microbiologists often specialize in one particular area, such as medical microbiology and food microbiology. Find out about the training and skills required to become a microbiologist.

44. Water and sewer authorities or departments, a part of most local governments, employ people who regularly check their systems for microorganisms. Find out about the training and licensing necessary to become employed by a water and sewer authority or department.

C32 Gram Staining of Bacteria

Skills
- using aseptic technique
- preparing a bacterial smear
- staining bacteria using the Gram-staining technique
- classifying bacteria

Objectives
- *Compare* and *contrast* Gram-positive and Gram-negative bacteria.
- *Relate* Gram staining to the selection of antibiotics for the treatment of disease.

Materials

PREPARATION
NOTES
Time Required: 2 days

Materials
Materials for this lab
activity can be pur-
chased from WARD'S.
See the *Master Materi-
als List* for ordering
instructions.

- safety goggles
- gloves
- lab apron
- disinfectant solution
- absorbent paper towels
- wax pencil
- glass slides (2)
- coverslips (2)
- distilled water
- inoculating loop
- Bunsen burner
- striker
- culture of Gram-positive
  bacteria, *Bacillus megaterium*
- culture of Gram-negative
  bacteria, *Aquaspirillum serpens*
- slide holder
- staining tray
- crystal violet stain
- clock or watch with a second hand
- 250 mL beakers (2)
- Gram's iodine stain
- 95% ethanol
- eyedropper
- Safranin O stain
- mounting medium
- lens paper
- compound microscope

Purpose

Your goal is to work in the medical field. You just landed a position as a part-
time lab technician in a doctor's office to see if you enjoy the work. You have been
directed to determine if cultures of bacteria are Gram-positive or Gram-negative.
Your first job is to learn the Gram-stain technique by testing known samples of
Gram-positive and Gram-negative bacteria. Then you will have the skills necessary to perform the Gram stain on cultures of unknown bacteria.

Background

Additional
Background
The protective layer,
lipopolysaccharide (LPS),
in Gram-negative cells
is an endotoxin. The
presence of LPS is what
actually causes people
to feel bad when they
have a Gram-negative
infection.

A **stain** gives color to cells that are colorless, such as bacterial cells. In 1884 the
Danish physician Hans Christian Gram developed the **Gram stain,** which is a
differential stain. Differential stains react differently to different types of bacteria. The Gram stain is used to classify bacteria into two large groups, **Gram positive** and **Gram negative.** After the Gram-staining procedure, Gram-positive
bacteria appear purple and Gram-negative bacteria appear pink.

To perform the Gram stain on bacteria, you must first prepare a **smear,** which is a
thin film of bacteria placed on a glass slide and air-dried. The slide is passed
through a Bunsen burner flame to **fix,** or attach, the smear to the slide. A stain
called crystal violet stains the bacterial cells purple. Gram's iodine, which is an

Other types of differential staining procedures include the negative stain and acid-fast stain. Cells appear colorless against a colored background in the negative-stain procedure. It is used to demonstrate the presence of a capsule. The capsule determines the degree to which a pathogenic organism can cause disease. The acid-fast stain procedure is used to help identify an organism that causes tuberculosis.

Safety Precautions

• Discuss all safety symbols and caution statements with students.

• Have students wear safety goggles, disposable gloves, and a lab apron.

Procedure

• Be sure students wash their hands thoroughly before leaving the laboratory.

Procedural Tips

• Demonstrate correct aseptic technique to students prior to conducting the lab.

• Model the procedure for making a bacterial smear. Remind students to take only a small amount of bacteria from the stock culture. Make sure students mix the bacteria thoroughly with the distilled water on the slide.

• Model the procedure for fixing a smear. Show students how to blot slides without damaging the bacteria.

• If an oil immersion objective is available, set up a station where students can observe their prepared slides using the oil immersion lens.

iodine solution, is then applied. The iodine acts as a **mordant,** which is a substance that intensifies the color of cyrstal violet and helps it adhere to the cells. Next, the smear is washed with ethanol. The ethanol **decolorizes,** or removes the color from, the cells of Gram-negative bacteria. A **counterstain,** Safranin O, which is pink in color, is then applied. It is called a counterstain because its color contrasts with that of crystal violet. After the smear is stained with Safranin O, Gram-negative bacteria appear pink.

Gram-positive and Gram-negative bacteria have differences in the chemical makeup of their cell walls, which determines how they are stained. Gram-positive bacterial cell walls contain a thick layer of peptidoglycan, which allows the crystal violet stain to penetrate and retains the purple color despite the decolorization step. Gram-negative bacterial cell walls have a thin layer of peptidoglycan and a protective layer of lipopolysaccharide. Crystal violet stains Gram-negative bacterial cells purple. But, when Gram-negative cells are decolorized, they do not retain the crystal violet because of their cell wall arrangement.

Identification of bacteria by Gram staining helps determine which drugs will be the most effective in the treatment of disease. Therefore, Gram staining is an important technique in medicine. Gram-positive bacteria tend to be killed by antibiotics such as penicillin and erythromycin. Gram-negative bacteria are resistant to these drugs but are sensitive to streptomycin and tetracycline.

1. 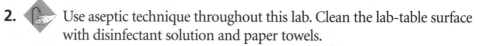 Put on safety goggles, gloves, and a lab apron.

2. Use aseptic technique throughout this lab. Clean the lab-table surface with disinfectant solution and paper towels.

3. Label one glass slide "Gram +" and one "Gram –" using a wax pencil.

4. Place one drop of distilled water in the center of the slide labeled "Gram +."

5. Sterilize an inoculating loop by holding the wire in the flame of a Bunsen burner until it glows red. **CAUTION: Do not touch hot objects with your bare hands.** Open the lid of the *B. megaterium* culture. *Note: Remember to slowly lift one side of the lid to a 45° angle.* Cool the loop by touching it on the agar of the stock culture in an area of no growth, and then transfer a colony of *B. megaterium* from the culture to the water on the slide.

6. Stir the bacteria in the drop of distilled water using the inoculating loop, as shown in the diagram at right. Allow the smear to air-dry. Resterilize the inoculating loop.

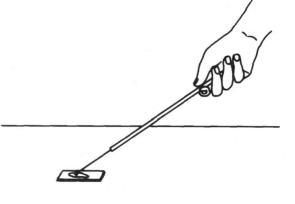

• You may want to furnish prepared slides of bacteria so that students can review the different shapes of bacteria.

Preparation Tips

• You may purchase live cultures of bacteria on agar plates or freeze-dried cultures, which must be rehydrated.

• To rehydrate freeze-dried cultures of bacteria, perform the following steps:

1. Lift the lid of the bacteria container, and remove the inner crystal. Using sterile technique, untwist the cap of the cryovial.

2. Using a sterile pipet, add 0.5 mL of nutrient broth to the pellet in the cryovial.

3. Allow 10 seconds for the pellet to soften; then mix by drawing the suspension up and down through the pipet 8–10 times.

4. Using aseptic technique, dip a sterile cotton swab into the suspension and streak the surface of labeled agar slants. Incubate at 37°C for 24 hours.

5. Autoclave all instruments used at 121°C and at 15 psi for 20 minutes. Dispose of contaminated swabs and paper towels in a biohazard bag.

6. After incubation of slant cultures, transfer them to agar plates.

• Purchase the stains ready-made. Place each stain in a labeled dropper bottle.

• Pour 95% ethanol into labeled dropper bottles. Limit the total amount of 95% ethanol in the lab to 100 mL.

7. After the smear has dried completely, pick up the slide with a slide holder, bacteria side up, and pass the slide quickly through the Bunsen burner flame three times, as shown in the diagram at right. *Note: Do not hold the slide directly in the flame. When finished, be sure to turn off the Bunsen burner.* **CAUTION: Do not touch hot objects with your bare hands. Use a slide holder to hold hot slides.** Place the slide into a staining tray, and allow it to cool.

8. Cover the smear with 10 drops of crystal violet for 60 seconds. **CAUTION: Crystal violet will stain your skin and clothing. Promptly wash off spills to minimize staining.**

9. After 60 seconds, pour off the excess stain. Gently rinse the smear in a beaker filled with distilled water, as shown in the diagram at right. Tilt the slide to remove any remaining water droplets. *Note: Be careful not to disturb the area where the smear is located.*

10. Place the slide back into the staining tray, and cover it with 10 drops of Gram's iodine for 60 seconds. **CAUTION: Gram's iodine will stain your skin and clothing. Promptly wash off spills to minimize staining.**

11. While holding the slide at an angle over an empty beaker, use an eyedropper filled with ethanol to gently rinse the smear, as shown in the diagram at right, until no stain rinses off. **CAUTION: Do not use alcohol when there are flames in the room. Do not light burners when others are using alcohol.**

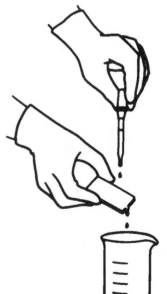

• Prepare several bottles of disinfectant solution. Pour fresh household bleach into squirt bottles labeled "household bleach disinfectant solution."

• Prepare separate containers for the disposal of broken glass, inoculating loops, and contaminated slides.

Disposal

• Never allow students to clean up bacteriological spills. Keep on hand a spill kit containing 500 mL of fresh bleach (full strength), biohazard bags (autoclavable), forceps, and paper towels. In the event of a bacterial spill, cover the area with a layer of paper towels, soak the towels with the bleach solution, and allow them to stand for 15–20 minutes. Wearing gloves and using forceps, place the residue in the biohazard bag. If broken glass is present, use a brush and dustpan to collect material, and place it in a suitably marked container.

• Used slides should be placed in a solution of fresh household bleach for 24 hours.

• All materials that come in contact with bacteria should be autoclaved.

Analysis

• All surfaces should be cleaned and decontaminated at the conclusion of each lab.

• Mix a small amount of fresh, full-strength household bleach with wastes containing Gram's iodine stain. When the mixture becomes colorless, rinse it down the drain with water.

12. Rinse the slide in a beaker filled with clean distilled water. Shake off the excess water.

13. Cover the smear with 10 drops of Safranin O for 15 seconds. **CAUTION: Safranin O will stain your skin and clothing. Promptly wash off spills to minimize staining.**

14. Rinse the slide in a beaker filled with clean distilled water, and then carefully blot the slide dry using an absorbent paper towel. *Note: Be careful not to disturb the smear.*

15. Repeat steps 4–14 using the slide labeled "Gram –" and *A. serpens*.

16. Add a drop of mounting medium to the smear, and then place a coverslip on each slide. View each slide under low power, and then switch to high power. Observe the bacteria, and record your observations in the table below.

Observations

Bacterium	Description (shape and color)	Drawing (high power)
B. megaterium	purple rods with bulges called endospores	
A. serpens	pink spirals	

17. Dispose of your materials according to the directions from your teacher.

18. Clean the lab-table surface with disinfectant solution and paper towels. Clean up your work area and wash your hands before leaving the laboratory.

19. You prepared slides using bacteria classified as Gram positive and Gram negative. What color should your bacteria be? Do your results agree?

Student answers will vary depending on technique. *B. megaterium* is a Gram-positive bacterium

and should appear as purple rods. *A. serpens* is Gram negative and should appear as pink spirals.

- Dispose of all other stains by washing them down the sink with water.
- Disinfectant solutions, staining solutions, and ethanol can be stored for future use.
- Solid trash that has been contaminated by biological waste must be collected in a separate and specially marked disposal bag. Package all sharp instruments in separate metal containers for disposal. No pipets should protrude from the disposal bag. It is recommended that disposal bags be autoclaved and then placed into a sealed container, such as a plastic bucket with a lid. Liquids and gels must be absorbed by paper towels to minimize the risk of leakage.

20. Why is the Gram stain useful in classifying bacteria?

It is useful in identifying bacteria with different types of cell walls. Staining also makes bacterial

cells easier to observe. It is also useful in medicine for determining the correct treatment of disease.

21. Why is the inoculating loop heated in a flame and then placed on agar in an area of no growth?

The loop is flamed to kill any bacteria that might contaminate the stock culture. The loop is placed

on the agar to cool so that it won't kill the bacteria being transferred.

22. Why did you use ethanol and a counterstain during the procedure?

The cell walls of bacteria differ in composition. Ethanol is a decolorizer, which washes out the

crystal violet from Gram-negative bacteria while Gram-positive bacteria are not affected because

their cell wall retains the crystal violet. Cells that lose the crystal violet are colorless, so a

counterstain with a contrasting color is added to make them easier to see.

Conclusions

- For further investigation, you may wish to provide unknown species for students to test using the Gram-staining technique. Bacterial species you may wish to order include *Bacillus cereus* and *Staphylococcus epidermidis*.

23. If you were a doctor, and a lab technician told you that the results of a patient's lab test were Gram positive, what antibiotic would you prescribe for the patient? Why?

Penicillin or erythromycin could be prescribed depending on the patient's allergy history. These

two antibiotics kill Gram-positive bacteria.

24. What would happen if a doctor prescribed penicillin for a Gram-negative bacterial infection? Why?

The patient would not get better because penicillin is ineffective with Gram-negative bacteria. The

doctor would have to prescribe streptomycin or tetracyline to see results.

25. Review and compare the structural components of the cell walls of Gram-positive and Gram-negative bacteria. Why do you think Gram-negative bacteria are less sensitive to antibiotics?

Gram-negative bacteria are less sensitive to certain antibiotics because the protective

lipopolysaccharide layer in their cell walls makes it difficult for the antibiotics to enter the cells. This

protective layer does not exist in the cell walls of Gram-positive bacteria.

Extensions

26. *Clinical laboratory microbiologists* identify pathogens, or disease-causing organisms, present in the specimens sent to them by doctors. After isolating the microbes, microbiologists run tests to determine the proper antibiotics to kill the pathogens. Find out about the training and skills required to become a clinical laboratory microbiologist.

27. Do research to determine whether bacterial shape is related to the results of Gram staining.

Name _____

Date _____ Class _____

HOLT
BIOSOURCES
LAB PROGRAM
EXPERIMENTAL DESIGN

C33 Gram Staining of Bacteria— Treatment Options

Prerequisites — **Laboratory Techniques C32: Gram Staining of Bacteria** on pages 157–162

Review
- aseptic technique
- techniques for preparing a bacterial smear
- Gram-staining techniques
- identification of Gram-positive and Gram-negative bacteria

 **PETMED** ANIMAL HOSPITAL

Baton Rouge, Louisiana

October 17, 1997

Caitlin Noonan
Research and Development
BioLogical Resources, Inc.
101 Jonas Salk Dr.
Oakwood, MO 65432-1101

Dear Ms. Noonan,

I am a veterinarian at a clinic here in Baton Rouge. We have had problems recently with our computer system. Some of our records have been mysteriously altered or deleted, and many of our test results have been scrambled. I have been able to salvage and reorganize most of the information, but one case in particular has been a problem. The case involves a dog with a bacterial infection. The test results that identify the bacteria are missing.

Additionally, the animal in question, a champion English bulldog named Fitz, is allergic to tetracycline. I still have the bacterial culture, and I will be repeating the test myself. However, considering our recent computer problems and Fitz's allergies, I have decided to have this test verified elsewhere to avoid mistakes that could harm the animal and lead to a lawsuit.

I have sent a culture of the bacteria to your company. Please identify the bacteria as Gram-positive or Gram-negative and recommend an antibiotic. It is imperative that we resolve this situation as quickly as possible. The owner is planning to enter Fitz in a dog show next month and would like to start treatment as soon as possible. Thank you in advance for your participation.

Sincerely,

Karl Jefferies

Karl Jefferies, D.V.M.
PetMed Animal Hospital

BioLogical Resources, Inc. Oakwood, MO 65432-1101

M E M O R A N D U M

To: Team Leader Pathology Dept.

From: Caitlin Noonan, Director of Research and Development

Have your teams test the unknown bacterial culture mentioned in the attached letter by completing a Gram stain. The bacterial culture will be sent to your lab as soon as it arrives here. Remember to use aseptic technique throughout the test.

Dr. Jeffries called me and informed me that he will need to begin treatment by the end of next week in order for the dog to recover in time for the dog show. As always, please be as efficient with your time as possible.

Proposal Checklist

Before you start your work, you must submit a proposal for my approval. **Your proposal must include the following:**

_____ • the **question** you seek to answer

_____ • the **procedure** you will use

_____ • a complete, itemized list of proposed **materials** and **costs** (including use of facilities, labor, and amounts needed)

Proposal Approval: _____
(Supervisor's signature)

Report Procedures

When you finish your analysis, prepare a report in the form of a business letter to Dr. Jefferies. **Your report must include the following:**

_____ • a paragraph describing the **procedure** you followed to prepare, stain, and identify the bacteria

_____ • your **conclusions** about the identity (Gram-positive or Gram-negative) of the bacteria and your recommendations for treatment

_____ • a detailed **invoice** showing all materials, labor, and the total amount due

Safety Precautions

• ⬦ ⬦ ⬦ Wear safety goggles, disposable gloves, and a lab apron.

• ⬦ Glassware is fragile. Notify your teacher promptly of any broken glass or cuts. Do not clean up broken glass or spills unless your teacher tells you to do so.

• ⬦ Never use electrical equipment around water, nor with wet hands or clothing. Never use equipment with frayed cords.

• ⬦ Wash your hands before leaving the laboratory.

• If you are using a microscope with a mirror, do not use direct sunlight as a light source.

Disposal Methods

• ⬦ Dispose of waste materials according to instructions from your teacher.

• Place all solid, uncontaminated waste materials into a trash can.

• Place broken glass inoculating loops, and contaminated slides in the separate containers provided.

• Wash reusable materials such as glassware and lab utensils, and return them to the supply area.

FILE: PetMed Animal Hospital

MATERIALS AND COSTS (Select only what you will need. No refunds.)

I. Facilities and Equipment Use

Item	Rate	Number	Total
facilities	$480.00/day		
personal protective equipment	$10.00/day		
compound microscope	$30.00/day		
microscope slide with coverslip	$2.00/day		
scissors	$1.00/day		
Bunsen burner with striker	$10.00/day		
slide holder	$5.00/day		
staining tray	$5.00/day		
100 mL beaker	$5.00/day		
eyedropper	$2.00/day		
clock or watch with second hand	$10.00/day		
hot plate	$15.00/day		

II. Labor and Consumables

Item	Rate	Number	Total
labor	$40.00/hour		
Fitz's culture	provided		
disposable inoculating loop	$5.00/each		
crystal violet stain	$50.00/bottle		
Gram's iodine	$50.00/bottle		
Safranin stain	$50.00/bottle		
95% ethanol in dropper bottle	$20.00/bottle		
methylene blue	$50.00/bottle		
distilled water	$0.10/mL		
paper towel	$0.10/sheet		
wax pencil	$2.00 each		
disinfectant solution	$2.00/bottle		

Fines

OSHA safety violation	$2,000.00/incident		

Subtotal	
Profit Margin	
Total Amount Due	

Name _____

Date _____ Class _____

HOLT
BioSources
LAB PROGRAM
EXPERIMENTAL DESIGN

C34 *Limiting Fungal Growth*

Prerequisite • **Laboratory Techniques C31: Using Aseptic Technique** on pages 149–156
Review
• aseptic technique
• characteristics of yeast

ROUGH IT
TRAVEL COMPANY
VACAVILLE, CA 95687

April 20, 1997

Sam Ashike
Health Division
BioLogical Resources, Inc.
101 Jonas Salk Dr.
Oakwood, MO 65432-1101

Dear Mr. Ashike,

I am the director of Rough-It Travel Company, a business that specializes in adventure travel. Some of the most popular programs we offer are our guided backpacking trips. We run trips all over the country, in both dry and humid regions. In order to prevent problems such as poor nutrition or food contamination, we always consult closely with our clients about the food they pack in their backpacks.

We do the best we can with these consultations, but despite our efforts we have had recurring problems with mold, especially in certain areas of the country. Obviously we want to avoid such problems. In order to provide sound advice for our customers, we need to know which of several common food products are most likely to support the growth of mold and under what conditions (damp or dry) they do so.

I have included a sample of the most common mold we see, *Penicillium*. I will be contacting you with a list of the products we will need to have tested. Please let me know as soon as possible if you and your team are willing to take on this project.

Tom Swarner

Tom Swarner
Programming Director
Rough-It Travel Company

BioLogical Resources, Inc. Oakwood, MO 65432-1101

M E M O R A N D U M

To: Team Leader, Food Testing Dept.

From: Sam Ashike, Director of Health

Have your teams begin work on the attached request as soon as possible. Mr. Swarner has informed me that Rough-It, a rapidly growing business, is planning to increase its food research over the next few years. Our participation in this study could lead to future contracts with Rough-It.

Sample *Penicillium* cultures have been delivered to your lab. This morning I received a phone call from Mr. Swarner in which he gave me the list of items that he wants reviewed. They are potato flakes, cornstarch, raisins, and orange peels. He would like these products to be tested in damp and dry conditions. In order to ensure consistent results, I would suggest keeping all samples in a darkened area at a temperature of approximately 25°C.

Proposal Checklist

Before you start your work, you must submit a proposal for my approval. **Your proposal must include the following:**

_____ • the **question** you seek to answer

_____ • the **procedure** you will use

_____ • a detailed **data table** for recording observations

_____ • a complete, itemized list of proposed **materials** and **costs** (including use of facilities, labor, and amounts needed)

Proposal Approval: _____
(Supervisor's signature)

Report Procedures

When you finish your analysis, prepare a report in the form of a business letter to Mr. Swarner. **Your report must include the following:**

_____ • a paragraph describing the **procedure** you followed to determine which mediums experienced mold growth

_____ • a complete **data table** showing the number of colonies in and description of each medium over a period of days

_____ • your **conclusions** about which conditions are best for deterring mold growth

_____ • a detailed **invoice** showing all materials, labor, and the total amount due

Safety Precautions

• Wear safety goggles, disposable gloves, and a lab apron.

• Glassware is fragile. Notify your teacher promptly of any broken glass or cuts. Do not clean up broken glass or spills unless your teacher tells you to do so.

• Never use electrical equipment around water, nor with wet hands or clothing. Never use equipment with frayed cords.

• Wash your hands before leaving the laboratory.

• Do not taste or ingest any food items in the laboratory.

Disposal Methods

• Dispose of waste materials according to instructions from your teacher.

• Place used paper towels in a trash can.

• Place broken glass, used cotton swabs, and mold samples in the separate containers provided.

• Wash reusable materials such as glassware and lab utensils, and return them to the supply area.

FILE: Rough-It Travel Company

MATERIALS AND COSTS (Select only what you will need. No refunds.)

I. Facilities and Equipment Use

Item	Rate	Number	Total
facilities	$480.00/day	_____	_____
personal protective equipment	$10.00/day	_____	_____
compound microscope	$30.00/day	_____	_____
depression slide with coverslip	$2.00/day	_____	_____
hand lens	$5.00/day	_____	_____
eyedropper	$2.00/day	_____	_____
clock with second hand	$10.00/day	_____	_____

II. Labor and Consumables

Item	Rate	Number	Total
labor	$40.00/hour	_____	_____
cotton swabs	$0.10 each	_____	_____
Penicillium notatum	provided	_____	_____
potato flakes	$1.00/sample	_____	_____
cornstarch	$1.00/sample	_____	_____
raisins	$1.00/sample	_____	_____
orange peel	$1.00/sample	_____	_____
slowing agent	$1.00/drop	_____	_____
wax pencil	$2.00 each	_____	_____
small jar with lid	$2.00 each	_____	_____
disinfectant solution	$2.00/bottle	_____	_____
spray bottle of distilled water	$2.00/bottle	_____	_____
paper towel	$0.10/sheet	_____	_____

Fines

Item	Rate	Number	Total
OSHA safety violation	$2,000.00/incident	_____	_____

Subtotal	_____
Profit Margin	_____
Total Amount Due	_____

Name _____

Date _____ Class _____

HOLT
BIOSOURCES

LAB PROGRAM
LABORATORY TECHNIQUES

C35 Staining and Mounting Stem Cross Sections

Skills
- using a compound light microscope
- making a wet mount
- staining tissue

Objectives

PREPARATION NOTES

Time Required: one 50-minute period

- *Prepare* slides of stem cross sections for examination with a compound light microscope.
- *Compare* the arrangement of vascular tissue in monocots and dicots.
- *Identify* a plant as a monocot or a dicot based on the arrangement of its vascular tissue.

Materials

Materials
Materials for this activity can be purchased from WARD'S. See the *Master Materials List* for ordering instructions.

- safety goggles
- lab apron
- microscope slides (3)
- alcohol swab
- forceps
- vial of monocot stem cross sections
- treating dish
- trichrome stain in dropper bottle
- water
- stopwatch or clock with second hand
- tissue application stick
- mounting medium (Pyccolyte II) in dropper bottle
- coverslips (3)
- compound microscope
- vial of dicot stem cross sections
- prepared reference slide of monocot and dicot root cross sections
- vial of unknown stem cross sections
- prepared slide of unknown root cross section

Purpose

You are a plant ecologist who specializes in studying the biodiversity of the rain forest. On a recent collecting expedition to the Amazon rain forest, you discovered what you think is a new species of plant. It has characteristics similar to both monocots and dicots but is different from other plants in many of its characteristics. The first thing you need to do to identify the plant is to determine if it is a monocot or a dicot. To do this, you will mount and stain tissue from the plant's stem and observe it with a compound light microscope.

Background

Flowering plants, **angiosperms,** are divided into two subclasses: Monocotyledoneae, commonly called **monocots,** and Dicotyledoneae, commonly called **dicots.** Monocots include familiar plants such as orchids, irises, lilies, grasses, sedges, and palms. Typically, the embryo within the seed of monocot plants has one well developed cotyledon, or seed leaf. Monocot leaves are generally narrow and have parallel veins running from base to tip. In monocots, the vascular tissue in the stem is organized into separate bundles. When the stem is seen in cross section, these bundles are scattered throughout the stem. Monocot stems do not contain a cambium. A **cambium** is a ring of rapidly growing cells that separates the **xylem,** tissue that conducts water, from the **phloem,** tissue that conducts food.

The second group of flowering plants, the dicots, includes broadleaf trees such as ash, maple, and oak; shrubs and vines such as rose, poison ivy, and lilac; and other plants such as tomatoes, sunflowers, and morning glories. The embryo within a dicot seed contains two cotyledons. The leaves of dicot plants are typically broad, and the veins are usually arranged in a fanlike or netlike pattern. The vascular tissues of the stem are typically arranged in a circular pattern. Unlike monocot stems, dicot stems have a cambium. Many dicots live for more than one year.

Many of the common external characteristics used to distinguish a monocot from a dicot are not reliable. Some monocots, for example, have broad leaves. The arrangement of vascular tissue in a plant's stem, however, is a reliable way to tell if a plant is a monocot or dicot. The figure below compares the arrangement of vascular tissue in a monocot and dicot stem.

Monocot stems have vascular tissue arranged in randomly scattered bundles.

Dicot stems have vascular tissue arranged in rings.

Procedure

1. Put on safety goggles and a lab apron.

2. Swab three microscope slides with the supplied alcohol pad. Dispose of the alcohol pad as instructed by your teacher.
   CAUTION: Glassware is fragile. Notify your teacher promptly of any broken glass or cuts. Do not clean up broken glass or spills unless your teacher tells you to do so.

3. Use forceps to remove one monocot stem cross section from the vial, and place it in a treating dish. Place two drops of trichrome stain on the cross section. Let it stain for one to two minutes. Flood the dish with water and swirl.

Preparation Tips

• Provide an "unknown" plant for students to classify by relabeling one of the vials of monocot or dicot stem cross sections and relabeling a prepared slide of a monocot or dicot root cross section from the same type of plant.

• Provide separate containers for the disposal of stained plant tissue and broken glass.

Disposal

Any stained plant material that is on a slide can be disposed of in a waste glass container. Plant material not on a glass slide can be thrown in the trash.

PROCEDURAL NOTES

Safety Precautions

• Remind students to avoid any skin/eye contact with the stain, mounting medium, and holding fluid of the stems. Flush spills or splashes with water for 15 minutes.

• Discuss all safety symbols and caution statements with students.

• Instruct students to follow the safety rules for carrying and using the compound light microscope.

4. Remove the cross section with a wooden application stick, and place it on the center of a cleaned slide.

5. Add two or three drops of mounting medium on the stem cross section, and add a coverslip. *Note: Fill in any air spaces by adding more mounting medium to the side of the coverslip.*

6. Observe the prepared slide with a compound light microscope first under low power and then under high power. *Note: Be careful when you switch from low power to high power. Your slide may be too thick to observe under high power. You might damage the objective or break the slide.* In the spaces below, draw a representation of what you see under low power and one vascular bundle as it appears under high power. Carefully label the various stem features.

Monocot Stem Cross Section (low power)

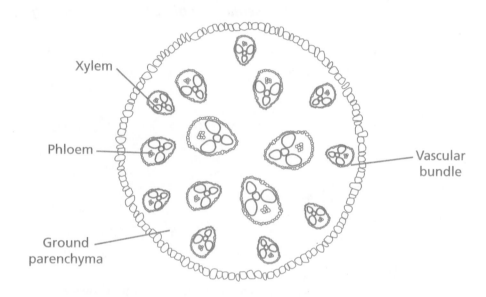

Vascular Bundle of Monocot Stem (high power)

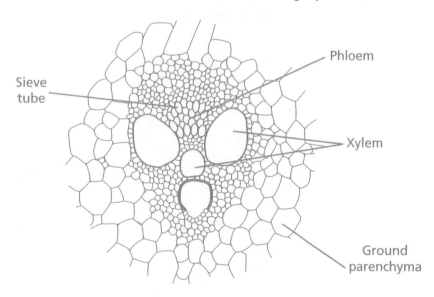

Procedural Tips
• The stem preparations that students make may be fairly thick. Recommend that the students take care not to force the objective down onto their preparation by mistake, thus crushing the slide and possibly damaging the microscope.
• You may want to display the prepared slides of monocot and dicot stem cross sections and the stem cross section from an unknown plant, under low power, on microscopes accessible to the whole class. Or you may want to project the slides on a projecting scope if one is available.

7. Repeat steps 3 through 6 for the dicot stem cross section. In the spaces below, draw a representation of what you see under low power and one vascular bundle as it appears under high power. Carefully label the various stem features.

Dicot Stem Cross Section (low power)

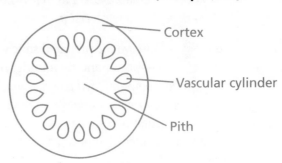

Vascular Bundle of Dicot Stem (high power)

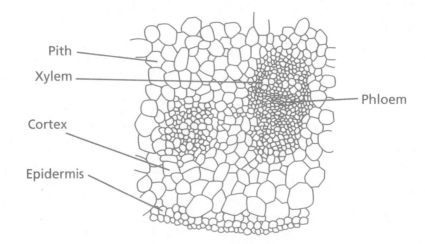

8. Observe a prepared reference slide that shows a side-by-side comparison of a monocot and dicot root. In the space below, draw and identify the various root structures.

Monocot Root   Dicot Root

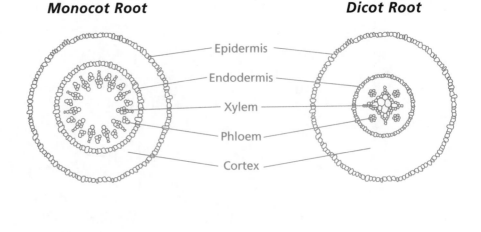

9. Repeat steps 3 through 6 for the unknown stem section. In the spaces below, draw a representation of what you see under low power and one vascular bundle as it appears under high power. Label the various stem features carefully.

Unknown Stem Cross Section (low power)

Vascular Bundle of Unknown Stem (high power)

10. Observe a prepared slide that shows a cross section of the root from the unknown plant. In the space below, draw and identify the various root structures.

Root Cross Section from Unknown Plant

11. Dispose of your materials according to the directions from your teacher.

12. Clean up your work area and wash your hands before leaving the lab.

Analysis **13.** Compare the arrangement of vascular tissue of the monocot and dicot stem cross sections in the two slides you made.

The monocot stem has vascular bundles scattered throughout the cortex, whereas the dicot stem

has well organized concentric rings of xylem and phloem with a vascular cambium region between

the two.

14. Compare monocot and dicot roots.

Monocot roots have vascular bundles arranged in a circular formation, whereas dicot roots have a

characteristic "X within a circle" formation.

Conclusions **15.** Is the unknown plant a monocot or a dicot? Explain your answer.

Answers will vary depending on the actual plant observed. Students should support their answers

with the proper arrangement of vascular bundles in the stem: scattered bundles for a monocot and

arranged in rings for a dicot.

16. Would it be safe to classify a plant as a monocot or a dicot on the basis of leaf, flower, and seed structure only? Explain or support your answer.

It would not be safe to classify a plant as a monocot or a dicot based on leaf, flower, and seed

structure only because these characteristics can be unreliable. The only sure way to classify a plant

as a monocot or a dicot is to study the arrangement of vascular tissue in stem and root cross

sections.

Extensions **17.** Collect a variety of plants from around your school or home. Identify each plant as either a monocot or a dicot. If you do not know the name of the plants, use a field guide or dichotomous key to find the plant's name.

• Be sure that students can identify any poisonous plants that grow in your area or that students are accompanied by an adult qualified to identify poisonous plants. Do not allow students to collect poisonous plants.

18. Collect a variety of plant stems from around your school or home. Identify the plant names if possible. Keep the plants in a plastic bag in the refrigerator until you are ready to study them. Using external features, separate the plants into monocot or dicot groups. Support your grouping by making very thin cross section cuts from the stems. Follow the procedure above to make wet mounts of each cross section. Identify each plant you observe as either a monocot or a dicot.

19. Research the training needed to become a *plant ecologist*. Identify the career possibilities and schools that offer this type of training.

HOLT
BIOSOURCES
LAB PROGRAM
LABORATORY TECHNIQUES

C36 *Using Paper Chromatography to Separate Pigments*

Skills
- using paper chromatography to study plant pigments
- interpreting a chromatogram
- making inferences

Objectives
- *Separate* simulated leaf pigments using paper chromatography.
- *Determine* the R_f value of each pigment.

Materials

PREPARATION NOTES

Time Required:
one 50-minute period

- safety goggles
- lab apron
- scissors
- chromatography-paper strip
- pencil
- simulated extract of plant pigment from desert plant
- micropipet
- chromatography reaction chamber
- graduated cylinder
- chromatography solvent (5 mL)
- stopwatch or clock with second hand
- metric ruler

Purpose

Materials

Materials for this lab activity can be purchased from WARD'S. See the *Master Materials List* for ordering instructions.

You are a plant physiologist working for a natural clothing company. All the garments manufactured by your company are made from natural plant fibers and are dyed with natural dyes. One of your company's designers wants you to develop a new yellow dye. Research shows that the leaves of a particular plant may contain the yellow pigment you need. You must separate the pigments in an extract from the leaves of this plant to determine whether the leaves contain a yellow pigment. The separation technique you decide to use is paper chromatography.

Background

Additional Background

One of the main tasks of chemists and biochemists is to unravel the complexities of chemical compounds and reduce them to their simple, individual components. Paper chromatography is an important technique used in the separation, isolation, and identification of chemical compounds. The term *chromatography* is derived from the Greek words *chromat,* which means color, and *graphon,* which means to write. Chromatography was originally used

The leaves of most plants contain many different **pigments,** or light-absorbing compounds. **Chlorophyll,** the main pigment of green plants, is needed for photosynthesis. The primary purpose of chlorophyll is to capture light energy, which is converted to chemical energy during photosynthesis. Chlorophyll is the most abundant and important photosynthetic plant pigment and exists in two forms: *chlorophyll a* and *chlorophyll b.* Both chlorophylls absorb blue and red light and reflect green light.

Chlorophyll often hides the other pigments present in leaves. In autumn, chlorophyll breaks down, allowing **xanthophyll,** which reflects yellow light, and **carotene,** which reflects orange light, to show their colors. Other pigments may also be present in leaves.

The individual pigments in a mixture of pigments from a leaf may be separated by the technique of **paper chromatography.** *Chromatography* means color writing. The separation takes place by *absorption* and *capillarity.* The paper holds substances by absorption. Capillarity pulls the substances up the paper at different rates. Pigments are separated on the paper and show up as colored streaks. The pattern of separated components on the paper is called a **chromatogram.**

The relative rate of migration, the R_f **value,** for each pigment can be determined from the chromatogram. The R_f value is the ratio of the distance a pigment moved on the chromatogram to the distance the solvent front moved.

Procedure

to separate pigments in leaves, berries, and natural dyes. Over the last 50 years, chromatography has been applied to the separation of a series of organic and biological compounds.

Preparation Tips

• This activity uses simulated plant pigments in nontoxic solutions. If you use real leaf pigments and organic solvents, you will need to make appropriate safety changes.

• The chromatography paper you purchase will be in sheets 7 cm × 10 cm. Cut each sheet into 10 strips 1 cm × 7 cm.

Disposal

• Because the plant pigments are simulated and the solvent nontoxic, all materials used in this lab can be flushed down the drain or placed in the trash.

• Provide separate containers for the disposal of any broken glass and other items used in the lab.

• Provide a container under the hood for solvent and pigment collection.

1. Put on safety goggles and a lab apron.

2. With scissors, cut the bottom of a strip of chromatography paper into a point as shown in the diagram at right. **CAUTION: Handle scissors carefully. Notify your teacher of any cuts.** Draw a faint pencil line a few millimeters above the pointed end of the paper strip.

3. With a micropipet, apply a very small drop of simulated plant pigment on the center of the pencil line. *Note: Do not allow the simulated plant pigment to dry on the strip of chromatography paper.*

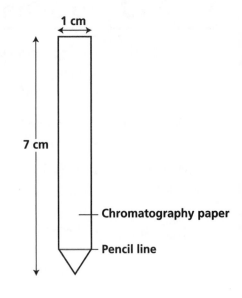

4. Pour about 5 mL of chromatography solvent into a chromatography reaction chamber. Pull the paper-chromatography strip through the opening of the cap as shown in the diagram at right. Adjust the paper strip so that the tip just touches the solvent. *Note: Do not immerse the pigment spot in the solvent.* **CAUTION: If you get a chemical on your skin or clothing, wash it off at the sink while calling to your teacher.**

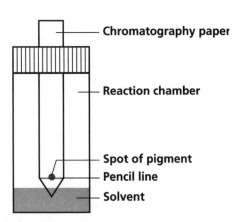

If you get a chemical in your eyes, promptly flush it out at the eyewash station while calling to your teacher. Notify your teacher in the event of a chemical spill.

5. Place the cap over the reaction chamber, and carefully bend the end of the paper strip over the cap. *Note: Make sure the strip of paper is hanging straight down within the reaction chamber. Also make sure the paper strip does not touch the sides of the chamber.*

6. Watch the solvent rise up the paper, carrying and separating the pigments as it goes. After five to seven minutes, you will notice an obvious change in the paper within the reaction chamber. *At the instant* the solvent reaches the top of the chamber, remove the paper, and let it air dry.

PROCEDURAL NOTES

Safety Precautions
• Discuss all safety symbols and caution statements with students.
• Remind students not to touch or taste any chemicals used in the lab.

Procedural Tip
The simulated plant pigment is specially formulated and does not include the typical hazardous solvents needed to extract real plant pigments. Additionally, the developer is a nontoxic detergent solution rather than the petroleum-ether and acetone mixture that is typically used to separate real plant pigments. The chromatogram of the simulated plant pigments will be almost identical to the chromatogram of real plant pigment extracts in terms of appearance and R_f values. Simulated plant pigments are used to eliminate the use of hazardous developers and solvents normally associated with the use of real plant pigments for chromatography.

7. Observe the bands of pigment. Identify the pigments by their colors: carotene (orange), xanthophyll (yellow), chlorophyll a (yellow-green), and chlorophyll b (blue-green). With a pencil, mark the highest point the solvent traveled up the chromatography-paper strip. Also, make a pencil mark at the center of each of the separated pigments. Write your name on the strip as well.

8. Measure the distance in mm of the solvent from its starting point to the highest point it traveled on the paper strip. Then measure the distance traveled by the different pigments from the starting point to the center of each pigment band. Record your measurements in the data table below.

Simulated Plant Pigments

Band number	Pigment	Color	Migration distance in mm	R_f value
1 (top)	carotene	orange	37	0.92
2	xanthophyll	yellow	29	0.72
3	chlorophyll a	yellow-green	17	0.42
4	chlorophyll b	blue-green	8	0.20
Solvent	NA	NA	40	NA

9. In the space provided at right, make a sketch of the chromatogram. The chromatogram of the simulated plant pigments will fade over time, so it is necessary to record the data in order to preserve it.

10. Use the following formula to calculate the R_f values for each pigment as a decimal fraction. Record your answers in the data table above.

$$R_f = \frac{\text{distance substance's pigment traveled}}{\text{distance solvent traveled}}$$

11. Dispose of your materials according to the directions from your teacher.

12. Clean up your work area and wash your hands before leaving the lab.

Analysis 13. Describe what happened to the original spot of simulated plant pigment.

The original spot of simulated plant pigment separated into four bands.

14. Which of the pigments migrated the farthest? Why?

Carotene migrated the farthest of the four pigments because it is the most soluble and has the

least retention compared with the other pigments.

Conclusions

15. Would a plant containing the pigments shown in the data table on the previous page be a source for yellow dye? Explain your answer.

Yes, the plant contains the pigment xanthophyll, which could be a source of yellow dye.

16. What other colors of dye might be available from the same plant? Explain your answer.

The same plant might also be a source of orange dye because the pigment carotene is present and

a source of yellow-green dye from chlorophyll a and blue-green from chlorophyll b.

17. Which of the chlorophyll forms is more soluble? Why?

Chlorophyll a appears to be more soluble than chlorophyll b since it migrated farther on the

chromatogram than chlorophyll b.

Extensions

18. Do research in the library or via the Internet to find out what professions use the technique of paper chromatography.

19. *Plant physiologists* use a variety of laboratory techniques to study the chemical processes that make a plant function. Find out about the training and skills required to become a plant physiologist. In addition to chromatography, find out what other kinds of laboratory tests a plant physiologist might do.

Name _____

Date _____ Class _____

HOLT
BIOSOURCES
LAB PROGRAM
LABORATORY TECHNIQUES

C37 *Growing Plants in the Laboratory*

Skills
- growing plants from seeds in the laboratory
- hand pollinating flowers

Objectives
- *Observe* the germination of radish seeds and the growth and development of radish plants.
- *Grow* radish plants until seeds can be harvested from the plants.
- *Distinguish* between self-pollination and cross-pollination.
- *Recommend* effective techniques for growing radish plants used in research.

PREPARATION NOTES

Time Required: about 35 days total: 1 class period, initially; 5 minutes daily to check and

Materials

water plants for up to 35 days; 10 minutes once a week to fertilize; 10 minutes on day 2 or 3 to transplant seedlings; 20 minutes on day 10 to skate seedlings; 10 minutes each on three consecutive days (beginning on about day 15) for pollinations; 1 class period on about day 35 for counting and compiling class data

- growing container with 4 sections
- adhesive label
- waterproof pen or pencil
- safety goggles
- lab apron
- Rapid Radish seeds (8)
- transparent tape
- water
- metric ruler

- plastic container
- plant labels (4)
- fertilizer
- plastic pipet
- forceps
- pollinating brushes (3)
- stakes (4)
- twist ties (4)

(class materials)
- growing/watering system
- lighting system

- algae inhibitor

Purpose

Materials

Materials for this lab activity may be purchased from WARD'S. See the *Master Materials List* for ordering instructions.

You work as a lab assistant in the Plant Sciences department of the university in your town. Your job is to grow and care for plants that are used in research. One of the researchers is planning a series of experiments on rapid-cycling radishes. The researcher has asked you to find out whether these radish plants are self-pollinating when grown in the laboratory or whether they must be cross-pollinated in order to produce seeds. To do so, you grow the radishes until mature seeds can be harvested from them.

Background

Additional Background

Radishes are self-incompatible, as students will discover when they conduct the experiment. Self-incompatibility, which is a genetic trait, is the inability of pollen from one individual to fertilize the eggs of the same individual. This

Rapid-cycling plants are plants with a short life cycle. The **life cycle** of a plant begins with the germination of a seed and includes growth, flowering, development of fruits, and maturation of seeds. Rapid-cycling radishes complete their life cycle in about 35 days.

Radish plants are easily grown from seeds. A **seed** is a plant embryo surrounded by a protective coat. When a viable seed (one with a live embryo) takes in water, it **germinates,** or begins to grow. The young plant that grows from a seed is called a **seedling.** When a seed germinates, the seedling breaks through the surface of the soil and is said to **emerge.** The first leaflike structures to appear on a radish seedling are the **cotyledons,** or seed leaves.

trait is an adaptation that insures cross-pollination and helps maintain genetic variability.

Preparation Tips

• Review the instructions in the growing guide that comes with WARD'S Rapid Radish kits.

• You may wish to set up additional containers of growth medium and plant seeds in them in case students don't have good success growing plants.

Procedure

• Mix fertilizer according to the directions, and distribute 10 mL to each group weekly.

• If you do not purchase the growing/watering system, prepare an area in the lab where growing containers can be placed. Three-inch or four-inch plastic pots can be substituted for the growing containers. If water is not easily accessible, provide water containers.

• Prepare an area in the lab where plants can receive adequate light. "Gro-lites" and cool white fluorescent lights provide the best source of light.

• Provide a place for students to put plant containers when finished so they can be reused.

Disposal

• All excess growth medium should be stored for future use or thrown into a trash can.

• Return liquid fertilizer to a stock container.

Radish plants produce **perfect flowers,** which have both male and female reproductive parts. **Stamens** are the male reproductive parts of flowers. They consist of a pollen-producing sac called an **anther** and a **filament,** which supports the anther. **Pistils** are the female parts of flowers. They consist of an ovary, a **style** (stalk that connects the ovary to the stigma), and a **stigma** (sticky top of a pistil). Seeds develop within the **ovary,** which becomes a **fruit** as it matures.

Pollination, the transfer of pollen from an anther to a stigma, is necessary for the production of seeds. **Self-pollination** is the transfer of pollen from an anther to the stigma of the same flower. **Cross-pollination** is the transfer of pollen from an anther of a flower on one plant to the stigma of a flower on a different plant. Researchers often **hand-pollinate** flowers, using a brush or some other means, to ensure that pollination and seed production occurs. After pollinations have been made, the researcher will **terminalize** (remove the tip of) the plants.

1. Using a waterproof pen or pencil, write the initials of all group members and the date on an adhesive label. Stick the label on one end of a growing container.

2. Put on safety goggles and a lab apron.

3. Fill each section in the growing container with growth medium. Gently tap the bottom of the container on your lab table to make the medium settle into each section. *Note: Do not press on the medium to pack it down.* Repeat this process until each section is filled to within 2 cm (0.8 in.) of the top.

4. Place eight Rapid Radish seeds on a piece of transparent tape, and transport them to your work area.

5. Place two seeds 2 cm apart in each section of the growing container. Fill a plastic container with water, and carefully water the seeds. Then cover the seeds with 2–3 mm of growth medium.

6. Using a waterproof pen or pencil, write the numbers 1, 2, 3, or 4 on four plant labels. Place a label in each section of the growing container.

7. Place the growing container on the mat on the platform of the growing/watering system. Be sure the mat is wet and that there is water in the watering tray.

8. Clean up your work area and wash your hands before leaving the lab.

9. Fertilize your plants once a week. To do so, use a plastic pipet to add about 2.5 mL (2 squirts) of fertilizer solution to each section. *Note: Be careful to place the fertilizer solution on the soil, not on the plants.*

10. If no seeds have germinated in a section by day 2 or 3, transplant a seedling from a section with 2 seedlings into the empty section. To transplant a seedling, first make a small hole in the growth medium using a pencil. Then, using forceps, loosen the growth medium under the seedling to be transplanted. Next, use the forceps to *carefully* and *gently* lift the seedling from its growth medium and place it gently in the hole. Gently push the medium around the roots, and gently press down on the medium to secure the seedling.

PROCEDURAL TIPS

Safety Precautions

• Discuss all safety symbols and caution statements with students.

• Caution students to not touch or taste any solution or chemical used in the laboratory.

Procedural Tips

• When students are transplanting plants, make sure each group has only one plant per pot for a total of four plants. Students may have to obtain a seedling from another group. Collect extra seedlings, and plant them in the stock pots.

• If a watering system is not available, have students use a watering can and water slowly.

• Make sure students check daily for moist soil. If soil is not moist, the problem needs to be investigated and solved.

• Prepare a class data table on the chalkboard or on an overhead transparency for students to use when compiling class data.

• You may want to have students graph their results. Make an overhead transparency of the graph with axes labeled and numbered. Guide students in the construction of the graph. Instruct students to make a key and create a title for their graph.

• Instruct students to empty all growing containers into a bucket, rinse out the growing containers, and turn them upside down to drain.

11. On day 10, stake each seedling. To do so, insert a stake in the soil and prop the seedling against it. Loop a twist tie tightly around the stake a couple of times, and then loosely wrap the twist tie around the tender stem of the plant at a node.

12. When plants begin to flower, pollinate the flowers on each plant for three consecutive days, according to the following treatments:
Treatment 1—Self-pollinate plant 1
Treatment 2—Cross-pollinate plant 2 with pollen from plant 3
Treatment 3—Cross-pollinate plant 3 with pollen from plant 2
Treatment 4—Control (no pollination on plant 4)

13. To self-pollinate plant 1, swirl a brush for a few seconds inside each open flower.

14. To cross-pollinate plants 2 and 3, swirl a different brush inside a flower on plant 2 to pick up some of its pollen. Then swirl the brush inside a flower on plant 3 to deliver pollen from plant 2 to the flower's stigma and to pick up some of plant 3's pollen. Using the same brush, return to the original flower on plant 2 and swirl the brush inside of that flower to deliver pollen from plant 3 to the flower's stigma. Using a new brush, repeat this procedure until you have pollinated an equal number of flowers on plants 2 and 3 each day. Remove any unpollinated flowers from the plants.

15. After making pollinations for 3 days, terminalize each plant by pinching off its tip just above the last open flower. Also pinch off any side branches that have developed on the plants.

16. After day 35, remove your plants from the growing/watering system, and allow the plants to dry. Count the number of seed pods per plant. After the pods have dried, open the pods and count the total number of seeds produced by each plant. Record your data in the table below. Share your data with the other groups in your class by recording your data in a class data table (on the chalkboard or overhead projector). Using the class data, determine the class mean for the number of pods per plant and the number of seeds per plant. Record the means for each treatment in the table below.

Results of Plant Pollinations

Plant	1	2	3	4
Treatment	Self-pollination	Cross-pollination with plant 3	Cross-pollination with plant 2	Control (no pollination)
Number of flowers pollinated				
Number of pods per plant				
Total seeds per plant				
Class mean for the number of pods per plant				
Class mean for the total number of seeds per plant				

17. Dispose of your materials according to the directions from your teacher.

18. Clean up your work area and wash your hands before leaving the lab.

Analysis **19.** Why do you think you were asked to compile the data from your class? Hint: What does each group's data represent?

Answers will vary. Students should recognize that each group's data represents a repetition of the

experiment. By repeating an experiment, you can be more certain that your results are due to the

treatment and not to some unrelated factor.

20. What was the purpose of terminalizing the plants after your pollinations were completed?

Answers will vary. Students should recognize that terminalization prevents the plants from

continuing to grow and enables them to devote all energy to fruit and seed production.

Conclusions **21.** Do rapid-cycling radish plants need to be cross-pollinated in order to produce seeds when grown in a laboratory? What evidence supports your conclusion?

Students' data should support their conclusions. Students should find that rapid-cycling radishes

must be cross-pollinated to produce seed.

22. Based on your experiment, what recommendations would you make for growing rapid-cycling radishes in the laboratory?

Students should summarize the growing procedures they used in class and could include any

modifications that they found to be helpful.

Extensions **23.** *Plant physiologists* study the growth, development, and functioning of plants. Find out about the training and skills required to become a lab assistant for a plant physiologist, and discover where plant physiologists work.

24. Do research about self-incompatibility in plants. Discover some important food plants that are self-incompatible, and find out what growers must do to ensure that these plants produce fruit and seeds.

HOLT
BIOSOURCES
LAB PROGRAM
EXPERIMENTAL DESIGN

C38 Growing Plants in the Laboratory— Fertilizer Problem

Prerequisites	• **Laboratory Techniques C37: Growing Plants in the Laboratory** on pages 181–184
Review	• procedure for growing plants in the laboratory
	• role of nutrients in plant growth
	• N, P, and K

AGRICULTURAL OUTREACH

Washington D.C.

February 7, 1998

Lee Kwan
Macrobiology Division
BioLogical Resources, Inc.
101 Jonas Salk Dr.
Oakwood, MO 65432-1101

Dear Ms. Kwan,

Agricultural Outreach is a privately funded consulting firm that was established to assist with foreign agricultural issues. We work closely with the U. S. Department of Agriculture, acting as consultants in areas not covered by USDA departments. Our clients include farmers, developers, environmentalists, and governments.

Recently, we received a request from a small vegetable growers cooperative in Belize. The growers have had problems with their crops and have asked us to help them solve their problem. Our experts think their crops are growing poorly because of their fertilizer. We have subcontracted with a local chemistry lab to analyze the vegetable growers' fertilizer. We would like your firm to do the horticultural analysis.

Please conduct the proper tests to determine the correct fertilizer mixture for optimal growth of radish plants. We will send you a sample of the fertilizer used by the vegetable growers in Belize.

Sincerely,

Chester Pennick

Chester Pennick
Director
Agricultural Outreach

BioLogical Resources, Inc. Oakwood, MO 65432-1101

M E M O R A N D U M

To: Team Leader, Botany Dept.

From: Lee Kwan, Director of Macrobiology

Please have your research teams design and conduct an experiment to compare the effects of different fertilizers on the growth of vegetable plants. Since you have worked with rapid-cycling radishes before, use them as your experimental vegetable plants. You will need to spend about two weeks on this project, so please start as soon as possible.

A sample of the fertilizer used by the vegetable growers in Belize has been delivered to your lab. I suggest that you compare plants grown with the fertilizer from Belize with plants grown with several different types of fertilizers, such as the following:

fertilizer with N, P, and K

fertilizer with N and P only

fertilizer with N and K only

fertilizer with P and K only

Proposal Checklist

Before you start your work, you must submit a proposal for my approval. **Your proposal must include the following:**

_____ • the **question** you seek to answer

_____ • the **procedure** you will use

_____ • a detailed **data table** for recording results

_____ • a complete, itemized list of proposed **materials** and **costs** (including use of facilities, labor, and amounts needed)

Proposal Approval: _____

(Supervisor's signature)

Report Procedures

When you finish your analysis, prepare a report in the form of a business letter to Mr. Pennick. **Your report must include the following:**

_____ • a paragraph describing the **procedure** you followed to determine the best fertilizer for growing radish plants

_____ • a complete **data table** showing plant growth for each type of fertilizer

_____ • your **conclusions** about the fertilizer mixture that should be recommended to the client

_____ • a detailed **invoice** showing all materials, labor, and the total amount due

Safety Precautions

• Wear safety goggles and a lab apron.

• Wash your hands before leaving the laboratory.

Disposal Methods

• Dispose of all waste materials according to instructions from your teacher.

• Place broken glass, extra soil, used solutions, unused fertilizer, and radish plants in the separate containers provided.

• Wash reusable materials such as glassware and lab utensils, and return them to the supply area.

FILE: Agricultural Outreach

MATERIALS AND COSTS (Select only what you will need. No refunds.)

I. Facilities and Equipment Use

Item	Rate	Number	Total
facilities	$480.00/day		
personal protective equipment	$10.00/day		
growing containers	$2.00/day		
petri dish	$5.00/day		
growing/watering system	$10.00/day		
lighting system	$10.00/day		
metric ruler	$1.00/day		
scissors	$1.00/day		

II. Labor and Consumables

Item	Rate	Number	Total
labor	$40.00/hour		
Rapid Radish seeds	$0.50/seed		
growth medium	$5.00/kg		
NPK fertilizer solution	$0.30/mL		
NP fertilizer solution	$0.30/mL		
NK fertilizer solution	$0.30/mL		
PK fertilizer solution	$0.30/mL		
vegetable growers' fertilizer	provided		
pan, plate, or piece of plastic	$2.00 each		
stakes	$2.00 each		
string	$0.10/cm		
twist ties	$0.10 each		
pollinating brushes	$3.00 each		
adhesive labels	$0.10 each		
plant labels	$0.25 each		
plastic pipets	$2.00 each		
algae inhibitor	$5.00/bottle		

Fines

OSHA safety violation	$2,000.00/incident		

Subtotal	
Profit Margin	
Total Amount Due	

HOLT
BIOSOURCES
LAB PROGRAM
EXPERIMENTAL DESIGN

C39 | *Response in the Fruit Fly*

Prerequisites • none

Review • the term *taxis*

CheMystery Laboratories
Tallahassee, FL 90743-0356

May 16, 1997

Lee Kwan
Macrobiology Division
BioLogical Resources, Inc.
101 Jonas Salk Dr.
Oakwood, MO 65432-1101

Dear Ms. Kwan,

I am a chemist here at CheMystery Laboratories. I have recently been assigned the task of creating a new pesticide specifically for *Drosophila melanogaster*, or the common fruit fly. Before I can start my research, I need some behavioral information about these insects. I need to know if the flies are attracted, repelled, or unaffected by the stimuli light and odor.

We are willing to pay the normal fee for your services. Please send me a report that includes your raw data, the procedures you used to test the flies' responses to the stimuli you provided, and a summary of your results.

Sincerely,

David Nichols, Ph.D.
Researcher
CheMystery Laboratories

BioLogical Resources, Inc. Oakwood, MO 65432-1101

M E M O R A N D U M

To: Team Leader, Zoology Dept.

From: Lee Kwan, Director of Macrobiology

Please have several of your team members design two separate experiments, one that tests for chemotaxis and the other for phototaxis. As you may recall, chemotaxis is a response to an odor stimulus, and phototaxis is a response to a light stimulus.

I have spoken with Dr. Nichols, and it seems that his research team is operating on a rather tight schedule. Please be as efficient and as accurate as possible in completing this assignment. Please remember to make your experimental designs as simple as possible.

While researching similar studies of animal behavior, I found one study of fruit flies which used standard equipment in a unique way. Two flasks were connected at the mouths by wide, transparent tape, allowing the flies to move freely between the two flasks. Use this set-up in your experiment. Also, you should probably choose a very aromatic fruit to test for chemotaxis, and use the same fruit as a control in the test for phototaxis.

Proposal Checklist

Before you start your work, you must submit a proposal for my approval. **Your proposal must include the following:**

_____ • the **question** you seek to answer

_____ • the **procedure** you will use

_____ • two detailed **data tables** for recording results

_____ • a complete, itemized list of proposed **materials** and **costs** (including use of facilities, labor, and amounts needed)

Proposal Approval: _____
(Supervisor's signature)

Report Procedures

When you finish your analysis, prepare a report in the form of a business letter to Dr. Nichols. **Your report must include the following:**

_____ • a paragraph describing the **procedure** you followed to determine how the fruit flies respond to light and odor

_____ • two complete **data tables** showing the results for chemotaxis and phototaxis

_____ • your **conclusions** about fruit flies' response to the tested stimuli

_____ • a detailed **invoice** showing all materials, labor, and the total amount due

Safety Precautions

- Glassware is fragile. Notify your teacher promptly of any broken glass or cuts. Do not clean up broken glass or spills unless your teacher tells you to do so.

- Wash your hands before leaving the laboratory.

- Do not taste or ingest any food items in the laboratory.

Disposal Methods

- Dispose of waste materials according to instructions from your teacher.

- Place used banana slices into a trash can.

- Place broken glass and aluminum foil in the separate containers provided.

- Wash reusable materials such as glassware and lab utensils, and return them to the supply area.

FILE: CheMystery Laboratories

MATERIALS AND COSTS (Select only what you will need. No refunds.)

I. Facilities and Equipment Use

Item	Rate	Number	Total
facilities	$480.00/day	_____	_____
personal protective equipment	$10.00/day	_____	_____
compound microscope	$30.00/day	_____	_____
microscope slide with coverslip	$2.00/day	_____	_____
scissors	$1.00/day	_____	_____
ruler	$1.00/day	_____	_____
hot plate	$15.00/day	_____	_____
laboratory tray	$5.00/day	_____	_____
500 mL Florence flask	$10.00/day	_____	_____
stopper assembly	$15.00/day	_____	_____
light source	$15.00/day	_____	_____
test tube and test tube rack	$10.00/day	_____	_____

II. Labor and Consumables

Item	Rate	Number	Total
labor	$40.00/hour	_____	_____
wide transparent tape	$0.25/m	_____	_____
banana slices	$0.10/slice	_____	_____
aluminum foil	$0.25/m	_____	_____
container of ice	no charge	_____	_____
vial of fruit flies (20 in each)	$30.00/vial	_____	_____
clear packing tape	$2.00/m	_____	_____

Fines

OSHA safety violation	$2,000.00/incident	_____	_____
		Subtotal	_____
		Profit Margin	_____
		Total Amount Due	_____

C40 Conducting a Bird Survey

Prerequisites • none

Review • observation techniques

THE A U D U B O N SOCIETY

Washington, D.C.

April 17, 1998

Lee Kwan
Macrobiology Division
BioLogical Resources, Inc.
101 Jonas Salk Dr.
Oakwood, MO 65432-1101

Dear Ms. Kwan,

Here at the Audubon Society we have begun a project to update our bird identification books. The distribution of bird populations has been known to change in response to environmental factors such as climate changes and pollution. Our current statistics are very thorough, but we fear that they are out of date. We have contacted a variety of research groups similar to yours to help us update our statistics on the distribution of bird species in North America. We are hoping that your company will participate in our project by collecting data about the birds in your region.

Because of the tremendous variety of species, it is important that we standardize our information. Observations should include the following information: (1) name of researcher(s); (2) date of sighting; (3) location of sighting; (4) county or region in which the sighting occurred; (5) name of bird, including order, genus, and species; (6) physical description, including coloration, wing type, neck length, tail length, tail shape, and position of feet in flight; (7) whether the bird was sighted alone, with a group, or with a mate; (8) food acquisition techniques; (9) nest location and type, if available; and (10) diet.

I am looking forward to your findings. Thank you for your cooperation.

Sincerely,

Elaine Diflorio

Elaine Diflorio
Director of Research
The Audubon Society

BioLogical Resources, Inc. Oakwood, MO 65432-1101

M E M O R A N D U M

To: Team Leader, Zoology Dept.

From: Lee Kwan, Director of Macrobiology

Ms. Diflorio has sent a copy of a table that describes commonly observed bird traits. Use this table to record observations for each bird. For record-keeping purposes, write your observations on a 5 × 8 index card along with information about the sighting. Also, try to include a sketch, photo, or videotape of the birds you observe.

Proposal Checklist

Before you start your work, you must submit a proposal for my approval. **Your proposal must include the following:**

_____ • the **procedure** you will use

_____ • a complete, itemized list of proposed **materials** and **costs** (including use of facilities, labor, and amounts needed)

Proposal Approval: _____
(Supervisor's signature)

Report Procedures

When you finish your analysis, prepare a report in the form of a business letter to Ms. Diflorio. **Your report must include the following:**

_____ • a paragraph describing the **procedure** you followed to identify the birds

_____ • a complete **bird identification form** and **index card**

_____ • a **sketch or photo** of the bird or birds observed

_____ • your **conclusions** about the populations of birds in the area

_____ • a detailed **invoice** showing all materials, labor, and the total amount due

Safety Precautions

• Do not touch any plant or approach any animal in the wild.

• Wash your hands after leaving the field.

Disposal Methods

• Dispose of waste materials according to instructions from your teacher.

Bird Identification Form

Bird Name: _____   Number: _____

Type of trait	Term for trait	Description of trait	✔
Wing type	pointed	outermost feathers are the longest	
	rounded	middle primaries are the longest; remaining are graduated	
	slotted	every other feather stands out	
Neck length	long	as long as or longer than the body	
	short	shorter than the body	
Tail length	long	longer than the body	
	short	same length as or shorter than the body	
Tail shape	square	all tail feathers are the same length	
	rounded	feathers are successively longer toward the center	
	pointed	middle feathers are much longer than the others and come to a point	
	forked	feathers increase in length from the middle out	
Feet in flight	extended	extended beyond the body	
	retracted	drawn under the body	
Nest types	spherical	shaped like a hollow sphere	
	pendant	shaped like an elongated sac	
	platform	flat, built like a structure, large enough for the bird to land on	
	flat	shallow, may be on the ground	
	cavity	dug into a tree or a hole in a dead tree limb	
	burrow	on the ground at the end of a tunnel	
	crevice	in cracks in a manmade wall, in a crevice of a natural cliff, or in spaces between boulders	
Social behavior	solitary		
	in a group		
	with a mate		
Food acquisition	walking	picking up food wile walking on ground, branch, etc.	
	flying	picks up food by hovering, swooping, diving, or in flight	
Diet			
Coloration			

FILE: The Audubon Society

MATERIALS AND COSTS (Select only what you will need. No refunds.)

I. Facilities and Equipment Use

Item	Rate	Number	Total
facilities	$480.00/day	_____	_____
personal protective equipment	$10.00/day	_____	_____
compound microscope	$30.00/day	_____	_____
microscope slide with coverslip	$2.00/day	_____	_____
bird identification book	$5.00/day	_____	_____
scissors	$1.00/day	_____	_____
camera	$20.00/day	_____	_____
binoculars	$15.00/day	_____	_____
video camera	$30.00/day	_____	_____

II. Labor and Consumables

Item	Rate	Number	Total
labor	$40.00/hour	_____	_____
bird identification form	provided	_____	_____
5 × 8 index cards	$0.25 each	_____	_____
pencils	$2.00 each	_____	_____
paper	$0.10/sheet	_____	_____
magazines	$2.00 each	_____	_____
videotape	$30.00 each	_____	_____
film with developing	$15.00/roll	_____	_____

Fines

OSHA safety violation	$2,000.00/incident	_____	_____

Subtotal	_____
Profit Margin	_____
Total Amount Due	_____

Name _____

Date _____ Class _____

HOLT
BioSources
LAB PROGRAM
EXPERIMENTAL DESIGN

C41 *Evaluating Muscle Exhaustion*

Prerequisites • none

Review • none

 chester's cosmic candy
Pottstown, Pennsylvania

November 21, 1997

Sam Ashike
Health Division
BioLogical Resources, Inc.
101 Jonas Salk Dr.
Oakwood, MO 65432-1101

Dear Mr. Ashike,

I am the director of a small candy factory that makes hand-decorated gourmet chocolate candies. We recently received national attention in several cooking magazines. Consequently, orders for our candy have been pouring in, and we have all been working overtime to meet the increased demand.

Soon after business picked up, our decorators started experiencing muscle fatigue in their hands and arms. To decorate the candy, the decorators squeeze a pouch of frosting to create the specific pattern that identifies the filling inside the chocolate. They must apply a significant amount of pressure on the frosting pouch in order to control the amount of frosting squeezed out. Previously, the decorators worked at their own pace, taking breaks when needed. However, since we have been so busy, every decorator has been working almost continuously.

We need to know how to reduce muscle fatigue without sacrificing a great deal of productivity. Please research this situation, and provide any data and recommendations that you may have. We are willing to provide our employees with any training that you deem necessary. This matter is very important to us. We are willing to pay your normal fee. We are looking forward to your findings.

Sincerely,

Jonathan W. Bookout

Jonathan W. Bookout
Director
Chester's Cosmic Candy

Biological Resources, Inc. Oakwood, MO 65432-1101

M E M O R A N D U M

To: Team Leader, Physiology Dept.

From: Sam Ashike, Director of Health

The problem that Mr. Bookout describes in the attached letter is not an uncommon one. Muscle fatigue affects million of people laboring in factories and warehouses, driving trucks, and typing on keyboards. Although the exact cause of the fatigue is sometimes hard to identify, the problem usually results from repeated activity without sufficient rests.

I have spoken with Mr. Bookout, and he thinks that the problem at his factory may be caused by insufficient rests between squeezes of the frosting bag. However, he also suggested that other factors, such as repeated use of the same hand, individual muscle development, and the thickness of the frosting, could also contribute to the problem.

Please have your research teams design an experiment to study the muscle fatigue that occurs with different patterns of exercise. Use one of two methods to test for muscle fatigue. Either use the standard hand exerciser to measure the number of repetitions completed or use an exerciser that is linked to a computer or graphing calculator to measure the force exerted at each repetition. In either case, you should repeat your tests and record multiple sets of data.

Proposal Checklist

Before you start your work, you must submit a proposal for my approval. **Your proposal must include the following:**

_____ • the **question** you seek to answer

_____ • the **procedure** you will use

_____ • a detailed **data table** for recording results

_____ • a complete, itemized list of proposed **materials** and **costs** (including use of facilities, labor, and amounts needed)

Proposal Approval: _____
(Supervisor's signature)

Report Procedures

When you finish your analysis, prepare a report in the form of a business letter to Mr. Bookout. **Your report must include the following:**

_____ • a paragraph describing the **procedure** you followed to compare the effects of different patterns of rest and exercise on muscle fatigue

_____ • a complete **data table** showing the number of squeezes, the continuous minutes spent in exercise prior to exhaustion, and the number of squeezes per minute for each trial

_____ • your **conclusions** about how to prevent muscle fatigue during repeated use of the hand and arm

_____ • a detailed **invoice** showing all materials, labor, and the total amount due

Safety Precautions

• Never use electrical equipment around water, nor with wet hands or clothing. Never use equipment with frayed cords.

• Wash your hands before leaving the laboratory.

• Do not continue with this experiment if you experience any pain in your hand or arm.

Disposal Methods

• Dispose of waste materials according to instructions from your teacher.

FILE: Chester's Cosmic Candy

MATERIALS AND COSTS (Select only what you will need. No refunds.)

I. Facilities and Equipment Use

Item	Rate	Number	Total
facilities	$480.00/day	_____	_____
personal protective equipment	$10.00/day	_____	_____
computer	$40.00/day	_____	_____
graphing calculator	$30.00/day	_____	_____
hand exerciser probe	$20.00/day	_____	_____
pH probe	$10.00/day	_____	_____
hand exerciser, noncomputer	$5.00/day	_____	_____
thermometer	$5.00/day	_____	_____
clock or watch with second hand	$10.00/day	_____	_____
scissors	$1.00/day	_____	_____
ruler	$1.00/day	_____	_____

II. Labor and Consumables

Item	Rate	Number	Total
labor	$40.00/hour	_____	_____
wax pencil	$2.00 each	_____	_____

Fines

OSHA safety violation	$2,000.00/incident	_____	_____
		Subtotal	_____
		Profit Margin	_____
		Total Amount Due	_____

Name _____

Date _____ Class _____

HOLT
BIOSOURCES
LAB PROGRAM
LABORATORY TECHNIQUES

C42 | *Blood Typing*

Skills
- blood typing
- performing a cell count

Objectives
- *Determine* the ABO and Rh blood types of unknown simulated blood samples.
- *Observe* and *count* simulated red blood cells and white blood cells.

Materials

PREPARATION NOTES

Time Required: one 50-minute period

Materials

Materials for this lab activity can be purchased from WARD'S. See the *Master Materials List* for ordering instructions.

- safety goggles
- gloves
- lab apron
- wax pencil
- blood typing trays (4)
- simulated blood samples of unknown type from four subjects
- simulated Anti-A blood-typing serum
- simulated Anti-B blood-typing serum
- simulated Anti-Rh blood-typing serum
- toothpicks (12)
- ruled microscope slide
- coverslip
- compound light microscope

Purpose

Preparation Tips

Prepare separate containers for the disposal of broken plastic trays and toothpicks.

Disposal
- Simulated blood and typing sera may be washed down the drain.
- Broken typing trays and stirrers may be placed in the trash.

You are a new lab technician working in your city's blood bank. Before a person can donate blood, you must first determine his or her blood type. The blood type of the donor must match the blood type of the recipient before a blood transfusion can be given. Another part of your job is to take a person's blood count. A **blood count** provides the total number of blood cells and the percentage of red and white blood cells in a person's body. A blood count gives a physician valuable information about a person's health. If a person has a higher than normal count of white blood cells, for example, the physician will know that the person is fighting an infection. To make sure you know how to determine a person's blood type, your supervisor wants you to determine the blood type of four samples of simulated human blood. You are also asked to do a blood count.

Background

Occasionally an injury or disorder is serious enough that a person must receive a **blood transfusion,** or blood from another person. A blood transfusion can succeed only if the blood of the **recipient,** the person receiving the blood, matches the blood of the **donor,** the person giving the blood. One of the factors that must be considered in matching blood is **blood type.** Blood type is determined by the presence or absence of specific marker proteins called **antigens** found on the surfaces of red blood cells. The most familiar blood typing system uses the letters *A,* *B,* and *O* to label the different antigens. Under this system, the primary blood types are A, B, AB, and O.

PROCEDURAL NOTES

Safety Precautions

• Have students wear safety goggles, gloves, and a lab apron.

• DO NOT allow students to test any blood other than the simulated blood you provide.

• Tell students that gloves provide protection from blood-borne pathogens and are necessary when working with body fluids.

• Discuss all safety symbols and caution statements with students.

• Review rules for carrying and using the compound microscope.

Procedural Tips

• Remind students to use a new stirrer (or toothpick) for each test and to break used stirrers in two before placing them in the designated container.

• Blood cell counts may vary from student to student but should approximate the given results. Class counts may be combined for greater accuracy. The method given to the students is not the actual blood-cell counting procedure. A hemacytometer may be used to demonstrate to students the actual counting technique used in clinical settings. It is more accurate than a ruled microscope slide.

• Remind students to always begin using a microscope on low power before changing to high power.

Transfusions between people of different blood types usually are not successful because the immune system of the recipient attacks the transfused blood from the donor. People with type A blood have a marker protein called the A antigen on the surface of their red blood cells. A person with type B blood has a slightly different marker protein called the B antigen. People with type AB blood have both A and B antigens on their red blood cells. A person with type O blood has neither A nor B antigens on their red blood cells.

Antigen-antibody interactions are part of the immune system. Antigens label a cell as belonging to or not belonging to an organism. Antibodies produced in response to the presence of a foreign antigen attack the foreign antigen, defending the body against invasion.

Individuals produce antibodies, or **agglutinins,** to the marker proteins not found on their own cells. For example, individuals with type A blood produce antibodies against the B antigen, even if they have never been exposed to the antigen. The following table summarizes the transfusion capabilities of the different blood types.

Table 1 Transfusion Capabilities of Different Blood Types

Blood type	Antigens on red blood cells	Antibodies in plasma	Can receive blood from groups	Can give blood to groups
A	A	B	O, A	A, AB
B	B	A	O, B	B, AB
AB	A and B	none	O, A, B, AB	AB
O	none	A and B	O	O, A, B, AB

If type A blood is transfused into a person with type B or type O blood, antibodies against the A antigen will attack the foreign red blood cells, causing them to clump together, or **agglutinate.** Clumps of red blood cells can block capillaries and cut off blood flow, which may be fatal. Clumping also occurs if type B blood is transfused into individuals with type A or type O blood, or if type AB blood is given to people with type O blood.

Blood typing is performed using **antiserum,** blood serum that contains specific antibodies. For ABO blood typing, antibodies against the A and B antigens are used. These antibodies are called anti-A and anti-B agglutinins. If agglutination occurs in the test blood when it is exposed to anti-A serum, the blood contains the A antigen. The blood type of the test blood is A. If clumping occurs in the test blood when it is exposed to anti-B serum, the blood contains the B antigen. The blood type of the test blood is B. If clotting occurs with both anti-A and anti-B sera, the type is AB. If no clumping occurs with either serum type, the blood type is O. This information is summarized in the table on the next page.

Table 2 Agglutination Reaction of ABO Blood-Typing Sera

Reaction		Blood type
A antibodies (anti-A serum)	**B antibodies (anti-B serum)**	
agglutination	no agglutination	A
no agglutination	agglutination	B
agglutination	agglutination	AB
no agglutination	no agglutination	O

Another type of marker protein on the surface of red blood cells is the Rh factor, so named because it was originally identified in rhesus monkeys. People whose blood contains the Rh factor are said to be Rh positive (Rh^+). People whose blood does not contain the Rh factor are Rh negative (Rh^-). A person with Rh^- blood has no antibodies to Rh^+ blood unless the person was exposed to Rh^+ blood at an earlier age. No agglutination occurs the first time an Rh^- person receives a blood transfusion from an Rh^+ person. Agglutination can occur, however, the second time the Rh^- person receives Rh^+ blood. In addition to testing blood type, it is also important to test blood for transfusion for its Rh factor.

Procedure

Part 1—ABO and Rh Typing

1.   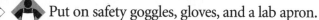 Put on safety goggles, gloves, and a lab apron.

2. With a wax pencil, label each of four blood-typing trays as follows: Tray 1—Mr. Smith, Tray 2—Ms. Jones, Tray 3—Mr. Green, Tray 4—Ms. Brown.

3. Place 3 to 4 drops of Mr. Smith's simulated blood in each of the A, B, and Rh wells of Tray 1.

4. Place 3 to 4 drops of Ms. Jones's simulated blood in each of the A, B, and Rh wells of Tray 2.

5. Place 3 to 4 drops of Mr. Green's simulated blood in each of the A, B, and Rh wells of Tray 3.

6. Place 3 to 4 drops of Ms. Brown's simulated blood in each of the A, B, and Rh wells of Tray 4.

7. Add 3 to 4 drops of the simulated anti-A serum to each A well on the four trays.

8. Add 3 to 4 drops of the simulated anti-B serum to each B well on the four trays.

9. Add 3 to 4 drops of the simulated anti-Rh serum to each Rh well on the four trays.

10. Use *separate* toothpicks to stir each sample of serum and blood. Record your observations in the data table below. Dispose of the toothpicks as directed by your teacher. Indicate an agglutination reaction with a + and no reaction with a −. *Note: A positive test is indicated by obvious clumping of the red blood cells.* Also record your observations of each test.

Table 3 Blood Typing

	A antibodies with type A blood	B antibodies with type B blood	Rh antibodies with Rh+ blood	Blood type	Observations
Tray 1: Mr. Smith	+	−	+	A+	Answers will vary.
Tray 2: Ms. Jones	−	+	−	B−	
Tray 3: Mr. Green	+	+	+	AB+	
Tray 4: Ms. Brown	−	−	−	O−	

Part 2—Blood Cell Count

11. Thoroughly shake a vial of simulated blood. Add a small drop of the simulated blood to the center grid of a ruled slide. Place a clean coverslip over the drop, and be careful not to trap any air bubbles.

12. To count the individual blood components, place the slide on the microscope stage under low power, find an area of the slide containing cells, and switch to high power. *Note: Red blood cells appear red, and white blood cells appear blue.* Count each cell in any clump separately. Count the red blood cells first, then count the white blood cells. Record your counts for each component in the data table below. Repeat this procedure for two other random areas on the slide.

Table 4 Blood Cell Count

Blood cell type	Cell count field no. 1	Cell count field no. 2	Cell count field no. 3	Average number of cells	Dilution factor	Total number of blood per mm^3 (average no. of cells × dilution factor)
Red blood cells	Answers will vary.				150,000	approximately 500,000
White blood cells	Answers will vary.				5,000	approximately 10,000

13. Add your counts for each kind of blood cell in the three fields, and divide by three to compute the average number of cells you counted. Record your answers in the data table above.

14. Multiply the average number of cells by the dilution factor to determine the number of red and white blood cells per cubic millimeter. Record your answers in the data table on the previous page.

15. Dispose of your materials according to the directions from your teacher.

16. Clean up your work area and wash your hands before leaving the lab.

Analysis **17.** If Ms. Brown were serving as a blood donor, what ABO blood type(s) could receive her blood safely?

Ms. Brown has type O⁻ blood. Therefore, she can give blood to groups O, A, B, and AB.

18. Which person among the four represented by the simulated blood samples can receive donated blood from Ms. Jones? Which blood type does each person have? Explain your answer.

Mr. Green can receive blood from Ms. Jones. Ms. Jones has blood type B, which can be received by

Mr. Green, who has blood type AB. A person with type AB blood has both A and B antigens and,

therefore, can receive blood from a person with type A, B, AB, or O.

19. What is the difference between an antigen and agglutinin?

Antigens are substances found on the surface of red blood cells; agglutinins are antibodies found in

plasma.

20. What factors determine the ABO blood types?

The ABO blood typing is based on the presence or absence of antigens A and B, which are encoded

by genes.

Conclusions

21. People with type O blood are commonly called universal donors. Explain this statement.

Type O blood carries no blood cell antigens to react with the recipient's agglutinins. Type O blood

can, therefore, be received by people with any blood type.

22. A person with what blood type would be considered a universal recipient? Explain your answer.

A person with type AB^+ blood would be considered a universal recipient because he or she can

receive types A, B, AB, and O blood and either Rh^+ or Rh^- blood.

23. Which type of blood cells are most numerous in blood?

red blood cells

Extensions

24. The technique you used in this lab to count blood cells is not the actual technique used in most laboratories and other clinical settings. Find out how lab technicians use a tool called a hemacytometer to accurately count blood cells.

25. The first baby with Rh^+ blood born to a woman with Rh^- blood usually has no health problems. The second Rh^+ child, however, is seriously threatened when the mother produces antibodies against the Rh antigens of her baby. Find out why this happens and what treatment is given to babies in this situation to save their life.

26. Find out what an emergency medical technician gives to a patient when he or she administers an emergency transfusion in the field. Why do they use this substance instead of blood for transfusing their patients?

Name _____

Date _____ Class _____

HOLT

BIOSOURCES
LAB PROGRAM
EXPERIMENTAL DESIGN

C43 | *Blood Typing—Whodunit?*

Prerequisites | • **Laboratory Techniques C42: Blood Typing** on pages 201–206
Review | • procedure for blood typing
| • the terms *antigen*, *antibody*, and *agglutination*
| • the possible blood types and what they stand for

CITY OF OAKWOOD
POLICE DEPARTMENT
Oakwood, Missouri 65432-1221

December 18, 1997

Caitlin Noonan
Research and Development Division
BioLogical Resources, Inc.
101 Jonas Salk Dr.
Oakwood, MO 65432-1101

Dear Ms. Noonan,

We have recently had yet another series of burglaries. This time, all the victims have been residents of the same neighborhood. We were able to obtain some stained cloth samples from the last crime scene. According to the victim, he came home to find the burglar in his living room. The burglar broke a window to escape and was apparently cut by the glass. We believe that the cloth samples, which were found among the broken glass, are stained with the burglar's blood.

We are trying solve this case as quickly as possible. Many of the residents in the area are concerned for their property and their safety. So far we have four suspects. All of them were seen near the crime scene, and all of them have cuts that could have been made by broken glass. We have blood samples from all four suspects and from the latest victim as well. Unfortunately, our forensics expert called in sick early this week and will not be able to complete blood-typing tests until next week. We need your research company to complete the necessary blood tests to help us narrow down our list of suspects. We will provide you with the stained cloth samples and the blood samples that we have collected. Please let me know what you find.

Sincerely,

Roberto Morales

Roberto Morales
Chief of Police
City of Oakwood Police Department

BioLogical Resources, Inc. Oakwood, MO 65432-1101

M E M O R A N D U M

To: Team Leader, Forensics Dept.

From: Caitlin Noonan, Director of Research and Development

Chief Morales has sent the blood samples he described in his letter, along with a list of tests that need to be completed. Please have your research teams examine the stain on the cloth to determine whether it is indeed blood. Then complete a blood-typing test for the crime-scene stain samples, the victim, and each of the four suspects.

Chief Morales told me that he was very pleased with the work your team did on the last burglary case. I congratulate you on your previous success and encourage you to complete this project with the same exemplary skill and determination that you have demonstrated thus far.

Proposal Checklist

Before you start your work, you must submit a proposal for my approval. **Your proposal must include the following:**

_____ • the **question** you seek to answer

_____ • the **procedure** you will use

_____ • a detailed **data table** for recording results

_____ • a complete, itemized list of proposed **materials** and **costs** (including use of facilities, labor, and amounts needed)

Proposal Approval: _____
(Supervisor's signature)

Report Procedures

When you finish your analysis, prepare a report in the form of a business letter to Chief Morales. **Your report must include the following:**

_____ • a paragraph describing the **procedure** you followed to examine the crime-scene blood sample and to complete blood-typing tests on all six blood samples

_____ • a complete **data table** showing blood-typing results for each sample

_____ • your **conclusions** about whether the crime-scene sample is indeed blood, and which, if any, of the other samples matches the blood type of the crime-scene sample

_____ • a detailed **invoice** showing all materials, labor, and the total amount due

Safety Precautions

• Wear safety goggles, disposable gloves, and a lab apron.

• Glassware is fragile. Notify your teacher promptly of any broken glass or cuts. Do not clean up broken glass or spills unless your teacher tells you to do so.

• Never use electrical equipment around water, nor with wet hands or clothing. Never use equipment with frayed cords.

• Wash your hands before leaving the laboratory.

• If you are using a microscope with a mirror, do not use direct sunlight as a light source.

• Under no circumstances are you to test any blood other than the simulated blood samples provided by your teacher.

Disposal Methods

• Dispose of waste materials according to instructions from your teacher.

• Place used toothpicks in a trash can.

• Place broken glass, unused simulated blood, unused antiserums, and used cloth squares in the separate containers provided.

• Wash reusable materials such as glassware and lab utensils, and return them to the supply area.

FILE: City of Oakwood Police Department

MATERIALS AND COSTS (Select only what you will need. No refunds.)

I. Facilities and Equipment Use

Item	Rate	Number	Total
facilities	$480.00/day		
personal protective equipment	$10.00/day		
compound microscope	$30.00/day		
microscope slide with coverslip	$2.00/day		
scissors	$1.00/day		
ruler	$1.00/day		
hot plate	$15.00/day		
blood-typing tray	$5.00/day		
test tube	$2.00/day		
test-tube rack	$5.00/day		

II. Labor and Consumables

Item	Rate	Number	Total
labor	$40.00/hour		
4 stained cloth samples	provided		
vial of simulated "Victim" blood	provided		
vial of simulated "Suspect 1" blood	provided		
vial of simulated "Suspect 2" blood	provided		
vial of simulated "Suspect 3" blood	provided		
vial of simulated "Suspect 4" blood	provided		
vial of anti-A typing serum	$20.00 each		
vial of anti-B typing serum	$20.00 each		
vial of anti-Rh typing serum	$20.00 each		
toothpicks	$0.10 each		
distilled water	$0.10/mL		
wax pencil	$2.00 each		

Fines

OSHA safety violation	$2,000.00/incident		

Subtotal		
Profit Margin		
Total Amount Due		

Name _____

Date _____ Class _____

HOLT
BIOSOURCES
LAB PROGRAM
EXPERIMENTAL DESIGN

C44 Blood Typing—Pregnancy and Hemolytic Disease

Prerequisites • **Laboratory Techniques C42: Blood Typing** on pages 201–206
Review • procedures for testing simulated blood
• antigen/antibody reactions
• ABO and Rh blood typing

ST. DAVID'S MEMORIAL HOSPITAL

Emporia, Virgina

October 22, 1997

Caitlin Noonan
Research and Development Division
BioLogical Resources, Inc.
101 Jonas Salk Dr.
Oakwood, MO 65432-1101

Dear Ms. Noonan,

St. David's is currently conducting a study on a new, injectable serum used to prevent Rh hemolytic disease. Rh hemolytic disease can occur in pregnancies when a mother is Rh− and her baby is Rh+. When an Rh− individual is exposed to Rh+ blood, anti-Rh antibodies can form. If the mother has not yet been exposed to Rh+ blood, the first birth of an Rh+ child to an Rh− mother usually causes no problems. This is because no antibodies have formed. However, exposure to Rh+ blood during a first birth can cause the formation of antibodies that can endanger a second child if that child is also Rh+. These antibodies can attack the child's red blood cells, causing anemia, brain damage, and, in the worst cases, death.

This disease can be prevented if Rh− women are given a serum immediately following exposure to Rh+ blood, such as during childbirth, a transfusion, or amniocentesis. This serum eliminates the foreign Rh+ blood cells in the mother's blood stream before antibodies can form. We suspect that our new serum will provide better results than the one currently used. We are currently conducting a study of this new serum, and we need confirmation of our blood types. Please let me know if your services are available.

Sincerely,

Dr. Nancy Reimer
Research Administrator
St. David's Memorial Hospital

M E M O R A N D U M

To: Team Leader, Medical Testing Dept.

From: Caitlin Noonan, Director of Research and Development

Please read the attached letter from St. David's Memorial Hospital. I was thrilled to hear that St. David's has been trying to develop a better serum to prevent Rh hemolytic disease. The disease is a serious problem for many people, and I am glad that our company will be participating in the study.

The hospital has sent samples of blood taken from several Rh− women who are pregnant with their first child and fetal blood from their unborn children. Please have your research teams test the blood samples for the presence or absence of the Rh factor. The hospital has also included blood from a known Rh+ and an Rh− blood source so that you can compare your results with these controls. For each case, compare the Rh blood type of the mother with that of the first-born child, and make conclusions about the potential danger to future offspring.

Proposal Checklist

Before you start your work, you must submit a proposal for my approval. **Your proposal must include the following:**

_____ • the **question** you seek to answer

_____ • the **procedure** you will use

_____ • a detailed **data table** for recording results

_____ • a complete, itemized list of proposed **materials** and **costs** (including use of facilities, labor, and amounts needed)

Proposal Approval: _____
(Supervisor's signature)

Report Procedures

When you finish your analysis, prepare a report in the form of a business letter to Dr. Reimer. **Your report must include the following:**

_____ • a paragraph describing the **procedure** you followed to test the Rh blood types of the given samples

_____ • a complete **data table** showing observations and Rh blood types

_____ • your **conclusions** about the risk of Rh hemolytic disease in future pregnancies

_____ • a detailed **invoice** showing all materials, labor, and the total amount due

Safety Precautions

• ◇ ◇ ⚠ Wear safety goggles, disposable gloves, and a lab apron.

• ⚠ Glassware is fragile. Notify your teacher promptly of any broken glass or cuts. Do not clean up broken glass or spills unless your teacher tells you to do so.

• ⚠ Wash your hands before leaving the laboratory.

• Under no circumstances are you to test any blood other than the simulated blood provided by your teacher.

Disposal Methods

• ⚠ Dispose of waste materials according to instructions from your teacher.

• Break used toothpicks, and place them in a trash can.

• Place broken glass, simulated blood, and antiserums in the separate containers provided.

• Wash reusable materials such as glassware and lab utensils, and return them to the supply area.

FILE: St. David's Memorial Hospital

MATERIALS AND COSTS (Select only what you will need. No refunds.)

I. Facilities and Equipment Use

Item	Rate	Number	Total
facilities	$480.00/day	_____	_____
personal protective equipment	$10.00/day	_____	_____
compound microscope	$30.00/day	_____	_____
dissecting microscope	$30.00/day	_____	_____
microscope slide with coverslip	$2.00/day	_____	_____
blood-typing tray	$5.00/day	_____	_____
eyedropper	$2.00/day	_____	_____

II. Labor and Consumables

Item	Rate	Number	Total
labor	$40.00/hour	_____	_____
vial of anti-A serum	$20.00 each	_____	_____
vial of anti-B serum	$20.00 each	_____	_____
vial of anti-Rh serum	$20.00 each	_____	_____
blood sample: mother	provided	_____	_____
blood sample: first-born child	provided	_____	_____
blood sample: known Rh+ blood	provided	_____	_____
blood sample: known Rh− blood	provided	_____	_____
toothpicks	$0.10 each	_____	_____

Fines

OSHA safety violation	$2,000.00/incident	_____	_____
		Subtotal	_____
		Profit Margin	_____
		Total Amount Due	_____

Name _____

Date _____ Class _____

HOLT
BioSOURCES
LAB PROGRAM
EXPERIMENTAL DESIGN

C45 *Screening Sunscreens*

Prerequisites • none

Review • aseptic technique

 SOLAR FLARE MODELING AGENCY

Palm Beach, Florida

May 20, 1998

Sam Ashike
Health Division
BioLogical Resources, Inc.
101 Jonas Salk Dr.
Oakwood, MO 65432-1101

Dear Mr. Ashike,

Solar Flare Modeling Agency represents several hundred talented men and women who are specially trained for outdoor photo shoots. We market our talent mainly to companies that sell beachwear and other products used for outdoor recreation.

Due to the health risks associated with prolonged exposure to UV radiation, we have strict policies about sun protection. As a rule, we insist that our models get limited exposure to the sun, and we require that they wear maximum-strength sunscreen during all outdoor photo shoots. We provide all of our models with sunscreen expressly for this purpose.

Recently, a local vendor has presented us with a considerable discount on sunscreen. We are interested in taking advantage of good prices and supporting local business, but we want to make sure that the product works.

I have sent samples of this new product along with samples of the more expensive product we have used for over ten years. Both claim to have the same level of UV protection. Please test these products for UV protection, and compare the two brands of sunscreen.

Sincerely,

Sandra Winters

Sandra Winters
Owner and President
Solar Flare Modeling Agency

BioLogical Resources, Inc. Oakwood, MO 65432-1101

M E M O R A N D U M

To: Team Leader, Product Testing Dept.

From: Sam Ashike, Director of Health

Your teams will be testing these two sunscreens for their effectiveness in filtering UV radiation. I suggest that you use a UV-sensitive strain of yeast, which can be grown on agar in petri dishes. A culture of the yeast and the samples of sunscreen have been delivered to your lab. Be sure to design a controlled experiment, and use aseptic technique to prevent contamination that might alter the results.

Ms. Winters has informed me that her company has maintained an exceptionally low rate of work-related health problems. She attributes this largely to the efforts of her and her staff, who go to great lengths to ensure that the models are protected from potential dangers such as dehydration and sunburn. She will only consider using discount sunscreen if she can be sure that it works as well or better than the brand the models currently use. Please be as careful as possible in completing your experiment, making sure that your procedures are precise.

Proposal Checklist

Before you start your work, you must submit a proposal for my approval. **Your proposal must include the following:**

_____ • the **question** you seek to answer

_____ • the **procedure** you will use

_____ • a detailed **data table** for recording observations

_____ • a complete, itemized list of proposed **materials** and **costs** (including use of facilities, labor, and amounts needed)

Proposal Approval: _____

(Supervisor's signature)

Report Procedures

When you finish your tests, prepare a report in the form of a business letter to Ms. Winters. **Your report must include the following:**

_____ • a paragraph describing the **procedure** you followed to test the effects of sunscreen on yeast growth

_____ • a complete **data table** showing the level of yeast growth for each treatment

_____ • your **conclusions** about the effectiveness of each sunscreen sample

_____ • a detailed **invoice** showing all materials, labor, and the total amount due

Safety Precautions

- Wear safety goggles, disposable gloves, and a lab apron.

- Glassware is fragile. Notify your teacher promptly of any broken glass or cuts. Do not clean up broken glass or spills unless your teacher tells you to do so.

- Never use electrical equipment around water, nor with wet hands or clothing. Never use equipment with frayed cords.

- Wash your hands before leaving the laboratory.

Disposal Methods

- Dispose of waste materials according to instructions from your teacher.

- Place all solid, uncontaminated waste materials in a trash can.

- Place broken glass and contaminated materials in the separate containers provided.

- Wash reusable materials such as glassware and lab utensils, and return them to the supply area.

FILE: Solar Flare Modeling Agency

MATERIALS AND COSTS (Select only what you will need. No refunds.)

I. Facilities and Equipment Use

Item	Rate	Number	Total
facilities	$480.00/day		
personal protective equipment	$10.00/day		
compound microscope	$30.00/day		
clock or watch with a second hand	$10.00/day		
inoculating loop	$2.00/day		
sterile test tube with cap	$3.00/day		
sterile 50 mL jar with lid	$2.00/day		
sterile 100 mL water bottle	$5.00/day		
incubator	$30.00/day		
sun lamp	$15.00/day		

II. Labor and Consumables

Item	Rate	Number	Total
labor	$40.00/hour		
petri dishes with agar	$6.00 each		
UV-sensitive yeast culture	$10.00 each		
sample of original sunscreen	provided		
test sample	provided		
sterile spreader	$2.00 each		
sterile, calibrated 1 mL pipette	$2.00 each		
toothpicks	$0.05 each		
transparent tape	$0.25/m		
sterile cotton swab	$0.10 each		
alcohol wipe	$0.10 each		
disinfectant solution	$2.00/bottle		
paper towels	$0.10/sheet		
thick paper	$0.10/sheet		
distilled water	$0.10/mL		
wax pencil	$2.00 each		

Fines

OSHA safety violation	$2,000.00/incident		

Subtotal	
Profit Margin	
Total Amount Due	

Name _____

Date _____ Class _____

HOLT
BIOSOURCES
LAB PROGRAM
LABORATORY TECHNIQUES

C46 *Identifying Food Nutrients*

Skills
• qualitatively analyzing food for organic compounds

Objectives
• *Perform* standard chemical identification tests for organic compounds.
• *Relate* indicator reactions to the presence of organic nutrients.
• *Recognize* a standard.

REPARATION
OTES
ne Required:
minutes

Materials

aterials
aterials for this lab
tivity can be pur-
ased from WARD'S.
e the *Master Materi-
List* for ordering
structions.

• safety goggles
• lab apron
• test tubes (10)
• wax pencil
• test-tube rack
• dropping pipets (5)
• glucose solution
• starch solution
• water

• Benedict's solution
• hot water bath
• test-tube holder
• Lugol's iodine solution
• albumin (protein) solution
• biuret solution
• vegetable oil
• Sudan III

Purpose

You have just started a job as a food-quality tester. Your task will be to develop a kit to test foods for sugar, starch, protein, and lipids. The test kit will be used by dietitians who will analyze foods to create nutritional menus for hospital patients. Your first assignment is to demonstrate proficiency using standard tests for nutrients in foods.

eparation Tips
Wear safety goggles
d a lab apron if you
epare the solutions
-low.

Background

Benedict's solution is
ailable and should be
rchased, not made.
owever, to make it
urself, dissolve 17.3 g
sodium citrate
a₃C₆H₅O₇•2H₂O) and
g of sodium carbon-
e (Na₂CO₃) in 80 mL
distilled water in a
aker. In a smaller
aker, dissolve 1.7 g
copper sulfate
uSO₄•5H₂O) in about
mL of distilled
ater. While stirring
e first solution,
owly pour the copper
lfate solution into it.
ur the combined
lutions into a 100 mL
aduated cylinder and
lute to 100

Substances or compounds that supply your body with energy and the building blocks of macro-molecules are called **nutrients.** The food you eat contains nutrients important to your body. **Sugars** and **starches** make up a group of organic compounds called carbohydrates, which are important in supplying your body with energy. Some starches provide your body with indigestible fiber, or roughage, which aids digestion. Organic compounds called **proteins** are important for growth and repair. **Lipids** are organic compounds that can supply as much as four times the amount of energy as carbohydrates or proteins.

You can perform **qualitative tests** to identify the presence of organic compounds in food using **indicators,** chemical substances that react in a certain way when a particular substance is present. **Benedict's solution** is used to identify the presence of reducing sugars, such as glucose. **Lugol's iodine solution** is used to indentify the presence of starch. **Biuret solution** is used to identify the presence of protein. **Sudan III** is used to identify the presence of lipids. A **standard** is a positive test for a known substance. Unknown substances can be tested and compared with the standard for positive identification of the substance.

Procedure

1.   Put on safety goggles and a lab apron.

Dispense the solution in dropper bottles labeled, "Benedict's solution." This should be enough for several classes. It can be stored for long periods of time.

• Lugol's iodine (or iodine-iodide solution) is available ready-made. NOTE: Iodine reacts with metal, skin, and many other substances. To make Lugol's iodine, dissolve 1.0 g of potassium iodide (KI) in 15 mL of distilled water in a 250 mL beaker. Add 0.7 g of iodine (I₂) and stir until dissolved, adding no more than 50 mL of water, if desired, to facilitate dissolution. Dilute to 100 mL while stirring well. Dispense in amber-colored dropper bottles labeled, "Lugol's iodine solution" because sunlight or strong lights can cause the solution to deteriorate.

• Biuret solution is available ready-made. NOTE: Biuret reagent is a caustic solution. Avoid spilling it on yourself or others. Dissolve 8.0 g of sodium hydroxide (NaOH) in 100 mL of water in a flask. Add 1.0 g of copper sulfate. Stopper and shake the flask. Dispense in dropper bottles labeled, "Biuret reagent." This should provide enough for several classes. It can be stored for long periods of time.

• Sudan III is available ready-made. To prepare a 2% stock solution of Sudan III, add 2 g of Sudan III to a flask and dilute to a final volume of 100 mL with undiluted ethyl alcohol. Just before using, mix the stock solution with an equal amount of 45% ethyl alcohol. Dispense in

2. To perform the Benedict's test, select three clean test tubes. With a wax pencil, label the tops of the test tubes "1," "2," and "3." To test tube 1, add 40 drops of glucose. To test tube 2, add 40 drops of starch. To test tube 3, add 40 drops of water.

3. Add 10 drops of Benedict's solution to each test tube. **CAUTION: If you get Benedict's solution on your skin or clothing, wash it off while calling to your teacher.** Heat the test tubes in a hot water bath with a temperature range of 40–50°C for five minutes.

◆ In which test tube do you see a reaction?

—————————— 1, glucose ——————————

◆ What color change occurs when Benedict's solution is heated in the presence of this substance?

—————————— blue to orange-yellow or red ——————————

In the table below, write the name of this indicator under the substance that showed a reaction, record the color change that occurred, and identify the nutritioal role of this substance.

Substance	Glucose	Starch	Protein	Lipid
Nutritional role	energy	energy, fiber	maintenance and repair	energy
Indicator	Benedict's solution	Lugol's iodine	Biuret solution	Sudan III
Positive result	orange-yellow to red	blue-black	pink	red-orange

4. To perform the Lugol's iodine test, select three clean test tubes. With a wax pencil, label the tops of the test tubes "1," "2," and "3." To test tube 1, add 40 drops of glucose. To test tube 2, add 40 drops of starch. To test tube 3, add 40 drops of water.

5. Add 2 drops of Lugol's iodine solution to each test tube. **CAUTION: Lugol's solution will stain your skin and clothing. Promptly wash off spills to minimize staining.**

◆ In which test tube do you see a reaction?

—————————— 2, starch ——————————

◆ What color change occurs when Lugol's iodine solution is in the presence of this substance?

—————————— yellow-orange to blue-black ——————————

In the table above, write the name of this indicator under the substance that showed a positive reaction, record the color change that occurred, and identify the nutritional role of this substance.

dropper bottles labeled, "Sudan III." Be sure there are no sources of ignition in the room when preparing Sudan III. Restrict the amount of Sudan III in each dispensing bottle to a maximum of 100 mL.

• Provide hot, soapy water, test-tube brushes, and test-tube racks for students to clean their test tubes after performing the tests.

• Provide separate containers for the disposal of wastes containing each type of indicator solution.

Disposal

• Combine all wastes containing Benedict's solution and biuret solution. Slowly add 1 M sulfuric acid with stirring until the pH is between 5 and 6. Scour six 6d iron nails with steel wool until they are bright and shiny. Immerse the nails in the acidified solution. Let the nails remain immersed until all of the copper has precipitated (overnight should be sufficient). Remove the nails, and filter the solution. Heat the nails and any precipitate obtained from the filtrate sufficiently to convert the copper precipitate and copper on the nails into copper oxide. Let the nails and the precipitate cool, and then place them in the trash. Treat the filtrate with sufficient 1 M NaOH to bring the pH to from 8 to 10. Filter the liquid again. Let

Analysis

the precipitate dry, and place it in the trash. Pour the filtrate down the drain.

6. To perform the biuret test, select two clean test tubes. With a wax pencil, label the tops of the test tubes "1" and "2." To test tube 1, add 40 drops of albumin solution (a protein). To test tube 2, add 40 drops of water.

7. Add 3 drops of biuret solution to each test tube. **CAUTION: If you get biuret solution on your skin or clothing, wash it off while calling to your teacher.**

♦ In which test tube do you see a reaction?

1, protein

♦ What color change occurs when biuret solution is in the presence of this substance?

from blue to pink

In the table on the previous page, write the name of this indicator under the substance that showed a positive reaction, record the color change that occurred, and identify the nutritional role of this substance.

8. To perform the Sudan III test, select two clean test tubes. With a wax pencil, label the tops of the test tubes "1" and "2." To test tube "1," add 5 drops of vegetable oil. To test tube "2," add 5 drops of water.

9. Add 3 drops of Sudan III to each test tube. **CAUTION: Sudan III will stain your skin and clothing. Promptly wash off spills to minimize staining. Do not use Sudan III when there are flames in the room.**

♦ In which test tube do you see a positive reaction?

1, vegetable oil

♦ What color change occurs when Sudan III is in the presence of this substance?

from red to red-orange

In the table on the previous page, write the name of this indicator under the substance that showed a positive reaction, record the color change that occurred, and identify the nutritional role of this substance.

10. Dispose of your materials according to the directions from your teacher.

11. Clean up your work area and wash your hands before leaving the lab.

12. What are the controls in this lab?

The test tubes filled with water are the controls.

13. Which indicator detects the presence of glucose? starch? protein? lipid?

glucose—Benedict's solution; starch—Lugol's iodine; protein—Biuret test; lipid—Sudan III

14. Describe the color change of the indicators in the presence of glucose, starch, protein, and lipids.

glucose—blue-green to orange-yellow to red; starch—brown-yellow to blue-black; protein—royal

blue to pink; lipid—stains the lipids red-orange

Conclusions
PROCEDURAL NOTES

15. Which test tubes contain standards that you could use for comparing tests on unknown substances?

after step 3, test tube 1 (glucose and Benedict's solution); after step 5, test tube 2 (starch and

Lugol's solution); after step 7, test tube 1 (protein and Biuret solution); after step 9, test tube 1

(vegetable oil and Sudan III solution)

Safety Precautions
• Discuss all safety symbols and caution statements with students.
• Remind students to inform you of any problems.

Procedural Tips
• When students finish each test, instruct them to wash out their test tubes and drain them upside down.
• Model the correct technique for placing test tubes in a hot water bath.
• Regulate the temperature so that the hot water bath is in a 40–50°C range.
• Review qualitative and quantitative research methodology. Provide definitions of both.

16. You are asked to analyze and compare a food substance with standards for organic compounds. You observe a positive response with Lugol's iodine solution and biuret solution. What can you conclude about this food?

The food contains both starch and proteins.

17. How would tests such as these help a nutritionist plan a balanced diet?

To plan a diet, a nutritionist needs to know which nutrients are present in each food.

18. What additional information would a nutritionist require about food to plan a balanced diet?

A nutritionist would also need to know the caloric value and quantity of each nutrient in the food.

Combine all wastes containing Lugol's solution. Add sufficient 1 M sodium thiosulfate solution to reduce all iodine to iodide, and pour down the drain.
• Combine all other wastes, and pour them down the drain.

Extensions

19. *Dietetic technicians* assist therapeutic dietitians in planning the diets of hospital patients and overseeing food-tray preparation. Find out about the skills and training required to become a dietetic technician. Investigate job opportunities and the types of businesses that employ dietetic technicians.

20. Design an experiment using standards to test for organic compounds in a variety of food, such as honey, pasta, oats, and butter. Do not attempt to carry out the experiment unless your teacher has approved your procedures.

HOLT
BIOSOURCES
LAB PROGRAM
EXPERIMENTAL DESIGN

C47 Identifying Food Nutrients—Food Labeling

Prerequisites · **Laboratory Techniques C46: Identifying Food Nutrients** on pages 219–222

Review · procedure for testing nutrient content

The
Kathy O'Brian
Cincinnati, Ohio SHOW

April 10, 1998

Sam Ashike
Health Division
BioLogical Resources, Inc.
101 Jonas Salk Dr.
Oakwood, MO 65432-1101

Dear Mr. Ashike,

I am the head of the research department for the Kathy O'Brian show, a morning talk show based here in Cincinnati. As you may know, we have a segment called the Critical Consumer. In this segment we investigate the claims that manufacturers put on their product packages.

We have recently taken an interest in a frozen breakfast product called the Good Morning Burrito. The package specifically reads, "This product contains no fat and no sugar," but we suspect that this is not entirely true. Last month, an observant viewer reported finding oily stains on her napkin, indicating that the product contains fat. After further investigation, we received an anonymous tip from an employee of the company that makes the burrito, claiming that sugar and cheese were added to the product.

I will send you a sample of the burrito filling. I would like your company to test the contents of this food product for sugar, starch, fat, and protein. I will call soon to discuss details. Thank you for your time.

Sincerely,

Mark Rapasarda

Mark Rapasarda
Research Department
The Kathy O'Brian Show

BioLogical Resources, Inc. Oakwood, MO 65432-1101

M E M O R A N D U M

To: Team Leader, Food Testing Dept.

From: Sam Ashike, Director of Health

Have your research team conduct a series of tests to determine which nutrients are found in the food product described in the attached letter. Along with the sample of burrito filling, Mr. Rapasarda has sent us a copy of the package ingredients. The ingredients listing reads, "Filling Ingredients: protein, soluble starch, water."

Use glucose test strips when testing for sugar. These can be used simply by inserting a test strip into the food substance. Mr. Rapasarda informed me that this particular story segment is scheduled to air in the next few months and they need plenty of time to follow up on our research. Please keep this time constraint in mind as you plan your experiment.

Proposal Checklist

Before you start your work, you must submit a proposal for my approval. **Your proposal must include the following:**

_____ • the **question** you seek to answer

_____ • the **procedure** you will use

_____ • a detailed **data table** for recording observations

_____ • a complete, itemized list of proposed **materials** and **costs** (including use of facilities, labor, and amounts needed)

Proposal Approval: _____
(Supervisor's signature)

Report Procedures

When you finish your analysis, prepare a report in the form of a business letter to Mr. Rapasarda. **Your report must include the following:**

_____ • a paragraph describing the **procedure** you followed to determine which nutrients are found in the food product

_____ • a complete **data table** showing the presence or absence of a nutrient and the observation that supports it

_____ • your **conclusions** about the accuracy of the claims made on the product package

_____ • a detailed **invoice** showing all materials, labor, and the total amount due

Safety Precautions

- Wear safety goggles, disposable gloves, and a lab apron.

- Glassware is fragile. Notify your teacher promptly of any broken glass or cuts. Do not clean up broken glass or spills unless your teacher tells you to do so.

- Wash your hands before leaving the laboratory.

- Do not taste or ingest any food item in the laboratory.

Disposal Methods

- Dispose of waste materials according to instructions from your teacher.

- Place broken glass in the container provided.

- Wash reusable materials such as glassware and lab utensils, and return them to the supply area.

FILE: The Kathy O'Brian Show

MATERIALS AND COSTS (Select only what you will need. No refunds.)

I. Facilities and Equipment Use

Item	Rate	Number	Total
facilities	$480.00/day	_____	_____
personal protective equipment	$10.00/day	_____	_____
compound microscope	$30.00/day	_____	_____
microscope slide with coverslip	$2.00/day	_____	_____
10 mL graduated cylinder	$5.00/day	_____	_____
25 mL graduated cylinder	$5.00/day	_____	_____
100 mL graduated cylinder	$5.00/day	_____	_____
test tubes	$3.00/day	_____	_____
test tube rack	$5.00/day	_____	_____
thermometer	$5.00/day	_____	_____
hot plate	$10.00/day	_____	_____
forceps	$5.00/day	_____	_____
chemical spatula	$2.00/day	_____	_____
watch or clock with second hand	$10.00/day	_____	_____

II. Labor and Consumables

Item	Rate	Number	Total
labor	$40.00/hour	_____	_____
burrito filling	provided	_____	_____
invertase	$0.50/g	_____	_____
Benedict's solution	$0.40/mL	_____	_____
Lugol's iodine solution	$0.40/mL	_____	_____
Biuret reagent	$0.40/mL	_____	_____
Sudan III reagent	$0.40/mL	_____	_____
glucose test strips	$2.00/strip	_____	_____
brown paper bag	$0.05/piece	_____	_____
distilled H_2O	$0.10/mL	_____	_____
plastic stirring rod	$1.00 each	_____	_____
filter paper	$0.10/sheet	_____	_____
wax pencil	$2.00 each	_____	_____
adhesive labels	$0.10 each	_____	_____

Fines

OSHA safety violation	$2,000.00/incident	_____	_____
		Subtotal	_____
		Profit Margin	_____
		Total Amount Due	_____

Name _____

Date _____ Class _____

HOLT
BIOSOURCES
LAB PROGRAM
LABORATORY TECHNIQUES

C48 *Urinalysis Testing*

Skills

PREPARATION NOTES

Time Required: two 50-minute periods

- observing and comparing physical characteristics of simulated urine samples
- testing simulated urine samples for the presence of sugar and protein
- using a compound microscope
- preparing a wet mount

Objectives

- *Understand* the basis of urinalysis and its application to the diagnosis of medical disorders.
- *Perform* a simulated urinalysis on simulated urine samples.
- *Diagnose* three different medical disorders using simulated urine samples, a control, and physical descriptions.

Materials

Materials

Materials for this lab activity can be purchased from WARD'S. See the *Master Materials List* for ordering instructions.

- safety goggles
- lab apron
- gloves
- medicine cups (4)
- wax pencil
- simulated urine samples (10 mL each of 4 samples)
- pH strips (4)
- stirring rod
- 400 mL beaker
- 250 mL of water
- hot plate
- test tubes (8) and test-tube rack

- graduated plastic pipets (6)
- pipet pump
- Benedict's solution
- test-tube holder
- stopwatch or clock with second hand
- biuret solution
- medicine dropper (4)
- microscope slide (4)
- coverslip (4)
- compound microscope
- prepared reference slide identifying urine crystals

Purpose

You are a physician's assistant working in a rural county health clinic. One of your jobs is to conduct laboratory tests to help the doctor make a diagnosis. One test you will need to do frequently is urinalysis. Today, the physician has given you four case studies and four corresponding simulated urine samples on which to practice the technique of urinalysis.

Case Studies

Patient 1—Mr. Jones is 19 years old. He notices that he has increased urine output, increased appetite, and great thirst. He has also experienced unexplained weight loss.

Patient 2—Mr. Thompson is 60 years old and has been unusually tired for several weeks. He occasionally feels dizzy. Lately he finds it increasingly difficult to sleep at night. He has swollen ankles and feet, and his face looks puffy. He experiences a burning pain in his lower back, just below the rib cage. He also notices that his urine is dark in color. He goes to see his physician, who finds that he has elevated blood pressure and that the kidney region is sensitive to pressure.

Patient 3—Ms. Smith is 27 years old and has been experiencing painful urination, frequency of urination, and urgency. Her urine has a milky color. She also has fever and malaise, which is evidence of infection. Upon seeking treatment, she is given antibiotic therapy. After a few days of antibiotics, her symptoms disappear.

Patient 4—Mrs. Martinez is 46 years old. She has regular yearly checkups and has no visible symptoms.

Background

Additional Background

Another important use for urinalysis is pregnancy testing. When a woman becomes pregnant, the hormone chorionic gonadotropin (HCG) begins to be secreted by the embryonic tissues shortly after fertilization. HCG secretion then increases until it reaches a peak in about 50 to 60 days. Thereafter, the HCG concentration drops to a much lower level and remains relatively stable throughout the pregnancy. Because HCG is excreted in the urine, urinalysis is used to detect the hormone and indicate the presence of an embryo. Such a pregnancy test may give positive results as early as 8 to 10 days after fertilization.

Many diseases are identified and diagnosed based on recognizable signs and abnormalities called **symptoms.** Laboratory tests, such as X rays and blood tests, help a physician make a correct diagnosis and also rule out any other disorders that may have some of the same symptoms.

Many different tissues and fluids from the body—including blood, feces, and cerebrospinal fluids—are tested for symptoms of disease. Urine is just one of many body fluids and tissues that give a physician clues about a disease. Urine testing, or **urinalysis,** is a very important diagnostic tool. It involves the physical, chemical, and visual examination of a patient's urine sample. A thorough analysis of the urine may provide more information about the general condition of the body than any other set of tests. Urinary tract infections, kidney malfunction, diabetes, and liver disease are just some of the medical problems that can be diagnosed through urinalysis. Several factors are examined when analyzing a urine sample: color and appearance, pH, microscopic appearance, and presence of glucose and proteins.

The color of normal urine ranges from pale yellow to amber, depending on the concentration of the pigment **urochrome,** which is the end product of the breakdown of the blood protein hemoglobin. The appearance of the urine may serve as an indication of a disorder. Pale yellow urine, for example, may indicate diabetes insipidus or granular kidney, or the urine may simply be very dilute due to the ingestion of large amounts of water. A milky color might signify fat globules or pus, the latter possibly indicating a urogenital tract infection. Reddish colors may be due to food pigments (such as those in beets), certain drugs or foods, or blood in the urine. Greenish colors indicate either bile pigment (jaundice) or certain bacterial infections. Brown-black urine can indicate phenol or metallic poisonings or hemorrhages due to kidney injury or malaria.

The pH of normal urine ranges from 4.5 to 8.0. The pH of urine fluctuates depending on the type of food a person has eaten. Fevers and acidosis lower the pH. Anemia, vomiting, and urine retention raise the pH.

The microscopic observation of urine is a vital aspect of routine urinalysis. Urine is made up primarily of water, with some salts and organic materials dissolved in it. Inorganic substances normally found in the urine include sulfates, chlorides, phosphates, and ammonia. Crystals, cells, and microorganisms are among the significant elements found in the urine sediments (the solids that precipitate out of urine).

Disposal

- Simulated urine samples can be flushed down the drain using copious amounts of water.
- Place pH test strips in the trash.
- Always wear personal protective equipment (goggles, apron, and nitrile gloves) when disposing of Benedict's solution and biuret reagent. Make sure an eyewash station is nearby.
- Combine all wastes containing Benedict's solution and biuret solution. Slowly add 1 M sulfuric acid with stirring until the pH is between 5 and 6. Scour six 6d iron nails with steel wool until they are bright and shiny. Immerse the nails in the acidified solution. Let the nails remain immersed until all of the copper has precipitated (overnight should be sufficient). Remove the nails, and filter the solution. Heat the nails and any precipitate obtained from the filtrate sufficiently to convert the copper precipitate and copper on the nails into copper oxide. Let the nails and the precipitate cool, and then place them in the trash. Treat the filtrate with sufficient 1 M NaOH to bring the pH to from 8 to 10. Filter the liquid again. Let the precipitate dry, and place it in the trash. Pour the filtrate down the drain.
- Clean glassware, plastic medicine cups, and graduated pipets with soap and water.

The variety of crystals and shapeless compounds found in normal urinary sediment may represent both the end product of tissue metabolism and the excessive consumption of certain foods or drugs. The type of crystals and other compounds found in urine depends to some extent on the pH of the urine. The presence of some types of crystals is of little or no significance while others indicate a positive symptom for a disease. Common crystals are normally present in acidic, neutral, or alkaline urine. Abnormal crystals, however, are almost always associated only with acidic or neutral urine. Urine pH and the types of urine crystals and the possible disorders they indicate are shown in the table below.

Table 1 Urine pH and Crystal Type

pH	Crystal type	Possible disorders
Alkaline	Calcium phosphate	Calculi (kidney stone) formation
Alkaline	Triphosphate	Calculi formation Urinary tract infection
Alkaline	Calcium carbonate	Calculi formation
Acid	Calcium oxalate	Excessive intake of oxalate-rich foods such as spinach, garlic, tomatoes, oranges Diabetes mellitus
Acid	Uric acid	Gout Leukemia Chronic nephritis

Cells are exfoliated, or sloughed off, from different parts of the urinary tract for various reasons, including normal "wear and tear," the presence of degenerative and inflammatory disorders, or the formation of tumors. Microscopic examination of cells in urinary sediment may detect certain types of cancer and other types of urinary tract diseases.

The presence of various substances in the urine that are not normally there can, in some cases, be an indication of a disorder. The presence of albumin (a protein that helps regulate osmotic concentration of the blood) may indicate a kidney malfunction. One of the kidneys' functions is to filter out albumin and glucose from waste materials and return them to the body. The presence of glucose in the urine may indicate diabetes mellitus, which is caused by a deficiency of insulin. The hallmark of diabetes is an increase in the concentration of blood sugar. When the blood sugar reaches a certain high concentration, it exceeds the renal (kidney) threshold, and the kidneys begin to excrete the excess. At this point, glucose appears in the urine. Other symptoms of diabetes mellitus include excessive urine output, dehydration accompanied by great thirst, and increased appetite. The person is also likely to lose weight.

Procedure

Part 1—Observing the Physical Characteristics of Urine

1. Put on safety goggles and a lab apron.

PROCEDURAL NOTES

Safety Precautions
• Have students wear safety goggles and lab aprons.
• Discuss all safety symbols and caution statements with students.
• Review the rules for carrying and using the compound microscope.
• Caution students to be very careful when working with chemicals such as acids and stains. Remind them to notify you immediately of any chemical spills. Also caution them to never taste, touch, or smell any substance or bring it close to their eyes.

2. Use a wax pencil to mark four medicine cups as follows: Patient 1, Patient 2, Patient 3, and Patient 4. Obtain 10 mL of each patient's simulated urine sample in the correctly labeled cup.

3. Observe the color and clarity of the four simulated urine samples. Record your findings in the data table below.

Table 2 Physical Characteristics of Urine

Sample	Color	Clarity	pH
Patient 1	yellow	clear	acidic
Patient 2	yellow/orange	cloudy	acidic
Patient 3	yellow	cloudy	alkaline
Patient 4	yellow	clear	normal

Part 2—Testing the pH of Urine

4. Dip the tip of a glass stirring rod into the simulated urine from Patient 1. Touch the rod to a pH test strip. Compare the color of the test strip to the pH color chart within 30 seconds of sampling the urine. Record the pH in the data table above.

5. Repeat step 4 for the remaining samples. *Note: Rinse off the stirring rod between each test. Use a new pH test strip for each test.*

Part 3—Testing Urine for Glucose (Benedict's Test)

6. Use a wax pencil to label four test tubes 1, 2, 3, and 4. **CAUTION: Glassware is fragile. Notify your teacher promptly of any broken glass or cuts. Do not clean up broken glass or spills unless your teacher tells you to do so.**

7. Place 250 mL of water into a 400 mL beaker. Place the beaker on a hot plate, and heat to boiling. **CAUTION: Do not touch hot objects with your bare hands. Use tongs, test tube holders, and padded gloves as appropriate.**

8. Use a clean, plastic graduated pipet and pipet pump to add 3 mL of the simulated urine sample from Patient 1 to test tube 1. Set the pipet aside; you will be using it again in step 13. Use another clean, plastic graduated pipet to add 3 mL of Benedict's solution to the test tube. **CAUTION: If you get a chemical on your skin or clothing, wash it off at the sink while calling to your teacher. If you get a chemical in your eyes, promptly flush it out at the eyewash station while calling to your teacher. Notify your teacher in the event of a chemical spill.** Immediately record the color of the solution in the test tube in the data table at the top of the next page. A positive reaction will result in a yellow to red color.

Table 3 Benedict's Test

Sample	Color before heating	Color after heating	Result (+ or −)
Patient 1	blue	red precipitate	+
Patient 2	blue	blue	−
Patient 3	blue	blue	−
Patient 4	blue	blue	−

9. Using a test-tube holder, place the test tube in a hot-water bath and leave it for two minutes. Remove the tube from the hot-water bath, and record any color change in the data table above.

10. Repeat steps 8 and 9 for each of the remaining simulated urine samples. *Note: Be sure to use a clean pipet for each urine sample.*

Part 4—Testing for Protein (Biuret Test)

11. Use a wax pencil to label four clean test tubes 1, 2, 3, and 4.

12. Use a plastic graduated pipet (the same one you used in step 8 above) to add 3 mL of the simulated urine sample from Patient 1 to test tube 1.

13. With a clean, plastic graduated pipet, add 1 mL of biuret solution drop by drop to the test tube. Swirl the tube.

14. Record the initial color of the solution in the data table below. After just a few seconds, record the final color. A positive reaction results in an orange-red color. A negative reaction results in a green color. Record your findings in the data table below.

Table 4 Biuret Test

Sample	Initial color	Final color	Result (+ or −)
Patient 1	green	green	−
Patient 2	green	orange-red	+
Patient 3	green	green	−
Patient 4	green	green	−

15. Repeat steps 12 through 14 for the remaining simulated urine samples.

Part 5—Microscopic Observations

16. With a wax pencil, label four microscope slides 1, 2, 3, and 4. Make a wet mount by placing one drop of the simulated urine from Patient 1 onto slide 1 and adding a coverslip.

17. Observe the slide with both low power and high power of a compound microscope. Scan the slide for any simulated red blood cells (visible as small red spheres), or simulated leukocytes (visible as small blue spheres) that may be present. Also look for any crystals that may be present in the urine. Record your findings in the data table below.

Table 5 Microscopic Observations

Sample	Red blood cells	White blood cells	Crystals
Patient 1	none	none	calcium oxalate
Patient 2	present	none	none
Patient 3	none	present	triphosphate
Patient 4	none	none	none

18. Repeat steps 16 and 17 for the remaining simulated urine samples.

19. Observe a prepared slide that identifies urine crystals. Compare any crystals you find in the samples with those on the slide. Identify the crystals in the urine samples. Record the names of the crystals in the table above.

20. Dispose of your materials according to the directions from your teacher.

21. Clean up your work area and wash your hands before leaving the lab.

Analysis

22. Summarize the results of the glucose test.

    Patient 1 has glucose in the urine. Patients 2, 3, and 4 show no glucose in the urine.

23. Summarize the results of the protein test.

    Patient 2 has protein in the urine. Test results for patients 1, 3, and 4 show no protein in the urine.

24. Summarize the results of the microscopic observations.

    Patients 1 and 3 showed crystals of calcium oxalate and triphosphate respectively. Patient 3 showed

    white blood cells. Patient 2 showed the presence of red blood cells in the urine.

25. Which patient is the "control" patient? Support your answer.

    Patient 4, Mrs. Ramirez, is the control patient. She shows no visible symptoms in regular yearly

    testing.

26. What disorder does Mr. Jones probably have? What evidence supports your diagnosis?

He probably has diabetes mellitus. Glucose and calcium oxalate crystals were found in the urine sample. He has reported increased urine output, increased appetite, and great thirst, all symptoms of diabetes mellitus.

27. What disorder does Mr. Thompson probably have? What evidence supports your diagnosis?

He probably has chronic kidney failure. The presence of red blood cells found in the microscopic observation relates to the dark color of his urine. He also shows protein in his urine, which may indicate kidney malfunction. All his physical symptoms indicate the diagnosis.

28. What disorder does Ms. Smith probably have? What evidence supports your diagnosis?

The white blood cells, the triphosphate crystals, and the cloudy or milky color of the urine indicate a urinary tract infection. The milky color might be due to white blood cells because of her infection.

Conclusions

29. Why is it important to perform various laboratory tests to diagnose a disorder?

Certain conditions share the same signs and/or symptoms. Laboratory tests can give additional important information and help a physician rule out disorders to arrive at a correct diagnosis.

30. What causes the chemical content of urine to change throughout the day?

Eating and physical activity can cause changes in the chemical content of urine throughout the day.

31. Why is it important to perform tests on a control sample of urine?

The control sample of urine does not contain any chemical substances; it simulates the urine of a

normal, healthy person. The control sample will show how simulated urine reacts with the test

chemicals if certain substances (e.g., glucose) are not present.

32. If you compare abnormal and normal urine, how can this help you determine the function of the kidneys?

If substances such as albumin and glucose are found in the abnormal urine sample, one can

speculate that the kidneys are not functioning properly.

33. The presence of blood in the urine can indicate a serious kidney problem. Why are kidney problems so serious?

The kidneys function to filter the blood, removing nitrogenous waste materials so that the person

not poisoned by his or her own metabolism. The kidneys also excrete excess water and certain salt

from the blood. These functions are important in maintaining homeostasis in the body.

34. Suppose a urine sample revealed abnormal results, such as protein in the urine. Should a physician always make an immediate diagnosis of a disorder based on abnormalities of the urine? Why or why not?

Changes in urine, such as the presence of certain substances, can sometimes be due to temporary

chemical imbalances, heavy exercise, or diet. After several days the substance may no longer be

present in the person's urine. Therefore, in some cases the physician may test the urine several

times over a certain time period before drawing any conclusion.

Extensions

35. Research what kidney dialysis is and how it works. Find out who needs kidney dialysis, how often people need it, and where they receive treatment.

36. Find out how urinalysis is used to determine if a woman is pregnant.

37. Find out how urinalysis is used to determine if a person has used drugs. Research your school's policy on drugs, and compare it to that of a major employer in your area. How are the policies similar? How are they different?